CHAPMAN & HALL/CRC FINANCIAL MATHEMATICS SERIES

Structured Credit Portfolio Analysis, Baskets & CDOs

CHAPMAN & HALL/CRC
Financial Mathematics Series

Aims and scope:
The field of financial mathematics forms an ever-expanding slice of the financial sector. This series aims to capture new developments and summarize what is known over the whole spectrum of this field. It will include a broad range of textbooks, reference works and handbooks that are meant to appeal to both academics and practitioners. The inclusion of numerical code and concrete real-world examples is highly encouraged.

Series Editors

M.A.H. Dempster
*Centre for Financial Research
Judge Business School
University of Cambridge*

Dilip B. Madan
*Robert H. Smith School
of Business
University of Maryland*

Rama Cont
*CMAP
Ecole Polytechnique
Palaiseau, France*

Published Titles

An Introduction to Credit Risk Modeling, *Christian Bluhm, Ludger Overbeck, and Christoph Wagner*
Financial Modelling with Jump Processes, *Rama Cont and Peter Tankov*
Robust Libor Modelling and Pricing of Derivative Products, *John Schoenmakers*
American-Style Derivatives; Valuation and Computation, *Jerome Detemple*
Structured Credit Portfolio Analysis, Baskets & CDOs, *Christian Bluhm and Ludger Overbeck*

Proposals for the series should be submitted to one of the series editors above or directly to:
CRC Press, Taylor and Francis Group
24-25 Blades Court
Deodar Road
London SW15 2NU
UK

CHAPMAN & HALL/CRC FINANCIAL MATHEMATICS SERIES

Structured Credit Portfolio Analysis, Baskets & CDOs

**Christian Bluhm
Ludger Overbeck**

Chapman & Hall/CRC
Taylor & Francis Group
Boca Raton   London   New York

Chapman & Hall/CRC is an imprint of the
Taylor & Francis Group, an informa business

Chapman & Hall/CRC
Taylor & Francis Group
6000 Broken Sound Parkway NW, Suite 300
Boca Raton, FL 33487-2742

© 2007 by Taylor & Francis Group, LLC
Chapman & Hall/CRC is an imprint of Taylor & Francis Group, an Informa business

No claim to original U.S. Government works
Printed in the United States of America on acid-free paper
10 9 8 7 6 5 4 3 2

International Standard Book Number-10: 1-58488-647-1 (Hardcover)
International Standard Book Number-13: 978-1-58488-647-1 (Hardcover)

This book contains information obtained from authentic and highly regarded sources. Reprinted material is quoted with permission, and sources are indicated. A wide variety of references are listed. Reasonable efforts have been made to publish reliable data and information, but the author and the publisher cannot assume responsibility for the validity of all materials or for the consequences of their use.

No part of this book may be reprinted, reproduced, transmitted, or utilized in any form by any electronic, mechanical, or other means, now known or hereafter invented, including photocopying, microfilming, and recording, or in any information storage or retrieval system, without written permission from the publishers.

For permission to photocopy or use material electronically from this work, please access www. copyright.com (http://www.copyright.com/) or contact the Copyright Clearance Center, Inc. (CCC) 222 Rosewood Drive, Danvers, MA 01923, 978-750-8400. CCC is a not-for-profit organization that provides licenses and registration for a variety of users. For organizations that have been granted a photocopy license by the CCC, a separate system of payment has been arranged.

Trademark Notice: Product or corporate names may be trademarks or registered trademarks, and are used only for identification and explanation without intent to infringe.

Library of Congress Cataloging-in-Publication Data

Bluhm, Christian.
    Structured portfolio analysis, baskets and CDOs / Christian Bluhm, Ludger Overbeck.
        p. cm. -- (Chapman & Hall/CRC financial mathematics series ; 5)
    Includes bibliographical references and index.
    ISBN 1-58488-647-1 (alk. paper)
    1. Portfolio management. 2. Investment analysis. I. Overbeck, Ludger. II. Title.

HG4529.5.B577 2006
332.64'5--dc22   2006049133

Visit the Taylor & Francis Web site at
http://www.taylorandfrancis.com

and the CRC Press Web site at
http://www.crcpress.com

Preface

The financial industry is swamped by structured credit products whose economic performance is linked to the performance of some underlying portfolio of credit-risky instruments like loans, bonds, credit default swaps, asset-backed securities, mortgage-backed assets, etc. The market of such collateralized debt or synthetic obligations, respectively, is steadily growing and financial institutions continuously use these products for tailor-made risk selling and buying.

In this book, we discuss mathematical approaches for modeling structured products credit-linked to an underlying portfolio of credit-risky instruments. We keep our presentation mathematically precise but do not insist in always reaching the deepest possible level of mathematical sophistication. Also, we do not claim to present the full range of possible modeling approaches. Instead, *we focus on ideas and concepts we found useful for our own work and projects in our daily business*. Therefore, the book is written from the perspective of practitioners who apply mathematical concepts to structured credit products.

As pre-knowledge, we assume some facts from probability theory, stochastic processes, and credit risk modeling, but altogether we tried to keep the presentation as self-contained as possible. In the bibliography, the reader finds a collection of papers and books for further reading, either for catching up with facts required for an even deeper understanding of the model or for following up on further investigations in alternative approaches and concepts found useful by other colleagues and researchers working in the field of portfolio credit risk. A helpful pre-reading for the material contained in this book is the textbook [25], which introduces the reader to the world of credit risk models. However, there are many other suitable textbooks in the market providing introductory guidance to portfolio credit risk and we are sure that for every reader's taste there is at least one 'best matching' textbook out there in the book market.

It is our hope that after reading this book *the reader will find the challenge to model structured credit portfolios as fascinating as we find it*. We worked in this area for quite some years and still consider it as a great and intriguing field where application of mathematical ideas, concepts, and models lead to value-adding business concepts.

Zurich and Munich, August 2006

Christian Bluhm and Ludger Overbeck

Acknowledgments

Christian Bluhm thanks his wife, Tabea, and his children, Sarah Maria and Noa Rebeccah, for their continuous support and generous tolerance during the writing of the manuscript. Their love and encouragement is his most important source of energy.

Ludger Overbeck is most grateful to his wife, Bettina, and his children, Leonard, Daniel, Clara, and Benjamin for their ongoing support.

We highly appreciate feedback and comments on the manuscript by various colleagues and friends.

We owe special thanks to Christoff Goessl (HypoVereinsbank, London), Christopher Thorpe (FitchRatings, London), Walter Mussil (Mercer Oliver Wyman, Frankfurt), and Stefan Benvegnu (Credit Suisse, Zurich) with whom we had the honor and pleasure to work for quite some years.

Disclaimer

This book reflects the personal view of the authors and does not provide information about the opinion of Credit Suisse and HypoVereinsbank. The contents of this book have been written solely for educational purposes. The authors are not liable for any damage arising from any application of the contents of this book.

About the Authors

Christian Bluhm is a Managing Director at Credit Suisse in Zurich. He heads the Credit Portfolio Management unit within the Credit Risk Management Department of Credit Suisse. Before that, he headed the team Structured Finance Analytics in HypoVereinsbank's Group Credit Portfolio Management in Munich, where his team was responsible for the quantitative evaluation of asset-backed securities with a focus on collateralized debt and loan obligations, collateralized synthetic obligations, residential and commercial mortgage-backed securities, default baskets, and other structured credit instruments, from origination as well as from an investment perspective. His first professional position in risk management was with Deutsche Bank in Frankfurt. In 1996, he earned a Ph.D. in mathematics from the University of Erlangen-Nuremberg and, in 1997, he was a post-doctoral fellow at the mathematics department of Cornell University, Ithaca, New York. He has coauthored a book on credit risk modeling, together with Ludger Overbeck and Christoph Wagner, and coauthored a book on statistical foundations of credit risk modeling, together with Andreas Henking and Ludwig Fahrmeir. He regularly publishes papers and research articles on credit risk modeling and management in journals and books, including RISK magazine and RISK books. During his academic time, he published several research articles on harmonic and fractal analysis of random measures and stochastic processes in mathematical journals. He frequently speaks at conferences on risk management.

Ludger Overbeck holds a professorship in mathematics and its applications at the University of Giessen in Germany. His main interests are quantitative methods in finance, risk management, and stochastic analysis. Currently, he also is a Director at HypoVereinsbank in Munich, heading the Portfolio Analytics and Pricing Department within the Active Portfolio Management unit in HypoVereinsbank's Corporates & Markets Division. His main tasks are the pricing of structured credit products and the risk management of the investment credit portfolio. Until June 2003, he was Head of Risk Research & Development in Deutsche Bank's credit risk function located in Frankfurt. His main responsibilities included development and implementation of the internal group-wide credit portfolio model, the operational risk model, specific (market) risk modeling, the EC/RAROC-methodology, integration of risk types, backtesting of ratings and correlations, quantitative support for portfolio management and all other risk types, and risk assessment of credit derivatives and portfolio transactions like collateralized debt and loan obligations. Prior to that, he worked for the Banking Supervision Department in the main office of Deutsche Bundesbank in Duesseldorf, Germany, mainly concerned with internal market risk models and inspections of the banks in line with the so-called 'minimum requirements for trading businesses'. He frequently publishes articles in various academic and applied journals, including RISK, and is coauthor of a book on credit risk modeling, together with Christian Bluhm and Christoph Wagner. He is a regular speaker at academic and financial industry conferences. Ludger holds a Ph.D. in probability theory and habilitations in applied mathematics from the University of Bonn and in economics from the University of Frankfurt.

A Brief Guide to the Book

Before we send the reader on a journey through the different chapters of this book, it makes sense to briefly summarize the contents of the different sections so that readers can define their own individual path through the collected material.

Chapter 1 is intended to be an introduction for readers who want to have a brief overview on credit risk modeling before they move on to the core topics of the book. For experienced readers, Chapter 1 can serve as a 'warming-up' and an introduction to the notation and nomenclature used in this book. The following keywords are outlined:

- Single-name credit risk measures like ratings and scorings, default probabilities, exposures, and loss given default

- Modeling of default risk via latent variable and threshold models

- Decomposition of credit risks into systematic and idiosyncratic components by factor models

- Credit portfolio loss distributions and their summary statistics like expected loss, unexpected loss, and quantile-based and expected shortfall-based economic capital

- Comments and remarks regarding the trade-off between accuracy and practicability of credit risk methodologies

It is not necessary to read Chapter 1 in order to understand subsequent chapters. But experienced readers will need not more than an hour to flip through the pages, whereas non-experienced readers should take the time until all of the keywords are properly understood. For this purpose, it is recommended to additionally consult an introductory textbook on credit risk measurement where the mentioned keywords are not only indicated and outlined but fully described with all the technical details. There are many textbooks one can use for cross-references; see Chapter 5 for recommendations.

Chapter 2 treats the modeling of basket products. The difference between basket products and collateralized debt obligations (CDOs) covered in Chapter 3 is the enhanced cash flow richness in CDO products, whereas baskets are essentially credit derivatives referenced to a portfolio of credit-risky names. The content of Chapter 2 can be divided into two broad categories:

- *Modeling techniques:*
  Essential tools for credit risk modeling are introduced and discussed in detail, including term structures of default probabilities, joint default probabilities, dependent default times and hazard rates, copula functions with special emphasis on the Gaussian, Student-t, and Clayton copulas, and dependence measures like correlations, rank correlations, and tail dependence.

- *Examples and illustrations:*
  Chapter 2 always tries out the just-developed mathematical concepts in the context of fictitious but realistic examples like duo baskets, default baskets (first-to-default, second-to-default), and credit-linked notes. One section in the context of examples is dedicated to an illustration on scenario analysis showing how modeling results can be challenged w.r.t. their plausibility.

In general, we present *techniques rather than ready-to-go solutions*, because, based on our own experience, we found that general principles and techniques are more useful due to a broader applicability. In many cases, it is straightforward to adapt modeling principles and techniques to the particular problem the modeler has to solve. For instance, we do not spend much time with pricing concepts but provide many techniques in the context of the modeling of dependent default times, which can be applied in a straightforward manner to evaluate cash flows of credit-risky instruments. Such evaluations can then be used to derive a model price of credit-risky instruments like default baskets or tranches of collateralized debt obligations.

Chapter 3 is dedicated to collateralized debt obligations (CDOs). In a first part of our exposition, we focus on a non-technical description of reasons and motivations for CDOs as well as different types and applications of CDOs in the structured credit market. For the modeling of CDOs we can apply techniques already elaborated in Chapter 2. This is exercised by means of some CDO examples, which are sufficiently realistic to demonstrate the application of models in a way ready to be adopted by readers for their own work. In addition to the dependent default times concept, Chapter 3 also includes a discussion on alternative modeling approaches like multi-step models and diffusion-based first passage times. Techniques applicable for the reduction of modeling efforts like analytic and semi-analytic approximations as well as a very efficient modeling technique based on the comonotonic copula function conclude our discussion on modeling approaches.

At the end of Chapter 3, a comprehensive discussion on single-tranche CDOs (STCDOs) and index tranches as important examples for STCDOs is included. In this section, we also discuss pricing and hedging issues in such transactions.

The two last topics we briefly consider in Chapter 3 are portfolios of CDOs and the application of securitization-based tranche spreads as building blocks in a cost-to-securitize pricing component.

Chapter 4 is a collection of literature remarks. Readers will find certain guidance regarding the access to a rich universe of research articles and books. The collection we present does by no means claim to be complete or exhaustive but will nevertheless provide suggestions for further reading.

The **Appendix** at the end of this book contains certain results from probability theory as well as certain credit risk modeling facts, which are intended to make the book a little more self-contained. Included are also certain side notes, not central enough for being placed in the main part of the book, but nevertheless interesting. One example for such side notes is a brief discussion on entropy-maximizing distributions like the Gaussian distribution and their role as standard choices under certain circumstances.

As mentioned in the preface of this book, the collection of material presented in this book is based on modeling techniques and modeling aspects we found useful for our daily work. *We very much hope that every reader finds at least some pieces of the presented material to be of some value for her or his own work.*

Contents

Preface	v
About the Authors	ix
A Brief Guide to the Book	xi

1 From Single Credit Risks to Credit Portfolios — 1
- 1.1 Modeling Single-Name Credit Risk 2
  - 1.1.1 Ratings and Default Probabilities 2
  - 1.1.2 Credit Exposure 10
  - 1.1.3 Loss Given Default 14
- 1.2 Modeling Portfolio Credit Risk 17
  - 1.2.1 Systematic and Idiosyncratic Credit Risk 17
  - 1.2.2 Loss Distribution of Credit Portfolios 20
  - 1.2.3 Practicability Versus Accuracy 24

2 Default Baskets — 27
- 2.1 Introductory Example: Duo Baskets 27
- 2.2 First- and Second-to-Default Modeling 34
- 2.3 Derivation of PD Term Structures 39
  - 2.3.1 A Time-Homogeneous Markov Chain Approach . 40
  - 2.3.2 A Non-Homogeneous Markov Chain Approach . 52
  - 2.3.3 Extrapolation Problems for PD Term Structures 57
- 2.4 Duo Basket Evaluation for Multi-Year Horizons 59
- 2.5 Dependent Default Times 67
  - 2.5.1 Default Times and PD Term Structures 67
  - 2.5.2 Survival Function and Hazard Rate 68
  - 2.5.3 Calculation of Default Time Densities and Hazard Rate Functions 69
  - 2.5.4 From Latent Variables to Default Times 78
  - 2.5.5 Dependence Modeling via Copula Functions . . . 85
  - 2.5.6 Copulas in Practice 93

          2.5.7   Visualization of Copula Differences and Mathe-
                  matical Description by Dependence Measures . .   99
          2.5.8   Impact of Copula Differences to the Duo Basket   113
          2.5.9   A Word of Caution   118
    2.6   Nth-to-Default Modeling 120
          2.6.1   Nth-to-Default Basket with the Gaussian Copula   121
          2.6.2   Nth-to-Default Basket with the Student-t Copula 127
          2.6.3   Nth-to-Default Basket with the Clayton Copula . 127
          2.6.4   Nth-to-Default Simulation Study 129
          2.6.5   Evaluation of Cash Flows in Default Baskets . . 136
          2.6.6   Scenario Analysis 140
    2.7   Example of a Basket Credit-Linked Note (CLN)   147

3   **Collateralized Debt and Synthetic Obligations**   **165**
    3.1   A General Perspective on CDO Modeling   166
          3.1.1   A Primer on CDOs   167
          3.1.2   Risk Transfer 172
          3.1.3   Spread and Rating Arbitrage 178
          3.1.4   Funding Benefits   184
          3.1.5   Regulatory Capital Relief 186
    3.2   CDO Modeling Principles   190
    3.3   CDO Modeling Approaches   194
          3.3.1   Introduction of a Sample CSO   194
          3.3.2   A First-Order Look at CSO Performance   199
          3.3.3   Monte Carlo Simulation of the CSO   202
          3.3.4   Implementing an Excess Cash Trap   210
          3.3.5   Multi-Step and First Passage Time Models . . .   213
          3.3.6   Analytic, Semi-Analytic, and Comonotonic CDO
                  Evaluation Approaches 220
    3.4   Single-Tranche CDOs (STCDOs)   250
          3.4.1   Basics of Single-Tranche CDOs 250
          3.4.2   CDS Indices as Reference Pool for STCDOs . . .   253
          3.4.3   ITraxx Europe Untranched   259
          3.4.4   ITraxx Europe Index Tranches: Pricing, Delta
                  Hedging, and Implied Correlations 271
    3.5   Tranche Risk Measures   287
          3.5.1   Expected Shortfall Contributions   288
          3.5.2   Tranche Hit Contributions of Single Names . . . 292
          3.5.3   Applications: Asset Selection, Cost-to-Securitize 294
          3.5.4   Remarks on Portfolios of CDOs 299

4	**Some Practical Remarks**	**303**
5	**Suggestions for Further Reading**	**307**
6	**Appendix**	**311**
	6.1 The Gamma Distribution	311
	6.2 The Chi-Square Distribution	312
	6.3 The Student-t Distribution	312
	6.4 A Natural Clayton-Like Copula Example	314
	6.5 Entropy-Based Rationale for Gaussian and Exponential Distributions as Natural Standard Choices	315
	6.6 Tail Orientation in Typical Latent Variable Credit Risk Models	318
	6.7 The Vasicek Limit Distribution	320
	6.8 One-Factor Versus Multi-Factor Models	322
	6.9 Description of the Sample Portfolio	329
	6.10 CDS Names in CDX.NA.IG and iTraxx Europe	332

References   **339**

Index   **349**

Chapter 1

From Single Credit Risks to Credit Portfolios

We begin our exposition by a brief non-informal tour through some credit risk modeling concepts recalling some basic facts we will need later in the book as well as making sure that we are all at the same page before Chapters 2 and 3 lead us to the 'heart' of our topic. A good starting point is a saying by ALBERT EINSTEIN who seemingly made the observation

> *As far as the laws of mathematics refer to reality they are not certain; and as far as they are certain they do not refer to reality.*[1]

We feel free to interpret his statement in the following way. Most of the time we cannot work with deterministic models in order to describe reality. Instead, our models need to reflect the *uncertainty* inherent in the evolution of economic cycles, customer behavior, market developments, and other random components driving the economic fortune of the banking business. Thanks to the efforts of probabilists over many centuries, we have a full range of tools from probability theory at our disposal today in order to model the uncertainties arising in the financial industry. By means of probability theory randomness can never be eliminated completely in order to overcome the uncertainty addressed in EINSTEIN's statement, but randomness can be 'tamed' and conclusions and forecasts with respect to a certain level of confidence can be made on which business and investments decisions can be based. It is the aim of the following sections to perform a quick but thorough walk through concepts exploiting probability theory to finally arrive at meaningful conclusions in structured finance.

---

[1] Albert Einstein in his address *Geometry and Experience* to the Prussian Academy of Sciences in Berlin on January 27, 1921.

As a reference for the sections in this chapter we refer to the book [25] where most of the introductory remarks made in this chapter are elaborated in much greater detail. In order to keep the exposition fluent, we do not interrupt the text too often to make bibliographic remarks. Instead, suggestions for further reading are provided in Chapter 5. The main purpose of Chapter 1 is to build up the basis of nomenclature, risk terms, and basic notions required for understanding the subsequent chapters of this book.

1.1 Modeling Single-Name Credit Risk

Based on EINSTEIN's observation regarding uncertainty in real world problems, we have to rely on probabilistic concepts already at the level of single borrowers. In the following brief exposition, we focus on ratings and default probabilities, loss quotes, and exposures.

1.1.1 Ratings and Default Probabilities

As a standard tool for the lending business, all banks have rating systems in place today ranking the creditworthiness of their clients by means of a *ranking*, e.g., AAA, AA, A, BBB, BB, B, CCC, or Aaa, Aa, A, Baa, Ba, B, Caa, respectively, for best, 2nd-best, 3rd-best, ..., and worst credit quality where the first row of letter combinations is used by the rating agencies Standard & Poor's[2] (S&P) and Fitch[3] and the second row of letter combinations is used by Moody's.[4] Banks and rating agencies assign *default probabilities* (short: PD for 'probability of default') with respect to certain time horizons to ratings in order to quantify the likelihood that rated obligors default on any of their payment obligations within the considered time period. In doing so, the ranking (with respect to creditworthiness) of obligors in terms of ratings (letter combinations) is mapped onto a metric scale of numbers (likelihoods) in the unit interval. Throughout the book we use one-year default probabilities assigned to ratings (in S&P notation) according to

---
[2] See www.standardandpoors.com
[3] See www.fitchratings.com
[4] See www.moodys.com

TABLE 1.1: One-year PDs for S&P ratings; see [108], Table 9

	D
AAA	0.00%
AA	0.01%
A	0.04%
BBB	0.29%
BB	1.28%
B	6.24%
CCC	32.35%

Table 1.1. Table 1.1 can be read as follows in an intuitive way. Given a portfolio of 10,000 B-rated obligors it can be expected that 624 of these obligors default on at least one of their payment obligations within a time period of one year.

Kind of 'non-acceptable' is the zero likelihood of default for AAA-rated borrowers. This would suggest that AAA-rated customers never default and, therefore, can be considered as risk-free asset for every investor. It is certainly very unlikely that a firm with a AAA-rating fails on a payment obligation but there always remains a small likelihood that even the most creditworthy client defaults for unexpected reasons. Moreover, as addressed[5] in research on *PD calibration for low default portfolios*, the zero default frequency for AAA is just due to a lack of observations of AAA-defaults. In order to assign a non-zero one-year PD to AAA-rated customers, we make a linear regression of PDs on a logarithmic scale in order to find a meaningful estimated PD for AAA-rated borrowers.

This yields a one-year PD for AAA-rated obligors of 0.2 bps, which we will assume to be the 'best guess PD' for AAA-rated clients from now on.

Default probabilities are the 'backbone' of credit risk management. In the new capital accord[6] (Basel II), the so-called *Internal Ratings Based* (IRB) approach (see [15]) allows banks to rely on their internal

---

[5] See PLUTO and TASCHE [100] and WILDE and JACKSON [114].

[6] Note that the Bank for International Settlements (BIS) is located in the Swiss city Basel such that the capital initiative, which led to the new regulatory framework is often addressed under the label 'Basel II'; see www.bis.org

estimates for default probabilities for calculating regulatory capital. Therefore, the better the quality of a bank's rating systems, the more appropriate are the bank's regulatory capital figures. But not only for regulatory purposes but, even more important, for economic reasons like competitive advantage in the lending market it is important that banks pay a lot of attention to their internal rating systems. It is beyond the scope of this book and beyond the purpose of this introductory chapter to go into great details here but the following remarks will at least provide a link to the practical side of ratings.

Internal ratings can be obtained by various methodologies. In this book, where the focus is on structured credit products like default baskets and collateralized debt obligations (CDOs; see Chapter 3), ratings and PDs are typically modeled via so-called *causal rating* methods. In such models, underlying drivers of the default event of a client or asset are explicitly modeled as well as the functional link between risk drivers and the default event. We will come back to this principle over and over again in subsequent chapters. For a first simple example, we refer to Figure 1.3. Causal rating models are a conceptually preferred type of ratings because nothing is more desirable for a bank than understanding the 'explicit mechanism' of default in order to define appropriate measures of default prevention. Unfortunately, it is not always possible to work with causal rating models. For example, private companies (Small and Medium Enterprises; SME) are difficult to capture with causal models because underlying default drivers are not given in up-to-date explicit form as it is in the case for listed corporate clients where public information, for instance, by means of stock prices, is available.

In cases where causal models are not implementable, we need to work with more indirect approaches like so-called *balance sheet scorings* where balance sheet and income statement informations (as 'explaining variables') are used to explain default events. In such rating systems, the bank optimizes the *discriminatory power* (i.e., the ability of the rating system to separate defaulting from non-defaulting clients w.r.t. a certain time horizon) by means of a scheme as illustrated in Figure 1.1. Starting with a long list of financial ratios, the rating modelers attempt to find the optimal selection of ratios and optimal weights for the combination of ratios to finally arrive at a rating score best possible identifying potential defaulters in the bank's credit approval process.

An important working step in the development of scoring models

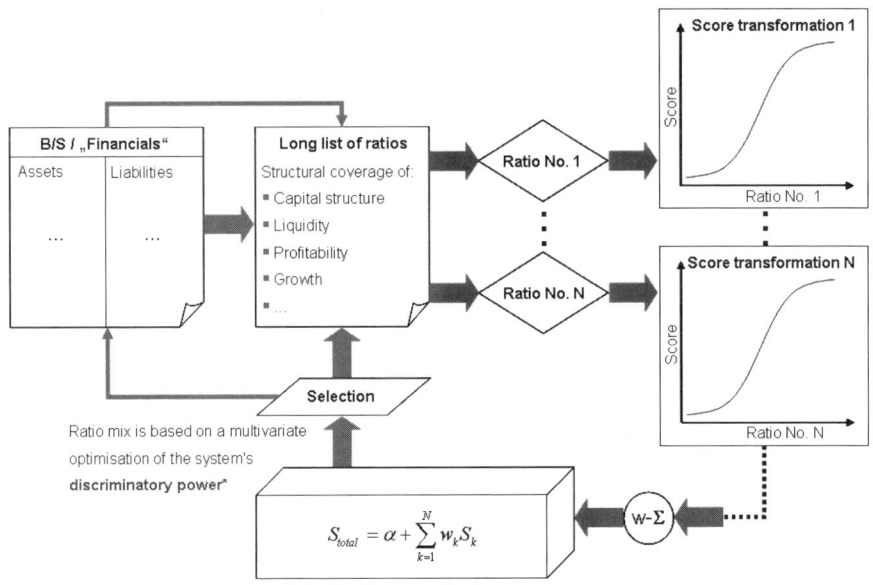

FIGURE 1.1: Rating/scoring system optimization (illustrative)

is the calibration of the scores to actual default probabilities. This procedure is briefly called the *PD calibration* of the model. A major challenge in PD calibration and rating/scoring model development is to find a healthy balance between the so-called error of first kind (α-error) and the error of second kind (β-error), where the first mentioned refers to a defaulting client approved by the scoring system and the latter-mentioned refers to a non-defaulting client erroneously rejected by the scoring system. The error of first kind typically contributes to the bank's P&L with the realized loss (lending amount multiplied by the realized loss quote) of the engagement, whereas the second error contributes to the bank's P&L in the form of a missed opportunity, which in monetary terms typically sums up to the lending amount multiplied by some margin. It is obvious that the first error typically can be expected to be more heavy in terms of P&L impact. However, banks with too conservative PDs will have a competitive disadvantage in the lending market. If too many clients are rejected based on ratings or scorings, the earnings situation of a bank can severely suffer, in terms of a negative P&L impact as well as in terms of lost market share.

There are other rating system categories, besides causal rating and scoring systems, which are common in credit risk management. For instance, in some cases ratings are based on a *hybrid* methodology combining a scoring component and a causal model component into a common framework. We will not comment further on ratings in general but conclude our discussion with an example illustrating the competitive advantage banks have from better rating systems. This remark also concerns our structured credit applications since banks with more sophisticated models for, e.g., the rating or evaluation of tranches in collateralized debt obligations (CDOs), have a chance of being more competitive in the structured credit market.

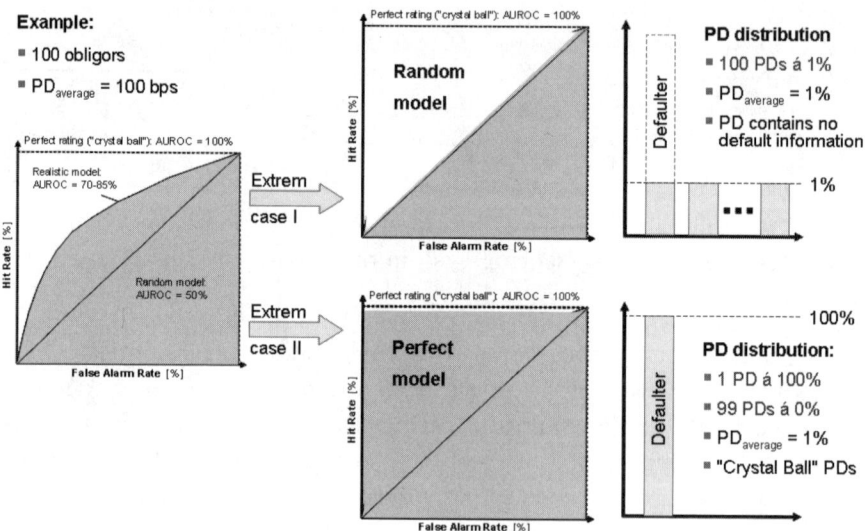

FIGURE 1.2: Illustration of the competitive advantage arising from discriminatory power in rating systems

For our example we consider a portfolio with 100 assets or clients. We assume that the average default quote in the portfolio is 1%. If the rating/scoring model of the bank has no discriminatory or prediction power at all, then the rating outcome for a client is purely based on chance. In Figure 1.2, we indicate such a worst case rating model by a so-called *receiver operating characteristic* (ROC)[7] on the diagonal of

---

[7]See [43] for an introduction to receiver operating characteristic (ROC) curves. In our chart, the x-axis shows the *false alarm rate* (FAR) and the y-axis shows the *hit rate* (HR) of the scoring system. Let us briefly explain the meaning of FAR and HR. Given that credit approval is based on a score S satisfying $S > c$, where c denotes

the unit square. The AUROC is a common measure for discriminatory power; see, e.g., [43]. The AUROC for the worst case rating system (see Figure 1.2) equals 0.5, which corresponds to the area under the diagonal in the unit square. This means that the credit score in the credit approval process of the bank does a job comparable to a coin-flipping machine with a fair coin: 'head' could mean approval of the loan and 'tail' could mean rejection of the loan. If the bank has such a rating/scoring system, then clients cannot be distinguished regarding their default remoteness. Therefore, given a portfolio default quote of 1%, every client will get assigned a PD of 1%. This is extreme case I in Figure 1.2. Extreme case II refers to a 'crystal ball' rating/scoring system where the model with certainty identifies the one defaulting client out of 100 clients in the portfolio who is responsible for the portfolio default quote of 1%. In such an unrealistically lucky case, it is natural to assign a zero PD to all non-defaulting clients and a PD of 100% to the one defaulting client.

In extreme case I as well as in extreme case II, the average of assigned PDs in the portfolio equals 1%. If the bank, starting from the worst case rating system in extreme case I, manages to improve their ratings toward the 'crystal ball' rating system from extreme case II, then 99 PDs experience a significant reduction (from 1% down to 0%) and one PD is significantly increased (from 1% to 100%). Our example is purely illustrative because typical rating/scoring systems have an AUROC between, e.g., 70% and 90%, but rating revisions, which improve the discriminatory power of the system exhibit the same behavior as our extreme case evolution in the example: *The benefit of improved predictive power in rating/scoring systems typically leads to lower PDs for assets/clients with higher credit quality and to higher PDs for assets/clients with worse credit quality.* In this way, improved ratings allow the bank to offer a more competitive or aggressive pricing for high quality assets and, at the same time, lead to an improved rejection mechanism for low quality assets. It can be observed in practice that better ratings have a potential to substantially contribute to an

---

a pre-specified critical cutoff, then a false alarm occurs if $S < c$ for a non-defaulting client. In contrast, the scoring system achieves a hit, if the score S shows $S < c$ for a defaulting client, because the comparison of score and cutoff values helped us indentifying a dangerous credit applicant. Now, the ROC curve is a plot of the curve $(\text{FAR}(c), \text{HR}(c))_{c \in \{\text{cutoffs}\}}$. A 'crystal ball' scoring system yields an area under the ROC curve (AUROC) of 100% because FAR=0 and HR=1 are best case situations.

improved P&L distribution.

After our brief excursion to the 'practical side' of ratings, we now come back to the development of the framework used in this book for modeling structured credit products. As already indicated, we are working in a *causal modeling* world when it comes to default baskets and CDOs; see Figure 3.8. But even at single-asset level, we can think of default events in a causal model context. For this purpose, we introduce for obligor i in a portfolio of m obligors a Bernoulli variable for indicating default within the time interval $[0, t]$,

$$L_i^{(t)} = \mathbf{1}_{\{\text{CWI}_i^{(t)} < c_i^{(t)}\}} \sim B(1; \mathbb{P}[\text{CWI}_i^{(t)} < c_i^{(t)}]) \qquad (1.1)$$

following a so-called *threshold model* in the spirit of MERTON [86], BLACK & SCHOLES [23], MOODYS KMV's PortfolioManager,[8] and CreditMetrics[9] from the RISK METRICS GROUP (RMG) for modeling default of credit customers. Here, $(\text{CWI}_i^{(t)})_{t\geq 0}$ denotes a *creditworthiness index* (CWI) of obligor i indexed by time t triggering a payment default of obligor i in case the obligor's CWI falls below a certain critical threshold $c_i^{(t)}$ within the time period $[0, t]$. The threshold $(c_i^{(t)})_{t\geq 0}$ is called the (time dependent) *default point* of obligor i. Equation (1.1) expresses the *causality* between the CWI of obligor i and its default event. Note that - despite its notation, *we do not consider* $(\text{CWI}_i^{(t)})_{t\geq 0}$ *as a* stochastic process *but consider it as a* time-indexed set of latent random variables *w.r.t. time horizons t where* $t \geq 0$ *denotes continuous time*. It is important to keep this in mind in order to avoid misunderstandings. It needs some work[10] (see Sections 3.3.5.1 and 3.3.5.2) to get from a sequence of latent variables to a stochastic process reflecting a certain *time dynamics*. For most of the applications discussed in this book, the 'fixed time horizon view' is completely sufficient. However, for the sake of completeness, we discuss in Section 3.3.5.2 a particular model based on the *first passage time* w.r.t. a *critical barrier* of a

---

[8] See www.kmv.com
[9] See www.riskmetrics.com
[10] The step from fixed horizon CWIs to a stochastic process can be compared with the construction of Brownian motion: just considering a sequence of normal random variables $(X_t)_{t\geq 0}$ with $X_t \sim N(0, t)$ does not mean that $(X_t)_{t\geq 0}$ already is a Brownian motion. It needs much more structure regarding the time dynamics and quite some work to choose $(X_t)_{t\geq 0}$ in a way making it a Brownian motion.

stochastic process instead of a sequence of (fixed horizon) latent variables w.r.t. a sequence of default points; see Figure 1.3 illustrating the concept of stochastic processes as default triggers. The process introduced in Section 3.3.5.2 will be called an *ability to pay process* (APP) because it reflects a stochastic time dynamics, whereas the sequence of CWIs can be seen as a *perfect credit score* indicating default with certainty according to Equation (1.1).

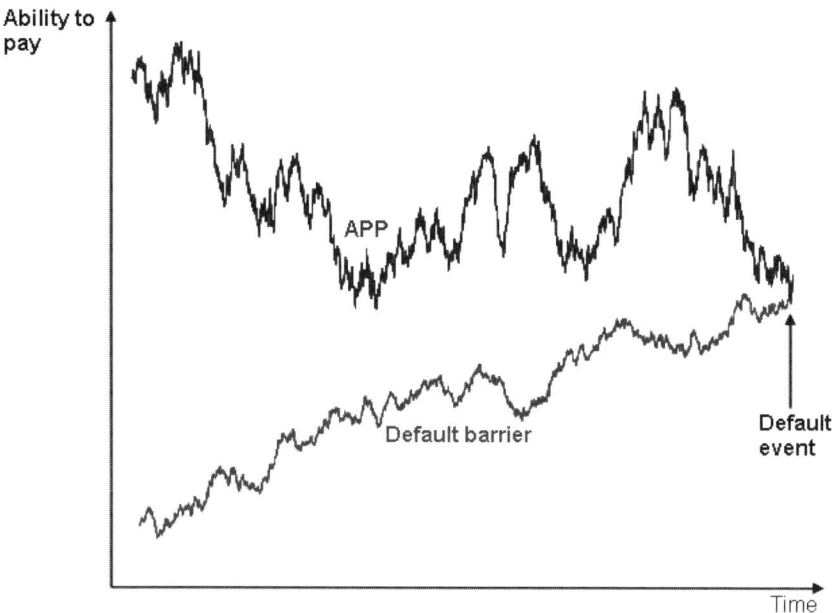

FIGURE 1.3: Ability to pay process (APP) as default trigger

CWIs are difficult or may be even impossible to observe in practice. In the classical MERTON model, the *ability to pay* or *distance to default* of a client is described as a function of *assets* and *liabilities.*. However, the concept of an underlying latent CWI triggering default or survival of borrowers is universally applicable: not only listed corporate clients but any obligor has its own individual situation of wealth and financial liabilities w.r.t. any given time horizon. If liabilities exceed the financial power of a client, no matter if the client represents a firm or a private individual, bankruptcy and payment failure will follow. In this way, the chosen approach of modeling default events by means of latent variables is a concept meaningful for corporate as well as for private clients and can be applied to the total credit portfolio of banks.

For the 'fixed time horizon' CWI approach, the time-dependent default points $(c_i^{(t)})_{t\geq 0}$ are determined by the obligor's *PD term structure* (see Section 2.3)

$$(p_i^{(t)})_{t\geq 0} = (\mathbb{P}[\text{CWI}_i^{(t)} < c_i^{(t)}])_{t\geq 0} \tag{1.2}$$

in a way such that

$$c_i^{(t)} = \mathbb{F}_{\text{CWI}_i^{(t)}}^{-1}(p_i^{(t)}) \tag{1.3}$$

where $\mathbb{F}_Z$ denotes the distribution function of any random variable Z and $\mathbb{F}_Z^{-1}$ denotes the respective (generalized) quantile function

$$\mathbb{F}_Z^{-1}(z) = \inf\{q \geq 0 \mid \mathbb{F}_Z(q) \geq z\}.$$

The derivation of PD term structures is elaborated in Section 2.3. Note that the before-mentioned term structures and CWIs so far reflect only the one-dimensional flow of the marginal distributions of single-name credit risks, not the joint distributions of CWIs in the sense of a multi-variate distribution. Later, we will catch-up in this point and spend a lot of time with multi-variate CWI vectors reflecting dependencies between single-name credit risks.

1.1.2 Credit Exposure

Despite default probabilities we need to know at least two additional informations for modeling single-name client risk, namely the *exposure at default* (EAD) outstanding with the considered obligor and the *loss given default* (LGD) reflecting the overall default quote taking place if obligors default. One could easily dedicate a separate chapter to each of both risk drivers, but again we have to restrict ourselves to a few remarks roughly indicating the underlying concepts.

Exposure measurement has several aspects. For drawn credit lines without further commitment, exposure equals the outstanding[11] *notional exposure* of the loan, whereas already for *committed undrawn*

---
[11] Taking amortizations into account.

lines exposure starts to involve a random component, in this case depending on the drawing behavior of the client. Another exposure notion appears in the context of counterparty credit risk in derivative products where exposure equals *potential exposure* (PE; typically applied in the context of limit setting, etc.) or *expected positive exposure* (EPE; exposure often used for economic capital calculation, etc.); see, e.g., [79] as well as [13] for an introductory paper[12] on exposure measurement. In this book we will not need PE or EPE techniques and can restrict our exposure notion to EAD; see also the new capital accord [15].

The addendum 'at default' in EAD refers to the fact that for the determination of realized loss we have to take into account all monetary amounts contribution to the realized loss. Hereby, it is important that we think of EAD as a conglomerate of *principal* and *interest streams*; see Figure 1.4 where the interest stream consists of coupon payments and the principal stream is made up by the repayment of capital at the maturity of the bond. In asset-backed transactions as discussed later in this book, we have to deal with EAD at two different levels, namely at the level of the underlying assets (e.g., some reference credit portfolio) as well as at the level of the considered structured security, e.g., some CDO tranche. Actually, the same two-level thinking has to take place for LGDs where *the LGD at structured security level typically depends on the LGDs of the underlying assets, loans, or bonds, respectively*. We will come back to this point later in the book.

Another aspect of EAD, mentioned for reasons of practical relevance in the same way as our excursion on ratings, is the potential uncertainty regarding the actual outstanding exposure at the default time of an asset. In CDOs, the amount of outstandings typically is well defined and controlled by the *offering circular* or the term sheet, respectively, describing the structure of the considered transaction. Nevertheless, one could easily think of situations where uncertainties in the outstanding exposure can occur, e.g., in case of *residential mortgage backed securities* (RMBS) where loans can be prepaid[13] or in cases where the

---

[12]Note that the nomenclature in exposure measurement exhibits some variability in the literature due to a certain lack of standardization of notions although the concepts discussed in different papers are most often identical or at least close to each other.

[13]In RMBS transactions, prepayed loans are typically replaced by other mortgage-backed loans matching the eligibility criteria of the transaction; in such cases, the offering circular defines clear rules how and in which cases replenishments can be

12   *Structured Credit Portfolio Analysis, Baskets & CDOs*

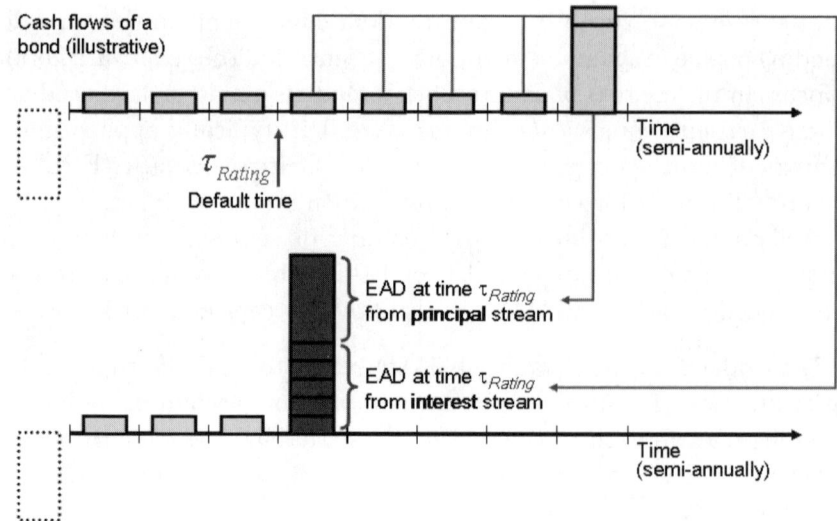

FIGURE 1.4: Exposure as a conglomerate of principal and interest streams

underlying assets are *PIKable* where 'PIK' stands for *payment in kind* addressing more or less the option to exchange current interest payments against principal or capital (e.g., in a bond where the issuer has the option to capitalize interest at any payment date by issuing additional par amount of the underlying security instead of making scheduled interest payments). The most common uncertainty in lending exposures at single loan level is due to unforeseen customer behavior in times of financial distress; see Figure 1.5 where the well-known fact that obligors tend to draw on their committed but so far undrawn lines in times of trouble. The most sophisticated approach to deal with exposure uncertainties is by means of *causal modeling* in a comparable way to causal rating models but focussing on exposure instead of default events, hereby taking the underlying drivers for changes in the outstanding exposure into account and modeling the relationship between these underlying exposure drivers and EAD, e.g., by means of Monte Carlo simulation techniques. As an example, prepayment behavior of obligors in RMBS portfolios is strongly coupled with interest rates such that the interest rate term structure as underlying driver is a reasonable starting point for a causal modeling of prepayments in an RMBS

---

done by the collateral or asset manager in charge of the underlying asset pool.

model. In a comparable way, prepayments in corporate lending can be tackled, taking into account that in corporate lending prepayments are a function of interest rates, borrower's credit quality, and prepayment penalties (depending on the domestic lending market); see, e.g., [68].

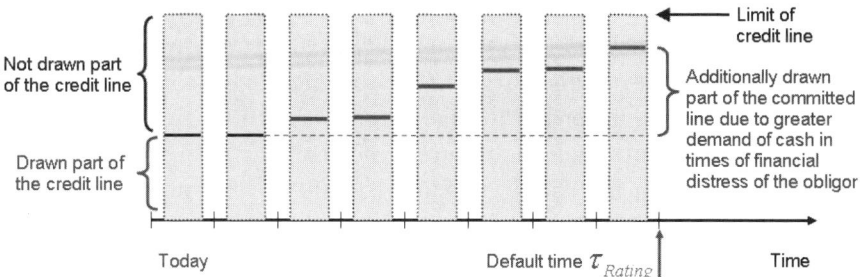

FIGURE 1.5: Exposure uncertainty due to unforeseen customer behavior

To mention another example, the new capital accord deals with exposure uncertainties for off-balance sheet positions by means of so-called *credit conversion factors* in the formula

$$\text{EAD} = \text{OUTSTANDINGS} + \text{CCF} \times \text{COMMITMENT}$$

where CCF denotes the credit conversion[14] factor determined w.r.t. the considered credit product; see [15], §82-89 and §310-316. For instance, in the so-called *Foundation Approach* a CCF of 75% has to be applied to credit commitments regardless of the maturity of the underlying facility, whereas a CCF of 0% applies to facilities, which, e.g., are uncommitted or unconditionally cancellable. In practical applications and in the sequel of this book, we will always consider EAD w.r.t. some time horizon t, written as $\text{EAD}^{(t)}$ (where necessary) addressing the outstanding exposure at time t.

---

[14] In general, credit conversion can address the conversion of non-cash exposures into cash equivalent amounts or the conversion of non-materialized exposures into expected exposures at default (for example, quantifying the potential draw down of committed unutilized credit lines), etc.

1.1.3 Loss Given Default

We now turn our attention to LGD. In general, LGD as a concept is simply explained but far from being trivial regarding modeling and calibration. In the early times of quantitative credit management, many banks defined 'standard LGDs' for broad loan categories, e.g., claiming that corporate loans on average admit, e.g., a 40% recovery quote such that the LGD of such products equals 60%. Today we see an increasing trend toward more sophisticated LGD approaches, to some extent due to the new capital accord [15] where banks approved for the already-mentioned IRB approach in its advanced form can base their capital figures not only on internally estimated PDs but also on their internal LGD estimates. In the sequel, we briefly summarize some basic facts about LGDs.

The most advanced LGD approach is, in the same way as mentioned in the context of PDs and EADs - the *causal modeling* approach again. We will later see that, e.g., the LGD of CDO tranches is a good example for an LGD derived by a causal modeling approach; see Section 3.1. Here, the loss of a CDO tranche is linked to the performance of the underlying asset pool such that in a Monte Carlo simulation one can model the economics of the underlying pool and the causality between the pool and the credit-linked securities at the liability side of the CDO structure. By transforming each asset pool scenario into a scenario at the CDO tranche level one obtains a loss distribution for the CDO tranche from which one can derive the LGD as the mean loss quote of the full distribution of possible *loss severities*.

At single loan level, a good compromise between sophistication and practicability is the following approach. Let us assume that the bank built up a collateral database containing comprehensive historic experience regarding the achieved value in selling collateral for loss mitigation in the context of defaulted credit-risky instruments. For example, such a database will contain information about the average market value decline of residential mortgages, different types of commercial mortgages, single stocks, bonds, structured products, cars for leasings, aircraft vessels for aircraft finance, ships for ship financing, etc. The database will also contain information about recovery quotes on unsecured exposures w.r.t. different seniorities, and so on. Based on such a database, *value quotes* (VQ) w.r.t. different types of collateral can be derived where the VQ incorporates the *expected market value decline* of the considered collateral category. In addition, the *collateral value volatility* can be derived, useful for the calibration of stochastic approaches to LGD modeling; see Section 3.3. Given such a database exists, AIRB[15] banks need to have such databases for regulatory approval of their internally calibrated LGDs, we can proceed as in the following example.

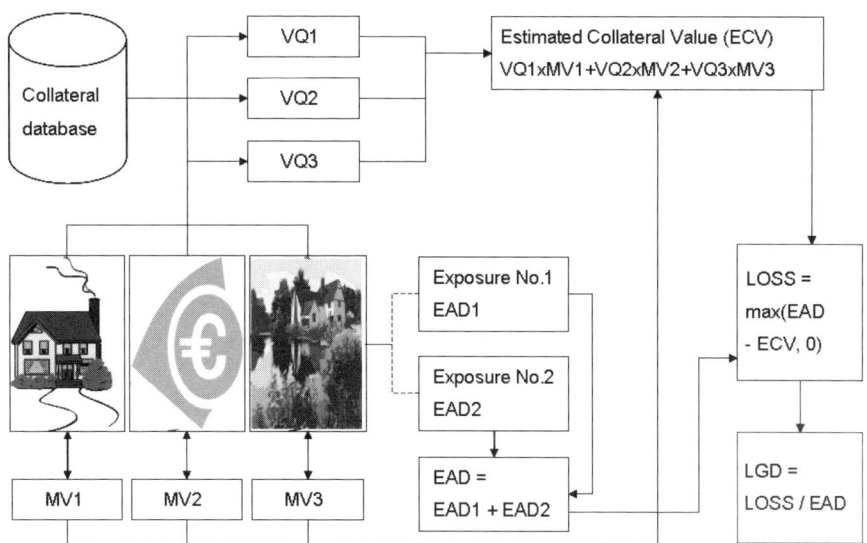

FIGURE 1.6: Illustration of LGD determination

Let us assume that we want to determine the LGD in a situation where some client has two loans secured by three types of collateral; see Figure 1.6. Out of the bank's collateral database we obtain value quotes VQ1, VQ2, VQ3 (in percentage) for the collateral values. Given the market value of the collaterals is given by MV1, MV2, and MV3, the overall, on average achievable, *estimated collateral value* (ECV) for loss mitigation is given by

$$\text{ECV} = \text{VQ1} \times \text{MV1} + \text{VQ2} \times \text{MV2} + \text{VQ3} \times \text{MV3}.$$

Note that in most of the cases, collateral selling to the market takes place quite some time after the actual default time of the obligor such

---
[15]Banks in the so-called *advanced internal ratings based* (AIRB) approach; see [15].

that an appropriate *discounting* of collateral values reflecting the time value of money has to be incorporated into the value quotes. Denoting the exposure of the two loans approved for the client by EAD1 and EAD2, the total exposure allocated to the client equals

$$EAD = EAD1 + EAD2.$$

Then, the *realized loss*, here denoted by LOSS, is given by

$$LOSS = \max\bigl[(1 - VQ_{unsecured}) \times (EAD - ECV), 0\bigr]$$

where $VQ_{unsecured}$ equals the percentage average recovery quote on unsecured exposures, also to be calibrated based on the collateral/loss database of the bank. Note that in Figure 1.6 we assume (for reasons of simplicity) that we lose 100% of the unsecured exposure in case of default, i.e., $VQ_{unsecured} = 0$. Because the value quotes already incorporate the potential market value decline of the considered collateral,[16] LOSS expresses an expectation regarding the realized loss amount after taking into account all available collateral securities for loss mitigation as well as the typical recovery rate on unsecured exposures. Then, the *severity of loss* in case the obligor defaults can be expressed as a percentage loss quote, namely the LGD, by

$$LGD = \frac{LOSS}{EAD}. \tag{1.4}$$

Because LOSS represents the realized loss amount in units of money, Equation (1.4) is 'compatible' with the well-known formula for the *expected loss* (EL) on single-names

$$EL[\$] = PD \times EAD \times LGD. \tag{1.5}$$

Note that the EL can be written in the simple form of Equation (1.5) only if the Bernoulli variable indicating default of the obligor and the two quantities EAD and LGD, e.g., considered as realizations of corresponding random variables due to certain uncertainties inherent in these quantities, are stochastically independent. If EAD and LGD are used as fixed values, they can be considered as expectations of corresponding random variables. Also note that, in general, in the same

---

[16] Equivalently to the chosen approach one could have defined LGDs for collateral securities and an LGD for the overall exposure net of recoveries and then aggregated both figures into an estimate for the 'realized loss' accordingly.

way as in the case of EAD, we have an additional degree of complexity in the time dimension. Because Definition 1.4 involves exposures, and exposures (EADs) are time dependent, we will write $LGD^{(t)}$ in the sequel where necessary and meaningful, in line with our time-dependent notion of exposures, $EAD^{(t)}$.

Under all circumstances it is essential that *LGD modeling clearly reflects the recovery and workout process of the bank*. Deviations from the bank-internal workout practice will lead to distortions in the LGD calibration and will cause unwanted long-term deviations of provisions or realized losses, respectively, from forecasts based on the expected loss of the bank's portfolio. Advanced IRB banks have to make sure that PDs and LGDs can be backtested and validated in order to achieve regulatory approval of their internal estimates.

1.2 Modeling Portfolio Credit Risk

Turning from single-name to portfolio credit risk is a challenging step in credit risk modeling. For the topics covered in this book, portfolio credit risk is not only one of several issues but rather the fundamental basis for modeling structured credit products. This is confirmed by the notion *correlation products* sometimes used as a headline for default baskets and CDOs, addressing the fact that these products *trade correlations* and interdependencies between single-name credit risks in a tailor-made way leading to interesting (portfolio-referenced) risk selling and buying opportunities.

1.2.1 Systematic and Idiosyncratic Credit Risk

The need for a sound modeling of correlations and 'tail dependencies' in the context of portfolio loss distributions is the main reason why for structured credit products we prefer latent variable models indicated in Equation (1.1) enabling an *explicit* modeling of correlations in contrast to other well-known models where dependencies between single-name credit risks are modeled *implicitly*, for example, by means of default rate volatilities in systematic sectors; see CreditRisk[+] [35] and Chapter 4 in [25]. According to Equation (1.1), every obligor i is represented by a default indicator

$$L_i^{(t)} = \mathbf{1}_{\{\text{CWI}_i^{(t)} < c_i^{(t)}\}}$$

such that the obligor defaults if and only if its CWI falls below its default point, always measured w.r.t. to some time interval $[0, t]$.

Now, the starting point of dependence modeling is a decomposition of the client's risk into a *systematic* and an *idiosyncratic* risk component; see Figure 1.7.

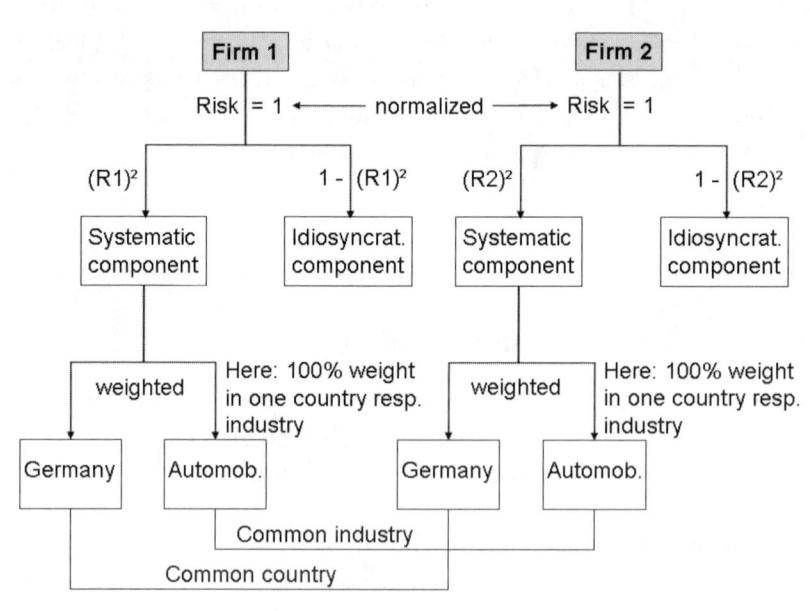

FIGURE 1.7: Decomposition of two firm's credit risk into systematic and idiosyncratic parts (separation approach; illustrative)

In Figure 1.7, the credit risk of two German automotive manufacturers is decomposed into systematic and idiosyncratic components, here with a 100% country weight in a factor representing Germany and a 100% industry weight in a factor representing the automobile industry. The basic assumption in such models is that *dependence takes place exclusively via the systematic components* of counterparties. After normalizing the overall firm risk (say, the volatility of the CWI) to 1, the so-called R^2 (R-squared) of the firm captures the *systematic risk*, whereas the quantity $1-R^2$ quantifies the *residual* or *idiosyncratic risk* of the firm, which cannot be explained by means of systematic risk

From Single Credit Risks to Credit Portfolios

drivers like countries and industries; see [25], Chapter 1 for a detailed discussion about systematic and idiosyncratic credit risk.

In more mathematical terms, Figure 1.7 suggests to decompose the CWI of obligor i in Equation (1.1) into systematic and idiosyncratic components,

$$\text{CWI}_i^{(t)} = \beta_i \Phi_i^{(t)} + \varepsilon_i^{(t)} \qquad (t \geq 0) \qquad (1.6)$$

where we assume that the residuals $(\varepsilon_1^{(t)})_{t\geq 0}, ..., (\varepsilon_m^{(t)})_{t\geq 0}$ are independent and for every fixed t identically distributed as well as independent of the systematic variables $\Phi_i = (\Phi_i^{(t)})_{t\geq 0}$. Here, m denotes the number of obligors in the portfolio and the index Φ_i is called the *composite factor* of obligor i because it typically can be represented by a weighted sum of indices,

$$\Phi_i^{(t)} = \sum_{n=1}^{N} w_{i,n} \Psi_n^{(t)} \qquad (i = 1, ..., m; \ t \geq 0), \qquad (1.7)$$

with positive weights $w_{i,n}$. Regarding the indices

$$\Psi_1 = (\Psi_1^{(t)})_{t\geq 0}, \ ..., \ \Psi_N = (\Psi_N^{(t)})_{t\geq 0}$$

we find two major approaches in the market and 'best practice' industry models, respectively.

- **The separation approach:** Here, $\Psi_1, ..., \Psi_{N_C}$ denote country indices and $\Psi_{N_C+1}, ..., \Psi_N$ are industry indices. The number of industry factors is then given by $N_I = N - N_C$, whereas N_C denotes the number of country factors. An industry example for such a model is the so-called *Global Correlation Model* by MOODYS KMV.[17]

- **The combined approach:** Here, every index Ψ_n refers to an industry within a country. *CreditMetrics* by the RISKMETRICS GROUP (RMG) follows such an approach[18] via MSCI indices.[19]

There are pros and cons for both approaches and it seems to us that both *factor models* are well accepted in the market. There are other differences in the way Moodys KMV and RMG model correlations, e.g.,

---

[17] See www.kmv.com
[18] See www.riskmetrics.com
[19] Morgan Stanley Capital International Inc.

KMV derives *asset value correlations* from a non-disclosed option-theoretic model transforming equity processes and related information into model-based asset value processes, whereas RMG works with *equity correlations* as a proxy for asset value correlations. For more information on factor modeling we refer to Chapter 1 in [25].

In this context, it is worthwhile to mention that correlations are subject to certain controversial discussions. One can find research papers in the market where people suddenly seem to discover that correlations are negligible. However, we are convinced from our practical experience that correlations are not only inherent/omnipresent in the credit market but play a fundamental role in the structuring and design of default baskets and CDOs. If correlations would be negligible, these kind of *correlation products* would not be as successful and growing as they are in the credit market. Moreover, in Section 3.4.4 we discuss implied correlations in index tranches. Here, investment banks actually quote correlations in the same way as they quote spreads and deltas. This provides strong support for the importance of correlations and dependencies in the structured credit market. For further study, we recommend to interested readers two very valuable discussion papers on the correlation topic. The first paper is from MOODY'S KMV [116] and the second paper is from FITCH RATINGS [6]. In these papers, the reader finds strong evidence that correlations are far from being negligible. Moreover, correlations can be measured within systematic sectors like industries (intra-industry) but also between different systematic sectors (e.g., inter-industry). The type of correlation considered in the two papers is the *asset correlation* referring to the correlation between underlying latent variables, in our case the CWIs of clients. We come back to this issue in Chapter 2 where we introduce correlations in a more formal way, distinguishing between *asset* or CWI *correlations* and the *default correlation*, which refers to the correlation of Bernoulli variables as in (1.1) indicating default. As we will see later, default correlations typically live at a smaller order of magnitude than CWI or asset correlations; see (2.5) and thereafter.

1.2.2 Loss Distribution of Credit Portfolios

In the early times of credit risk modeling, the 'end product' of credit risk models typically was the *portfolio loss distribution* w.r.t. some fixed time horizon. Today, and especially in structured finance, people are

From Single Credit Risks to Credit Portfolios 21

often more interested in looking at *dependent default times*. We will come back to this issue in Section 2.5. Obviously, the fixed time horizon point of view is a 'side product' of default time models just by restricting the default time τ of some obligor to a fixed time interval $[0, T]$ and considering the Bernoulli variable $\mathbf{1}_{\{\tau<T\}}$. For reasons of completeness, we illustrate in Figure 1.8 the loss distribution of a credit portfolio.

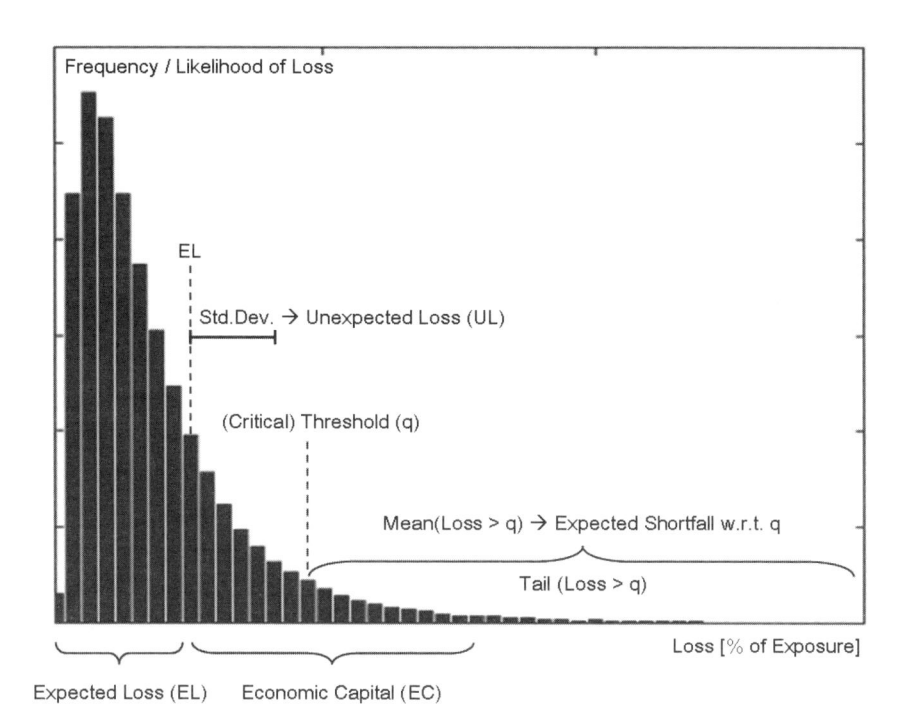

FIGURE 1.8: The loss distribution of a credit portfolio

Based on the loss distribution of the credit portfolio, all relevant risk quantities can be identified as 'summary statistics' of this distribution. For instance, the *expected loss* (EL) of the portfolio[20] (in percentage)

---

[20]Recall Equation (1.5) for the EL of single-name credit risk in case of independence of the default indicator, exposure, and severity of loss.

equals the sum of single-name ELs weighted by the total portfolio EAD,

$$\mathrm{EL}^{(t)}[\%] = \frac{1}{\sum_{j=1}^{m} \mathrm{EAD}_j^{(t)}} \sum_{i=1}^{m} \mathrm{EL}_i^{(t)} \qquad (1.8)$$

$$= \frac{1}{\sum_{j=1}^{m} \mathrm{EAD}_j^{(t)}} \sum_{i=1}^{m} \mathbb{E}[L_i^{(t)} \times \mathrm{LGD}_i^{(t)}] \times \mathrm{EAD}_i^{(t)}$$

where $L_i^{(t)}$ is defined in (1.1). Note that Equation (1.8) allows for *stochastic* (non-deterministic) LGDs but assumes EAD to be a deterministic (non-random) fixed quantity.

The *economic capital* (EC) of a credit portfolio typically is defined as the quantile of the loss distribution w.r.t. to some given target level of confidence, e.g., 99.9%, minus the EL of the portfolio, which is supposed to be fully priced-in by the front office of the bank,

$$\mathrm{EC}_\alpha^{(t)}[\%] = \frac{q_\alpha\left[\sum_{i=1}^{m} L_i^{(t)} \times \mathrm{EAD}_i^{(t)} \times \mathrm{LGD}_i^{(t)}\right]}{\sum_{j=1}^{m} \mathrm{EAD}_j^{(t)}} - \mathrm{EL}^{(t)}[\%] \qquad (1.9)$$

where $q_\alpha[X]$ denotes the α-quantile of some random variable X. Note that the quantile function $q_\alpha[X]$ in Equation (1.9) is a much more 'tricky' object than the expectation $\mathbb{E}[\cdot]$ arising in Equation (1.8). For instance, expectations are linear functions in their arguments whereas quantile functions are highly non-linear in general.

For capital allocation purposes better suitable than quantile-based risk measures are so-called *shortfall measures*; see Chapter 5 in [25] and the literature mentioned in the last chapter of this book for an explanation[21] why shortfall measures are superior to quantile-based measures. Figure 1.8 illustrates a typical shortfall measure. Let us say the senior management considers a certain loss threshold q as 'critical' for the economic future of the bank. Such a threshold typically will be much lower than the α-quantile of the loss distribution. Then, it makes sense to calculate the expectation of all losses exceeding q, weighted via the loss likelihoods illustrated by the loss distribution. Such a *conditional expectation* is called the *expected shortfall (ESF)* w.r.t. q of the credit portfolio. It is the mean loss or expected loss conditional

---
[21]Shortfall measures are the most prominent examples of so-called *coherent* risk measures.

on losses exceeding the threshold[22] q where q typically is chosen as a quantile of the portfolio's loss distribution,

$$\text{ESF}_q^{(t)}[\%] = \frac{1}{\sum_{j=1}^m \text{EAD}_j^{(t)}} \times \qquad (1.10)$$

$$\times \mathbb{E}\Big[\sum_{i=1}^m L_i^{(t)} \times \text{EAD}_i^{(t)} \times \text{LGD}_i^{(t)} \mid \sum_{i=1}^m L_i^{(t)} \times \text{EAD}_i^{(t)} \times \text{LGD}_i^{(t)} > q\Big].$$

Expected shortfall induces 'canonical' risk contributions at counterparty level just by application of the *linearity of conditional expectations*; see Chapter 5 in [25]. Banks can match the 'classical' quantile-based overall *capital determination* with expected shortfall-based *capital allocation* by choosing the critical threshold q in a way such that

$$\text{ESF}_q^{(t)}[\%] - \text{EL}^{(t)}[\%] \stackrel{!}{=} \text{EC}_\alpha^{(t)}[\%]. \qquad (1.11)$$

Then, capital allocation can be done in a coherent way for a quantile-based total EC w.r.t. the bank's preferred level of confidence α.

The last risk quantity derivable from the loss distribution in Figure 1.8 is the standard deviation of the loss distribution, often called the *unexpected loss* (UL) of the credit portfolio,

$$\text{UL}^{(t)}[\%] = \frac{\sqrt{\mathbb{V}\Big[\sum_{i=1}^m L_i^{(t)} \times \text{EAD}_i^{(t)} \times \text{LGD}_i^{(t)}\Big]}}{\sum_{j=1}^m \text{EAD}_j^{(t)}}. \qquad (1.12)$$

The unexpected loss provides some information about the deviation of losses from the EL, but based on the skewness of credit portfolio loss distributions *tail risk* measures like EC and ESF are more important than the UL of the credit portfolio. Nevertheless, it is a risk quantity useful to know. Especially in the early times of quantitative credit risk management, many banks applied UL contributions for economic capital allocation by means of the so-called *var/covar*-approach comparable to market risk methodologies; see any book on market risk, e.g., [63], to find an introduction to the var/covar approach.

---

[22]In Equation (1.10), q is stated in units of money. Also common is to write all quantities in percentage of the portfolio EAD. Equations have then to be adapted accordingly.

So far, we kept the presentation fairly generic without making explicit assumptions regarding the distribution of involved random variables. In the subsequent sections, we will be more explicit in tail risk modeling mainly involving four different so-called *copula* functions, namely the *Gaussian, Student-t, Clayton,* and *comonotonic copulas*. We will see that copula functions as *dependence modeling tools* have a major impact on portfolio-based risk measures like EC and ESF. We will also see how copula functions influence the distribution of joint default times in credit portfolios. The consequences for the modeling of structured credit products will be discussed in great detail later on.

1.2.3 Practicability Versus Accuracy

We close our brief walk through credit risk modeling aspects with Figure 1.9 illustrating the different trade-offs credit modelers are exposed to when balancing between *practicability* and *accuracy* of their models, parameterizations and estimates.

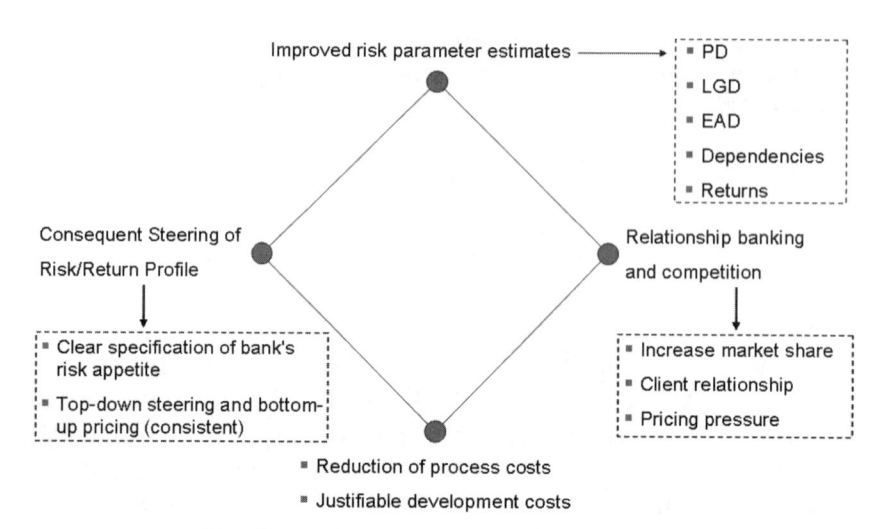

FIGURE 1.9: Finding a balance between practicability and accuracy in credit risk evaluation

In vertical direction, Figure 1.9 shows at the top corner the wish for steadily improved risk parameter estimates. For instance, in the previous section we discussed how positive an increase of the discriminatory power of a rating system can impact the bank's P&L. The more accurate the rating systems and PD estimates of the bank the more reliable are credit decisions. In opposite vertical direction we find drivers of efficiency in the bank's valuation and credit approval processes. For instance, reducing the number of qualitative factors in a rating system can significantly accelerate credit processes. Banks have to balance between efficiency and accuracy/sophistication in credit risk estimates, at single-name as well as at portfolio level.

In horizontal direction, Figure 1.9 shows some other diametral forces in banking. At the left-hand side we find the clear demand for consequent risk/return steering. As soon as a bank has recognized the potential of sound credit risk models and steering measures these intiatives should be applied to the day-to-day business of the bank. The classical 'relationship banking paradigm' points in the opposite direction. In times of over-banked loan markets, banks will think about it twice before they shock a long-term well-known client by an increased risk premium as credit price component due to an increased exposure concentration with this client. The way out of the dilemma of well-justified diametral demands is *active credit portfolio management* (ACPM). Creating tailor-made credit products for important clients and offering attractive conditions due to an active management of credit risks is a key factor of success in today's credit business.

Chapter 2

Default Baskets

A *default basket* is a portfolio of not too many obligors, e.g., not more than 10, although there is no 'officially agreed hard limit' regarding the number of obligors allowed in order to speak of a basket. At least it is clear that a basket contains more than one obligor or asset, respectively, such that in case of moderate or low correlation between the intruments in the basket *diversification effects* will reduce the overall portfolio or basket risk. In this chapter, we model default baskets and certain related products. This chapter also serves as a preparation for the slightly more complicated Chapter 3 on CDOs. As we will see later, default baskets and CDOs are 'close relatives'.

2.1 Introductory Example: Duo Baskets

We start our discussion by the most simple basket one can think of, namely, a *duo basket* consisting of two credit-risky instruments only. The purpose of the following introductory discussion is twofold. First, we want to review the idea of *diversification* and its interplay with correlation. Second, we want to elaborate a first example illustrating the necessary balance between *risk and return*, a fundamental way of thinking in portfolio management and especially in the field of structured credit products.

Let us assume that we consider the following two loans:

- Loan A has a one-year default probability of $p_A = 100$ bps, loan B has a one-year default probability of $p_B = 50$ bps.

- Both loans have a bullet-type exposure profile, where 'bullet' refers to a non-amortizing loan with 100% of the exposure outstanding from the first to the last day of the term of the loan.

28   *Structured Credit Portfolio Analysis, Baskets & CDOs*

- Both loans have an LGD of 100% such that in case of default the full outstanding exposure amount will be lost for the lending bank.

Let us restrict ourselves to the one-year horizon for the time being. According to our threshold model (see Equation (1.1)) explained in Chapter 1 we assume the existence of CWIs for obligors A and B and corresponding Bernoulli variables

$$L_A^{(1)} = \mathbf{1}_{\{CWI_A^{(1)} < c_A^{(1)}\}} \quad \text{and} \quad L_B^{(1)} = \mathbf{1}_{\{CWI_B^{(1)} < c_B^{(1)}\}} \tag{2.1}$$

indicating default or survival within one year. In line with Equation (1.3), the default points of the obligors are determined by

$$c_A^{(1)} = \mathbb{F}^{-1}_{CWI_A^{(1)}}(p_A^{(1)}) \quad \text{and} \quad c_B^{(1)} = \mathbb{F}^{-1}_{CWI_B^{(1)}}(p_B^{(1)}). \tag{2.2}$$

In order to come up with explicit numbers, let us assume that the CWIs of obligors A and B are standard normally distributed with a CWI correlation of 10%. If we want to embed the CWI correlation of two assets into a factor model as outlined in Section 1.2 then we can introduce a standard normally distributed random variable $Y \sim N(0,1)$ and two independent variables $\varepsilon_A, \varepsilon_B \sim N(0,1)$, independent of Y, and write (dropping the time index for a moment for reasons of a simplified notation)

$$CWI_A = \sqrt{\varrho}\, Y + \sqrt{1-\varrho}\, \varepsilon_A \tag{2.3}$$

$$CWI_B = \sqrt{\varrho}\, Y + \sqrt{1-\varrho}\, \varepsilon_B$$

such that $CWI_A, CWI_B \in N(0,1)$, and

$$\text{Corr}[CWI_A, CWI_B] = \varrho_{A,B} = \varrho$$

where $\text{Corr}[\cdot,\cdot]$ denotes correlation. A CWI correlation of 10% determines $\varrho = 0.1$ for our working example. The default points of A and B can then easily be calculated by application of the standard normal quantile function $N^{-1}[\cdot]$,

$$c_A = N^{-1}[p_A] = -2.33 \quad \text{and} \quad c_B = N^{-1}[p_B] = -2.58$$

confirming the natural expectation that the obligor with lower PD should be more bankruptcy remote; see also Appendix 6.6.

Default Baskets

Now we are ready for studying the interplay of the two loans in our duo basket regarding default behavior. Let's say the bank wants to allocate a certain amount of money to the duo basket, namely w percent of the available amount to obligor A and $(1-w)$ percent to obligor B. The question comes up which breakdown of capital by means of *exposure weights* $w_A = w$ and $w_B = (1-w)$ will lead to a maximum benefit for the lending institute. This question is the classical starting point for balancing between *risk and return* of an investment. Regarding returns we have to make a reasonable working assumption[1]. For the moment let us assume that the gross margin the bank earns with each loan is given by the formulas

$$\text{Margin}_A = w \times (p_A + 0.2 \times p_A(1-p_A) + 0.005)$$

$$\text{Margin}_B = (1-w) \times (p_B + 0.2 \times p_B(1-p_B) + 0.005).$$

At the one-year horizon, p_A and p_B are the percentage ELs and $p_A(1-p_A)$ and $p_B(1-p_B)$ are the percentage default variances of loans A and B, assuming an LGD of 100%. Therefore, we assume that loans earn a gross margin incorporating a 100% charge on the expected loss plus a 20% charge on the variance of losses[2] plus a uniform margin of 50 bps. For $w=1$ we obtain the gross margin of loan A

$$\text{Margin}_A = p_A + 0.2 \times p_A(1-p_A) + 0.005 = 0.017$$

and for $w=0$ we obtain the gross margin for loan B

$$\text{Margin}_B = p_B + 0.2 \times p_B(1-p_B) + 0.005 = 0.011.$$

Due to the low default probabilities of the assets, the gross margin formula approximately yields the EL scaled by 1.2 plus an absolute offset of 50 bps. We can now calculate the portfolio profit μ as a weighted sum of margins,

$$\begin{aligned}\mu = \mu(w) &= \text{Margin}_A + \text{Margin}_B \\ &= w \times (p_A + 0.2 \times p_A(1-p_A) + 0.005) \\ &+ (1-w) \times (p_B + 0.2 \times p_B(1-p_B) + 0.005).\end{aligned}$$

---

[1] In the credit market, margins or spreads are not only determined by the magnitude of credit risk inherent in the considered credit instrument but also by its liquidity complemented, in case of structured investments, by some 'complexity premium'.

[2] In a portfolio context, one would rather price-in the economic capital cost based on the EC-contribution of the loan. Comparable approaches are used in RAROC (risk-adjusted return over capital) and transfer pricing concepts.

For measuring diversification effects in our duo basket we have to choose a risk measure sensitive to portfolio effects. The three standard measures matching this condition are EC, ESF, and UL where UL stands for *unexpected loss* as a synonym for the standard deviation of the portfolio's loss distribution. For a two-name portfolio, EC and ESF are not the most useful measures so that in this particular case we decide in favor of the portfolio UL (here denotes by σ) given by the square root of the portfolio variance σ^2 w.r.t. the chosen weight w,

$$\sigma^2 = \sigma^2(w) = w^2 p_A(1-p_A) + (1-w)^2 p_B(1-p_B) \qquad (2.4)$$
$$+ 2\,r\,w(1-w)\sqrt{p_A(1-p_A)p_B(1-p_B)}$$

where $r = r_{A,B}$ denotes the *default correlation* of the two obligors,

$$r = r_{A,B} = \mathrm{Corr}\big[\mathbf{1}_{\{\mathrm{CWI}_A^{(1)} < c_A^{(1)}\}}, \mathbf{1}_{\{\mathrm{CWI}_B^{(1)} < c_B^{(1)}\}}\big] \qquad (2.5)$$
$$= \frac{N_2[N^{-1}[p_A], N^{-1}[p_B]; \varrho_{A,B}] - p_A p_B}{\sqrt{p_A(1-p_A)p_B(1-p_B)}}$$

based on the normal distribution assumption we made for the obligor's CWIs; see also Equation (2.3). In Equation (2.5), $N_2[\,\cdot\,,\cdot\,;\varrho]$ denotes the standard bi-variate normal distribution function with a correlation of ϱ. For $\varrho = 10\%$, $p_A = 0.01$ and $p_B = 0.005$ we obtain a default correlation of 74 bps. The order of magnitude of the default correlation compared to the CWI correlation is not unusual. Note that in typical credit model approaches, *default correlations are much lower than the correlation between underlying latent variables.*

Figure 2.1 illustrates the dependence of the default correlation $r = r_{A,B}$ on the PDs p_A and p_B and on the CWI correlation $\varrho = \varrho_{A,B}$. Table 2.1 shows the corresponding numbers.

For every $w \in [0,1]$, we obtain a duo basket with a mix of a w-portion of risk/return due to loan A and a $(1-w)$-portion of risk/return arising from loan B. The portfolio's risk/return can graphically be illustrated by plotting the curve of points

$$\big(\sigma(w), \mu(w)\big)_{0 \leq w \leq 1}$$

in 2-dimensional risk/return space; see Figure 2.2.

The case $w = 0$ leads to a duo basket consisting of loan B only, the case $w = 1$ is a basket with loan A only. Obviously, these cases are

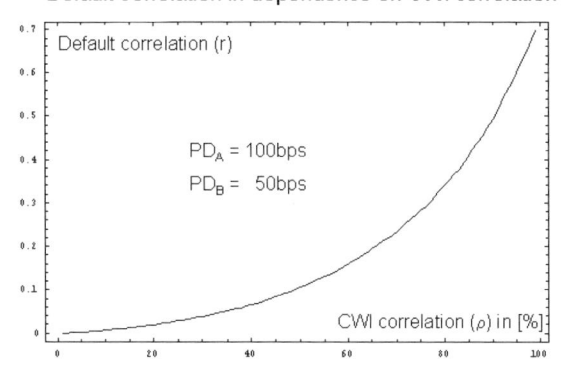

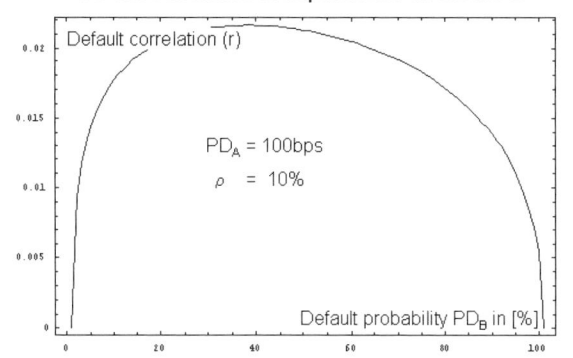

FIGURE 2.1: Default correlation for different levels of CWI correlation and pairs of PDs

not interesting because one can hardly call a single asset a basket or portfolio. For $0 < w < 1$, both assets are contributing to the duo basket's performance. With w increasing from 0 to 1 we move from the B-only portfolio to the A-only portfolio along the curve plotted in Figure 2.2. An interesting point on this curve is the duo basket where the portfolio UL attains its minimum. In line with classical portfolio theory we call this portfolio the *minimum variance portfolio*.

Calculating the derivative of the portfolio variance (i.e., UL^2) w.r.t. the weight w based on Equation (2.4),

$$\frac{\partial}{\partial w}\sigma^2(w) = 2\,w\,p_A(1-p_A) - 2\,(1-w)\,p_B(1-p_B) +$$

TABLE 2.1: Default correlation as a function of CWI correlation and default probabilities

PD_A	PD_B	CWI corr.	Def. corr	PD_A	PD_B	CWI corr.	Def. corr
1.00%	0.50%	1%	0.06%	1.00%	0.10%	10%	0.41%
1.00%	0.50%	2%	0.12%	1.00%	0.20%	10%	0.53%
1.00%	0.50%	3%	0.18%	1.00%	0.30%	10%	0.61%
1.00%	0.50%	4%	0.25%	1.00%	0.40%	10%	0.68%
1.00%	0.50%	5%	0.32%	1.00%	0.50%	10%	0.74%
1.00%	0.50%	6%	0.39%	1.00%	1.00%	10%	0.93%
1.00%	0.50%	7%	0.47%	1.00%	2.00%	10%	1.17%
1.00%	0.50%	8%	0.56%	1.00%	3.00%	10%	1.32%
1.00%	0.50%	9%	0.64%	1.00%	4.00%	10%	1.43%
1.00%	0.50%	10%	0.74%	1.00%	5.00%	10%	1.52%
1.00%	0.50%	12%	0.94%	1.00%	6.00%	10%	1.60%
1.00%	0.50%	15%	1.27%	1.00%	7.00%	10%	1.66%
1.00%	0.50%	20%	1.96%	1.00%	8.00%	10%	1.72%
1.00%	0.50%	25%	2.80%	1.00%	9.00%	10%	1.76%
1.00%	0.50%	30%	3.84%	1.00%	10.00%	10%	1.81%
1.00%	0.50%	35%	5.10%	1.00%	12.00%	10%	1.88%
1.00%	0.50%	40%	6.61%	1.00%	15.00%	10%	1.96%
1.00%	0.50%	45%	8.41%	1.00%	20.00%	10%	2.05%
1.00%	0.50%	50%	10.54%	1.00%	25.00%	10%	2.11%
1.00%	0.50%	55%	13.05%	1.00%	30.00%	10%	2.14%
1.00%	0.50%	60%	16.00%	1.00%	50.00%	10%	2.12%

$$+ 2\,r\,(1-2w)\sqrt{p_A(1-p_A)p_B(1-p_B)},$$

we obtain $(\partial/\partial w)\sigma^2(w) = 0$ for the *minimum variance weight*

$$w_{\min} = \frac{p_B(1-p_B) - r\sqrt{p_A(1-p_A)p_B(1-p_B)}}{p_A(1-p_A) + p_B(1-p_B) - 2\,r\sqrt{p_A(1-p_A)p_B(1-p_B)}} \quad (2.6)$$

yielding $w_{\min} = 0.33$ if we insert p_A, p_B, and r into Equation (2.6).

In Figure 2.2, one can see that the minimum variance portfolio corresponding to $w = 0.33$ offers *more return at less risk* (if risk is identified with volatility) than the single asset B. The benefit of $w = 0.33$ (basket) compared to $w = 0$ (single asset B) is due to the *diversification* effect arising from investing in two instead of in one single asset only. We also see that the diversification potential depends on the correlation of the assets in the basket. In Figure 2.2, we draw a return level (horizontal line) at $\mu = 150$ bps. The lower the correlation of the two obligors the lower the risk (in terms of UL) of the 2-asset portfolio and the higher the diversification benefit. If we consider the basket exclusively from a portfolio-UL perspective, then the perfect correlation case ($\varrho = 1$) is the worst situation one can obtain because the risk corresponding to a return of 150 bps attains its maximum. Increasing

Default Baskets 33

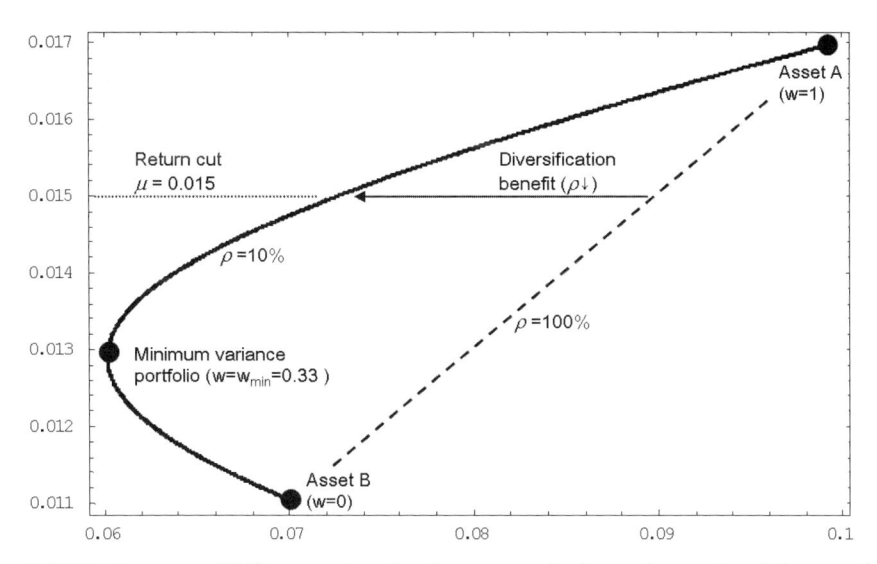

FIGURE 2.2: Different duo baskets consisting of w-mixed loans A and B in risk/return space

the weight w from 0 to 1 yields the dashed ($\varrho = 1$) or solid ($\varrho = 0.1$) line or curve, respectively, in Figure 2.2 starting in asset B ($w = 0$) and ending in asset A ($w = 1$). In the perfect correlation case, we have no diversification benefit at all in the basket. Due to the same argument, the best possible diversification benefit can be achieved if we combine independent[3] assets in a basket.

Figure 2.2 is well-known in portfolio theory and everyone involved in risk management has seen comparable pictures in the literature. Nevertheless, the duo basket case study will help us in the sequel to explore techniques for basket analysis in an easy 'test environment' before we turn our attention to more complicated portfolios.

---

[3]Note that based on our model setup we always assume non-negative CWI correlations. We do not allow for negative correlations leading to offsets in terms of hedging.

2.2 First- and Second-to-Default Modeling

Based on our duo basket example from Section 2.1, we can make some first steps in typical basket products. In Chapter 3, we will see that so-called *equity investors* in a CDO take the *first loss* of some reference portfolio consisting of credit-risky instruments. In the same way, the *first-to-default* buyer/taker in a basket structure bears the loss arising from the first default in the basket. Analogously, the *second-to-default* refers to the loss caused by the second default in the basket. Later, we will investigate first- and second-to-defaults regarding their dependence on the chosen time horizon, hereby relying on *dependent default times*. For the moment, we restrict ourselves to the one-year horizon in order to keep the exposition simple and reduced to certain aspects of these products. As a consequence, we suppress the time index for CWIs.

For the duo basket, the question arises what first- and second-to-default means in terms of event probabilities and corresponding losses in the context of our example. In the sequel, we systematically investigate this question. First of all, we define some notation by writing

$$p_{1st}^{(1)} = \mathbb{P}\big[\mathbf{1}_{\{CWI_A < c_A\}} + \mathbf{1}_{\{CWI_B < c_B\}} > 0\big] \qquad (2.7)$$

for the one-year first-to-default probability and

$$p_{2nd}^{(1)} = \mathbb{P}\big[\mathbf{1}_{\{CWI_A < c_A\}} \times \mathbf{1}_{\{CWI_B < c_B\}} > 0\big] \qquad (2.8)$$

for the one-year second-to-default probability. Note that $p_{1st}^{(1)}$ is the *hitting probability* of the basket, whereas $p_{2nd}^{(1)}$ is the *wipe-out probability* of the basket. Equations (2.7) and (2.8) can also be written in terms of sets (i.e., 'events' is the language of probability theory) as

$$p_{1st}^{(1)} = \mathbb{P}\big[\{CWI_A < c_A\} \cup \{CWI_B < c_B\}\big] \qquad (2.9)$$

$$p_{2nd}^{(1)} = \mathbb{P}\big[\{CWI_A < c_A\} \cap \{CWI_B < c_B\}\big] \qquad (2.10)$$

Let us now calculate $p_{1st}^{(1)}$ and $p_{2nd}^{(1)}$ for our duo basket example.

2.2.1 Proposition *The one-year first-to-default probability equals*

$$p_{1st}^{(1)} = \int_{-\infty}^{\infty} \big(g_{p_A,\varrho}(y) + g_{p_B,\varrho}(y)[1 - g_{p_A,\varrho}(y)]\big)\, dN(y) \qquad (2.11)$$

where the conditional one-year PD of A (analogously for B) is given by

$$g_{p_A,\varrho}(y) = N\left[\frac{c_A - \sqrt{\varrho}\,y}{\sqrt{1-\varrho}}\right]. \tag{2.12}$$

Here, $N[\cdot]$ denotes the standard normal distribution function and

$$c_A = N^{-1}[p_A]$$

represents the default point or default threshold of asset A.

Proof. First, let us recall the derivation of the conditional PD $g_{p_A,\varrho}(y)$,

$$\begin{aligned}
g_{p_A,\varrho}(y) &= \mathbb{P}[\mathbf{1}_{\{\mathrm{CWI}_A < c_A\}} = 1 \mid Y = y] \\
&= \mathbb{P}[\sqrt{\varrho}\,Y + \sqrt{1-\varrho}\,\varepsilon_A < c_A \mid Y = y] \\
&= \mathbb{P}\left[\varepsilon_A < \frac{c_A - \sqrt{\varrho}\,Y}{\sqrt{1-\varrho}} \,\bigg|\, Y = y\right] \\
&= N\left[\frac{c_A - \sqrt{\varrho}\,y}{\sqrt{1-\varrho}}\right]
\end{aligned}$$

(see also Appendices 6.6 and 6.7). Then, Equation (2.11) can be explained as follows. The first summand is the likelihood that A defaults, the second summand is the likelihood that B defaults and A survives. In set-theoretic notation,[4] the formula for $p_{1\mathrm{st}}^{(1)}$ reflects the equation

$$\{\mathrm{CWI}_A < c_A\} \cup \{\mathrm{CWI}_B < c_B\} =$$
$$= \{\mathrm{CWI}_A < c_A\} \cup \left(\{\mathrm{CWI}_B < c_B\} \setminus \{\mathrm{CWI}_A < c_A\}\right)$$

where $M \setminus N$ addresses all elements in M which are not in N. □

Equation (2.12) can be more generally described in the context of *Bernoulli mixture models*; see [58], [51], Chapter 2 in [25] and Appendices 6.6 and 6.7. We will come back to this and related formulas again in the following sections.

2.2.2 Proposition *The one-year second-to-default probability equals*

$$p_{2\mathrm{nd}}^{(1)} = \int_{-\infty}^{\infty} g_{p_A,\varrho}(y) g_{p_B,\varrho}(y)\, dN(y). \tag{2.13}$$

---
[4]See also Figure 2.41 for a comparable situation with three events.

Proof. Conditional on $Y = y$ the default probabilities multiply for the joint default probability due to conditional independence in the mixture model. As usual in mixture models, we have to integrate the product of conditional PDs w.r.t. the mixing variable Y. Again, we could have worked with a set-theoretic (event-based) approach as in the proof of Proposition 2.2.1. □

2.2.3 Remark The one-year second-to-default probability coincides with the *joint default probability* (JDP) of the two assets defined by

$$\text{JDP}_{A,B} = \text{JDP}^{(1)}_{A,B} = \mathbb{P}[\mathbf{1}_{\{\text{CWI}_A < c_A\}} = 1, \mathbf{1}_{\{\text{CWI}_B < c_B\}} = 1] \quad (2.14)$$

$$= \frac{1}{2\pi\sqrt{1-\varrho^2}} \int_{-\infty}^{N^{-1}(p_A)} \int_{-\infty}^{N^{-1}(p_B)} e^{-\frac{1}{2}(x_A^2 - 2\varrho x_A x_B + x_B^2)/(1-\varrho^2)} dx_A dx_B$$

based on the joint normal distribution of the obligor's CWIs in line with Equation (2.3).

In explicit numbers, we obtain in our example

- $p^{(1)}_{1\text{st}} = 0.0149$ for the one-year first-to-default probability and
- $p^{(1)}_{2\text{nd}} = 0.0001$ for the one-year second-to-default probability.

As a cross-reference, we can check that

$$p^{(1)}_{1\text{st}} + p^{(1)}_{2\text{nd}} = p_A + p_B. \quad (2.15)$$

Given $p^{(1)}_{1\text{st}}$ and $p^{(1)}_{2\text{nd}}$ we can now easily calculate the one-year EL and other risk quantities, incorporating LGDs of the underlying assets and the exposure weight w in our duo basket.

Before we close this section we want to briefly comment on the influence of PDs and the correlation impact on $p^{(1)}_{1\text{st}}$ and $p^{(1)}_{2\text{nd}}$. Figures 2.3 and 2.4 illustrate the dependence of $p^{(1)}_{1\text{st}}$ and $p^{(1)}_{2\text{nd}}$ on the CWI correlation ϱ.

Here, we make an important structural observation based on Figures 2.3 and 2.4.

2.2.4 Remark *The first-to-default probability attains its maximum in case of zero CWI correlation, whereas the second-to-default probability attains its maximum in case of perfect correlation.*

Default Baskets

[Figure: One-year first-to-default probability plotted against CWI correlation, ranging from ~0.015 at zero correlation down to ~0.010 at perfect correlation.

Zero correlation case:
$$p_{1st}^{(1)} = p_A + p_B(1-p_A)$$

Perfect correlation case:
$$p_{1st}^{(1)} = \max[p_A, p_B]$$]

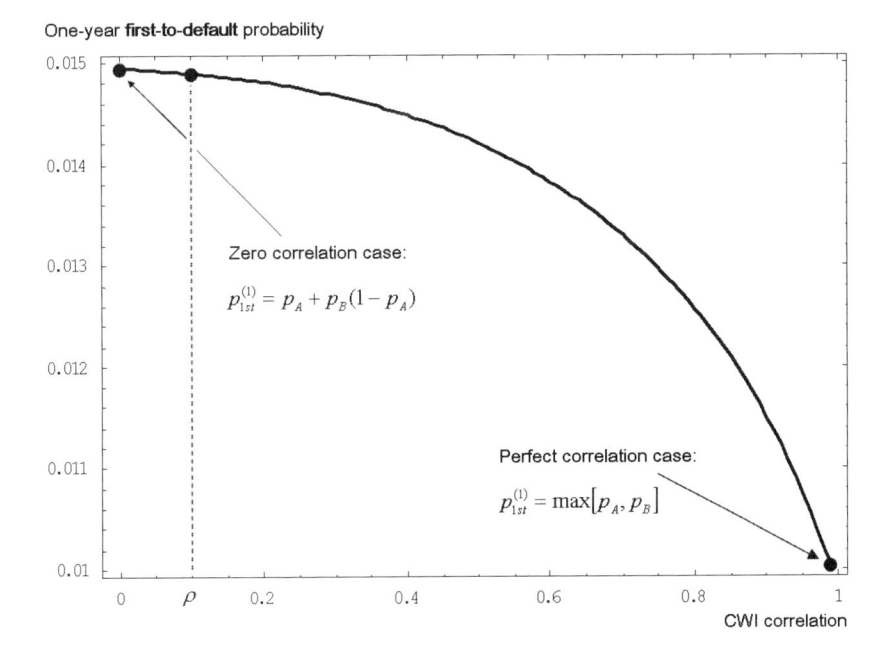

FIGURE 2.3: Influence of the CWI correlation on $p_{1st}^{(1)}$

Later in this book we will re-discover this fact and interpret it by saying that an investor taking the first default or loss in a basket or CDO is in a worst case scenario if assets default in a completely independent way because that makes defaults in a basket most unpredictable. In contrast, a second-to-default event will only take place if both assets default. In addition, the higher the CWI correlation the higher the probability of a joint default event; see Figure 2.5. In the extreme case of perfect correlation, both default events are linearly related leading to the highest possible value of $p_{2nd}^{(1)}$. Again we have parallels to CDOs: investors in senior tranches suffer from high correlations because they make tail events (joint defaults) more likely.

For reasons of completeness we briefly discuss the extreme cases $\varrho = 0$ and $\varrho = 1$ in terms of formulas.

- $\varrho = 0$: According to Equation (2.12), conditioning on Y has no effect in case of zero correlation. Equations (2.11) and (2.13) then yield
$$p_{1st}^{(1)} = p_A + p_B(1-p_A) = 0.01495$$
$$p_{2nd}^{(1)} = p_A \times p_B = 0.00005.$$

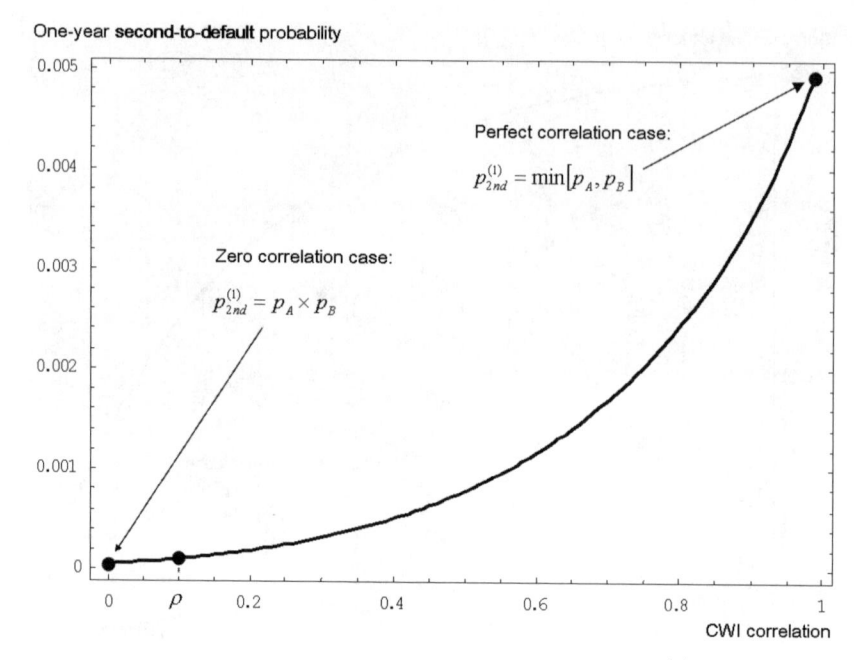

38   *Structured Credit Portfolio Analysis, Baskets & CDOs*

FIGURE 2.4: Influence of the CWI correlation on $p^{(1)}_{2nd}$

In other words, the zero correlation case corresponds to a situation where the systematic risk component is switched-off completely.

- $\varrho = 1$: In this case, idiosyncratic deviations from the systematic factor Y in Equation (2.3) are no longer possible. Then, we can replace CWI_A and CWI_B in Equation (2.14) by Y such that

$$\begin{aligned} JDP_{A,B} &= \mathbb{P}[\mathbf{1}_{\{Y<c_A\}} = 1, \mathbf{1}_{\{Y<c_B\}} = 1] \\ &= \mathbb{P}[Y < \min\{c_A, c_B\}] \\ &= N\big[\min\{N^{-1}[p_A], N^{-1}[p_B]\}\big] \\ &= \min[p_A, p_B]. \end{aligned}$$

Therefore, we have $p^{(1)}_{2nd} = \min[p_A, p_B]$. Since Equation (2.15) must be fulfilled, we can conclude that $p^{(1)}_{1st} = \max[p_A, p_B]$.

We will come back to these and related aspects later in the text when we consider baskets and CDOs over multi-year periods rather than with respect to the one-year horizon. Typically, baskets and CDOs have multi-year terms. A standard maturity for 'real life' baskets is 3 or 5

Joint default probability (JDP)

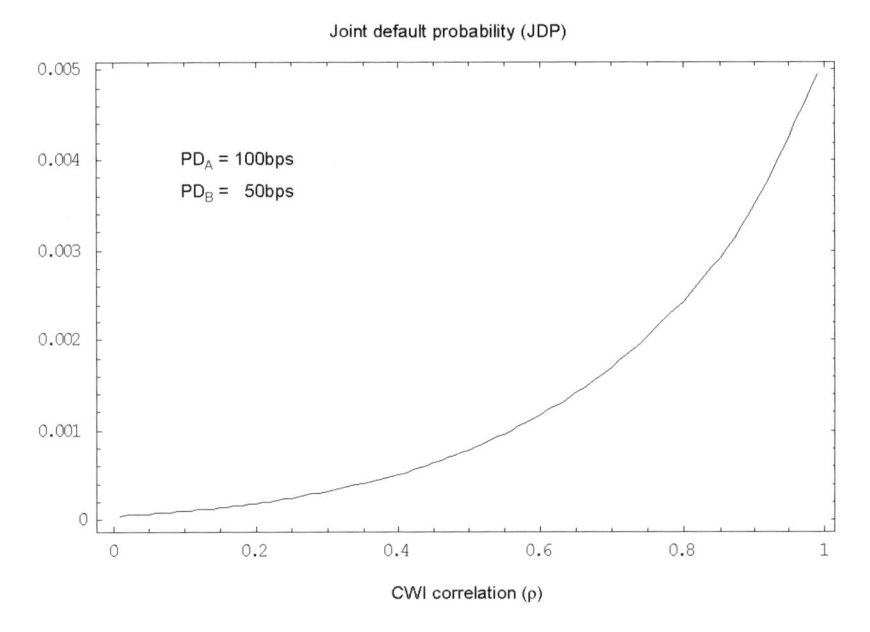

FIGURE 2.5: Joint default probability (JDP) as a function of CWI correlation

years so that we are motivated to leave the one-year horizon perspective behind and turn our attention to time horizons longer than a year. In order to incorporate the time dimension, we have to introduce a concept for modeling the term structure of default probabilities. This is exercised in the following section, which then directly will lead us to *dependent default times* in Section 2.5.

2.3 Derivation of PD Term Structures

In this section, we explain the calibration of *PD term structures* by means of an example based on published data from Standard and Poor's (S&P) [108]. Note that PD term structures sometimes are also called *credit curves*, just to mention another keyword for the same object in order to make any search in the literature easier for interested readers.

Note that, as long as nothing else is said, *time* in this section is measured in years.

Typically, every bank has its own way to calibrate credit curves. We proceed in our exposition in three steps. First, we calibrate a so-called *generator* or *Q-matrix* Q w.r.t. an average one-year migration matrix from S&P, describing migrations of a continuous-time Markov chain at 'infinitesimal small' time intervals. Next, we calculate the Markov PD term structure generated by Q and compare the result with empirical PD term structures from S&P. We will find what other people before us already discovered: although time-homogeneous Markov chains are very popular in credit risk modeling, their ability to fit empirical term structures is limited to some extent. Therefore, we leave the time-homogeneous Markov approach behind and turn our attention to non-homogeneous Markov chains in continuous time. We will see that this approach is more fruitful than our first attempt. The empirical term structures published by S&P can be approximated by a non-homogeneous Markov chain with surprising quality; see [29].

2.3.1 A Time-Homogeneous Markov Chain Approach

Time-homogeneous Markov chain approaches for the calibration of PD term structures are well-known in the literature and have been applied over and over again by financial practitioners; see ISRAEL, ROSENTHAL, and WEI [61], JARROW, LANDO, and TURNBULL [62], and KREININ and SIDELNIKOVA [70], just to mention a few examples in the literature. In these papers, one finds several alternatives for the calibration of continuous-time Markov chains to a given one-year migration matrix. For our examples in the sequel, we work with the so-called *diagonal adjustment* approach discussed in [70] and apply it to S&P data; see also [26]. Alternative approaches can be worked out analogously.

In Section 1.1 we introduced the notion for PD term structures in (1.2) for some obligor i by the curve

$$(p_i^{(t)})_{t \geq 0} = (\mathbb{P}[\mathrm{CWI}_i^{(t)} < c_i^{(t)}])_{t \geq 0}.$$

In other words, $p_i^{(t)}$ is the probability that obligor i defaults within the time interval $[0, t]$, and the curve of these cumulative default probabilities, naturally increasing with increasing time, constitutes the PD term structure of the considered obligor. Going back to Table 1.1 in Section

1.1, we find that one-year default probabilities $p_i^{(t)}$ and $p_j^{(t)}$ for differently rated obligors i and j are naturally different so that we expect to see the same dependence of PD term structures on ratings of the corresponding obligors not only at the one-year horizon but over the whole time axis. In fact, we will define term structures in a *rating-monotonic* way such that the credit curve of obligor i is above the credit curve of obligor j at all times if obligor i has a higher one-year PD (corresponding to a worse rating) than obligor j; see also Figure 2.13 and the discussion in the corresponding section.

As mentioned at the beginning of this section, we now provide a step-by-step calibration of credit curves based on the so-called *diagonal adjustment* approach to time-continuous homogeneous Markov chains. In practical applications, one can, in principal, follow a comparable approach but has to make sure that calibrated credit curves *comply with the multi-year default history* of the bank's portfolio. We will come back to this point later in this section when we compare the calibrated credit curves with empirically observed cumulative default frequencies. Another difficulty we see in practice in this context are certain client segments, e.g., private clients, for which migration history cannot as easily build up as it is the case for corporate clients where balance sheet data flows into the bank at least once a year. For such segments, PD term structures are more difficult to obtain and have to be derived from historical multi-year defaults rather than from estimates based on credit migration data. A last argument worthwhile to be mentioned is the discussion about how realistic the Markov assumption for PD evolution over time possibly can be. Here, we clearly look at the problem through the 'glasses of practitioners'. During recent years our jobs forced us to look at various approaches in the context of PD term structure calibration. Without going into details, we found, although the Markov assumption is somewhat questionable, that *Markov approaches can yield fairly well approximations to historically observed cumulative default frequencies*. In Section 2.3.2, we will see an example of a Markov approach that fits empirical data with outstanding quality.

Rating agencies publish discrete credit curves based on cohorts of historically observed default frequencies annually, [108] is an example for such an agency report. The problem with historically observed cohorts is that the resulting multi-year PDs have a tendency to look kind of 'saturated' at longer horizons due to lack-of-data problems. Comparing corporate bond default reports from 5 years ago with reports as of

TABLE 2.2: Modified S&P average one-year migration matrix

	AAA	AA	A	BBB	BB	B	CCC	D
AAA	91,68%	7,69%	0,48%	0,09%	0,06%	0,00%	0,00%	0,00%
AA	0,62%	90,49%	8,10%	0,60%	0,05%	0,11%	0,02%	0,01%
A	0,05%	2,16%	91,34%	5,77%	0,44%	0,17%	0,03%	0,04%
BBB	0,02%	0,22%	4,07%	89,72%	4,68%	0,80%	0,20%	0,29%
BB	0,04%	0,08%	0,36%	5,78%	83,38%	8,05%	1,03%	1,28%
B	0,00%	0,07%	0,22%	0,32%	5,84%	82,53%	4,78%	6,24%
CCC	0,09%	0,00%	0,36%	0,45%	1,52%	11,17%	54,06%	32,35%
D	0,00%	0,00%	0,00%	0,00%	0,00%	0,00%	0,00%	100,00%

today, one certainly recognizes major improvements on the data side. For ratings containing many clients (like BBB for instance), curves are smoother and do not imply zero forward PDs after a few years. Nevertheless, data quality still needs improvements in order to rely on historical data purely without 'smoothing' by some suitable model. The following Markov chain approach is an elegant way to overcome this problem and to generate continuous-time credit curves.

Let us now start with the adjusted average one-year migration matrix $M = (m_{ij})_{i,j=1,...,8}$ shown in Table 2.2 from S&P (see [108], Table 9). As said, we overcome the zero default observation problem for AAA-ratings by assuming an 0.2 bps default probability for AAA-rated customers. In addition, the matrix in Table 2.2 has been row-normalized in order to force row sums to be equal to 1 such that M is a stochastic matrix.

Next, we need the following theorem based on a condition, which will be fulfilled in most of the cases, although we saw a few examples of migration matrices (typically calibrated 'point-in-time' instead of 'through-the-cycle') for which the condition was hurt for certain rating classes. We denote by I the identity matrix and by N the number of credit states.

2.3.1 Theorem *If a migration matrix $M = (m_{ij})_{i,j=1,...,N}$ is strictly diagonal dominant, i.e., $m_{ii} > \frac{1}{2}$ for every i, then the log-expansion*

$$\tilde{Q}_n = \sum_{k=1}^{n}(-1)^{k+1}\frac{(M-I)^k}{k} \quad (n \in \mathbb{N})$$

converges to a matrix $\tilde{Q} = (\tilde{q}_{ij})_{i,j=1,...,N}$ satisfying

1. $\sum_{j=1}^{N} \tilde{q}_{ij} = 0$ for every $i = 1, ..., N$;

2. $\exp(\tilde{Q}) = M$.

The convergence $\tilde{Q}_n \to \tilde{Q}$ is geometrically fast.

Proof. See ISRAEL ET AL. [61]. □

Theorem 2.3.1 leads us to Condition 1 of the following remark, which serves, complemented by two other conditions, as definition for a special type of square matrices known as *generators* or *Q-matrices* in Markov chain theory; see Chapter 6 in [25].

2.3.2 Remark *The generator of a time-continuous Markov chain is given by a so-called Q-matrix $Q = (q_{ij})_{1 \leq i,j \leq N}$ satisfying the following properties:*

1. $\sum_{j=1}^{N} q_{ij} = 0$ *for every* $i = 1, ..., N$;

2. $0 \leq -q_{ii} < \infty$ *for every* $i = 1, ..., N$;

3. $q_{ij} \geq 0$ *for all* $i, j = 1, ..., N$ *with* $i \neq j$.

For background on Markov chains we refer to the book by NORIS [94].

Before we continue, let us briefly recall Theorem 2.3.1. If we calculate the log-expansion of M only for $k = 1$, then we obtain

$$M - I$$

as a first-order approximation to the log-expansion of M. Because M is a stochastic matrix, it is immediately clear that the row sums of $M - I$ are zero such that Condition 1 in Remark 2.3.2 naturally is fulfilled. Conditions 2 and 3 are also obvious. Therefore, $M - I$ is the most simple generator we can obtain from a migration matrix M.

Now, how are generators related to migration matrices? The next theorem, well known in Markov theory, gives an answer to this question.

2.3.3 Theorem *The following conditions are equivalent (here we have the special situation $Q \in \mathbb{R}^{8 \times 8}$):*

- *Q satisfies Properties 1 to 3 in Remark 2.3.2.*

- $\exp(tQ)$ *is a stochastic matrix for every* $t \geq 0$.

TABLE 2.3: Calculation of 1st-order generator approximation $M - I$

	AAA	AA	A	BBB	BB	B	CCC	D
AAA	-8,32%	7,69%	0,48%	0,09%	0,06%	0,00%	0,00%	0,00%
AA	0,62%	-9,51%	8,10%	0,60%	0,05%	0,11%	0,02%	0,01%
A	0,05%	2,16%	-8,66%	5,77%	0,44%	0,17%	0,03%	0,04%
BBB	0,02%	0,22%	4,07%	-10,28%	4,68%	0,80%	0,20%	0,29%
BB	0,04%	0,08%	0,36%	5,78%	-16,62%	8,05%	1,03%	1,28%
B	0,00%	0,07%	0,22%	0,32%	5,84%	-17,47%	4,78%	6,24%
CCC	0,09%	0,00%	0,36%	0,45%	1,52%	11,17%	-45,94%	32,35%
D	0,00%	0,00%	0,00%	0,00%	0,00%	0,00%	0,00%	0,00%

TABLE 2.4: Exponential $\exp(M - I)$ of 1st-order generator approximation

	AAA	AA	A	BBB	BB	B	CCC	D
AAA	92,04%	7,04%	0,73%	0,12%	0,06%	0,01%	0,00%	0,00%
AA	0,57%	91,03%	7,42%	0,76%	0,08%	0,11%	0,02%	0,02%
A	0,05%	1,98%	91,89%	5,27%	0,51%	0,19%	0,03%	0,06%
BBB	0,02%	0,24%	3,72%	90,46%	4,13%	0,87%	0,19%	0,36%
BB	0,04%	0,08%	0,44%	5,09%	85,01%	6,87%	0,91%	1,57%
B	0,00%	0,07%	0,22%	0,44%	4,97%	84,38%	3,52%	6,40%
CCC	0,07%	0,01%	0,30%	0,40%	1,38%	8,22%	63,36%	26,26%
D	0,00%	0,00%	0,00%	0,00%	0,00%	0,00%	0,00%	100,00%

Proof. See NORIS [94], Theorem 2.1.2. □

Above we mentioned that $M - I$ is the most simple generator, which can be derived from M. Let us see what we get when we calculate the matrix exponential $\exp(M - I)$ corresponding to the situation $t = 1$ in Theorem 2.3.1. For this purpose, we first calculate $M - I$. Table 2.3 shows the result of this exercise. Next, we calculate the matrix exponential of $M - I$. For this, recall that the matrix exponential is given by the expansion

$$\exp(A) = \sum_{k=0}^{\infty} \frac{A^k}{k!}$$

for any square matrix A. Table 2.4 shows the result. We find that $\exp(M - I)$ already shows some similarities to the one-year migration matrix M we started with, although the approximation is very rough. Based on this little exercise, one can imagine that with increasing k in Theorem 2.3.1 the logarithmic expansion of M applied as argument in the matrix-valued exponential function approaches the original matrix M with better and better approximation quality. However, as we will see in a moment, this does not necessarily mean that, in contrast to $M - I$, the log-expansion yields a generator at all.

Default Baskets 45

TABLE 2.5: Log-expansion of one-year migration matrix M

	AAA	AA	A	BBB	BB	B	CCC	D
AAA	-8,72%	8,44%	0,15%	0,07%	0,06%	-0,01%	0,00%	0,00%
AA	0,68%	-10,13%	8,91%	0,38%	0,02%	0,12%	0,02%	0,00%
A	0,05%	2,37%	-9,31%	6,37%	0,33%	0,15%	0,03%	0,02%
BBB	0,02%	0,19%	4,49%	-11,17%	5,39%	0,65%	0,22%	0,21%
BB	0,04%	0,08%	0,24%	6,68%	-18,71%	9,63%	1,15%	0,88%
B	-0,01%	0,08%	0,22%	0,12%	7,01%	-20,06%	7,09%	5,55%
CCC	0,13%	-0,02%	0,47%	0,54%	1,61%	16,59%	-62,20%	42,88%
D	0,00%	0,00%	0,00%	0,00%	0,00%	0,00%	0,00%	0,00%

TABLE 2.6: Approximative generator (Q-matrix) for M

	AAA	AA	A	BBB	BB	B	CCC	D
AAA	-8,73%	8,44%	0,15%	0,07%	0,06%	0,00%	0,00%	0,00%
AA	0,68%	-10,13%	8,91%	0,38%	0,02%	0,12%	0,02%	0,00%
A	0,05%	2,37%	-9,31%	6,37%	0,33%	0,15%	0,03%	0,02%
BBB	0,02%	0,19%	4,49%	-11,17%	5,39%	0,65%	0,22%	0,21%
BB	0,04%	0,08%	0,24%	6,68%	-18,71%	9,63%	1,15%	0,88%
B	0,00%	0,08%	0,22%	0,12%	7,01%	-20,06%	7,09%	5,55%
CCC	0,13%	0,00%	0,47%	0,54%	1,61%	16,59%	-62,22%	42,88%
D	0,00%	0,00%	0,00%	0,00%	0,00%	0,00%	0,00%	0,00%

Let us now see what we get from the log-expansion ($n \to \infty$ in Theorem 2.3.1) rather than looking at $M - I$ (case $n = 1$) only. We calculate the log-expansion $\tilde{Q} = (\tilde{q}_{ij})_{i,j=1,\ldots,8}$ of the one-year migration matrix $M = (m_{ij})_{i,j=1,\ldots,8}$ according to Theorem 2.3.1. This can be done with a calculation program like Mathematica or Matlab, but can as easily also be implemented in Excel/VBA. Table 2.5 shows the resulting matrix $\tilde{Q}$.

Theorem 2.3.1 guarantees that $\tilde{Q}$ fulfills Condition 1 of generators listed in Remark 2.3.2. Condition 2 is also not hurt by $\tilde{Q}$, but Condition 3 is not fulfilled. So we see confirmed what we already indicated, namely, that (in contrast to $M - I$) the log-expansion of M not necessarily yields a Q-matrix. However, there are only three migration rates in Table 2.5 not in line with generator conditions:

- $\tilde{q}_{AAA,B} = -1$ bps,
- $\tilde{q}_{B,AAA} = -1$ bps,
- $\tilde{q}_{CCC,AA} = -2$ bps.

Only these three entries disable $\tilde{Q}$ from being a generator matrix. Because these three values are very small numbers, we can follow the *diagonal adjustment approach* announced already several times and set the three negative migration rates equal to zero. In order for still being in line with the zero row sum condition, we then need to decrease the *diagonal* elements of rows AAA, B and CCC by an amount compensating for the increased row sums by setting negative entries to a zero value. The evident name *diagonal adjustment* for this procedure is mentioned in [70]. As a result we obtain a generator matrix $Q = (q_{ij})_{i,j=1,\ldots,8}$ as shown in Table 2.6.

From Theorem 2.3.1 we know that we get back the original migration matrix M from $\tilde{Q}$ by $\exp(\tilde{Q})$. But what about getting M back from Q? Because we manipulated $\tilde{Q}$ in order to arrive at a generator Q, $\exp(Q)$ will not exactly equal M. What is the error distance between M and $\exp(Q)$? Because the manipulation we did is kind of negligible, we already expect the result described in the following proposition.

2.3.4 Proposition M *has an approximate Q-matrix representation by* Q. *The mean-square error can be calculated as*

$$\|M - \exp(Q)\|_2 = \sqrt{\sum_{i,j=1}^{8} (m_{ij} - (\exp(Q))_{ij})^2} \approx 0.00023.$$

Proof. Just calculate the distance. □

We conclude from Proposition 2.3.4 that we can work with Q instead of $\tilde{Q}$, hereby accepting the minor approximation error.

Finding a Q-matrix representation Q for a time-discrete Markov chain represented by a transition matrix M is called an *embedding of the time-discrete Markov chain represented by M into a time-continuous Markov chain represented by its generator or Q-matrix Q* in Markov theory. Of course, our embedding only holds in an approximate manner (see Proposition 2.3.4), but the error is negligibly small. Probabilists know that the existence of such embeddings is far from being obvious, and to some extent we have been very lucky that it worked so well with the S&P-based migration matrix M. In [62], [61], and [70] one finds other ways to manage such embeddings, but as already said we found that the diagonal adjustment approach is well working in many situations. Readers interested in actually applying such embeddings in the context of their credit risk engines should take the time to experiment with different approaches in order to find their own 'best way' to deal with the challenge of PD term structure calibrations.

For the time-homogeneous case we achieved our target. Our efforts have been rewarded by a generator Q such that $\exp(Q) \approx M$. The

credit curves can be read off from the collection of matrices $(\exp(tQ))_{t\geq 0}$
by looking at the default columns. More precisely, for any asset i in
the portfolio with rating $R = R(i)$ we obtain

$$p_i^{(t)} = (e^{tQ})_{row(R),8} \qquad (2.16)$$

where $row(R)$ denotes the transition matrix row corresponding to the
given rating R. Figure 2.6 shows a chart of credit curves from $t = 0$ to
$t = 50$ years on a quarterly base, calibrated as just described.

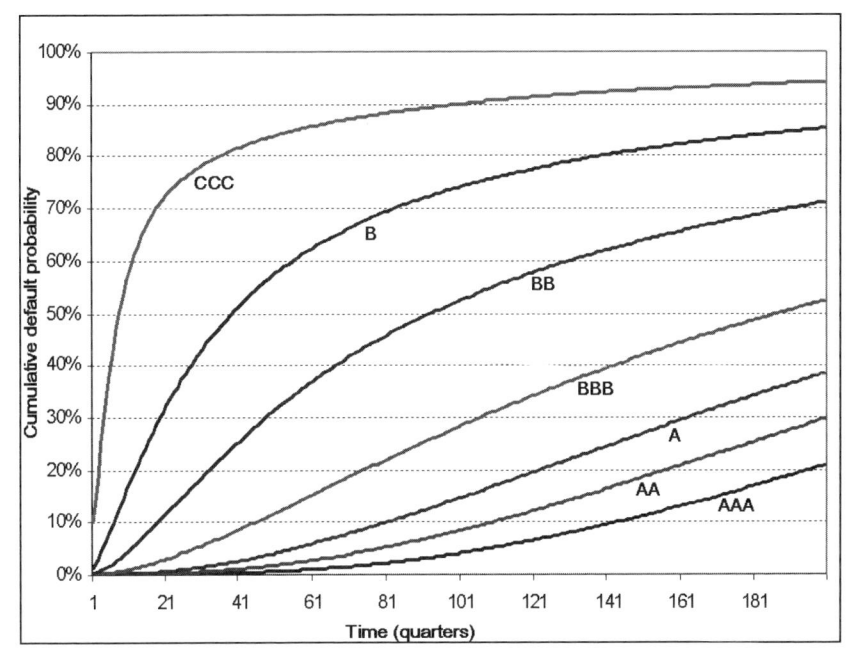

FIGURE 2.6: Calibrated credit curves based on a time-continuous
and time-homogeneous Markov chain approach

The curves in Figure 2.6 are kind of typical. Credit curves assigned to
subinvestment grade ratings have a tendency to slow down their growth,
because *conditional on having survived for some time the chances for
further survival improve*. For good ratings, we see the opposite effect.
For instance, assets with excellent credit quality have no upside potential because they are already outstanding but have a large downside
risk of potential deterioration of their credit quality over time.

If we calculate quarterly *forward PDs* via

$$\tilde{p}_i^{(t)} = \frac{p_i^{(t)} - p_i^{(t-0.25)}}{1 - p_i^{(t-0.25)}} \qquad (t \geq 0.25) \qquad (2.17)$$

we find a *mean reversion effect* in PD term structures. Moody's KMV writes in the documentation of their rating system RiskCalc for private companies the following statement:

> *We noticed that obligors appear to exhibit mean reversion in their credit quality. In other words, good credits today tend to become somewhat worse credits over time and bad credits (conditional upon survival) tend to become better credits over time. Default studies on rated bonds by Moodys Investors Services support this assertion, as do other studies. We found further evidence for mean reversion in both our proprietary public firm default databases and in the Credit Research Database data on private firms.* [85]

The homogeneous Markov approach, focussing on the aspect of mean reversion, is in line with findings of Moody's KMV. Mean reversion effects drive the forward or conditional PDs over time, respectively; see Figure 2.7. Therefore, we can hope to have at least a somewhat realistic model at disposal. However, if we compare PD term structures on a purely empirical base as published by S&P in [108] with the just calibrated term structure, we find as a disappointing surprise that curves do not match very well; see Figure 2.8.

When we compare the homogeneous continuous-time Markov chain (HCTMC-based) credit curves with observed multi-year default frequencies from S&P we immediately get the impression that HCTMC model-based PDs systematically overestimate empirically observed default frequencies (exception: AAA). Here are some remarks:

- For good rating classes, AAA is the extreme case, the number of observed defaults is too low to come up with reasonable empirical figures. For instance, we found a regression-implied one-year PD for AAA at the 0.2 bps level. If we increase the time observation window from one to several years, it still is exceptional to observe AAA-defaults. Figure 2.8 shows that there is some history of AAA-defaults but the curve clearly reflects the lack of data we have in this rating class. A comparable lack of data problem

Forward (conditional) default probability

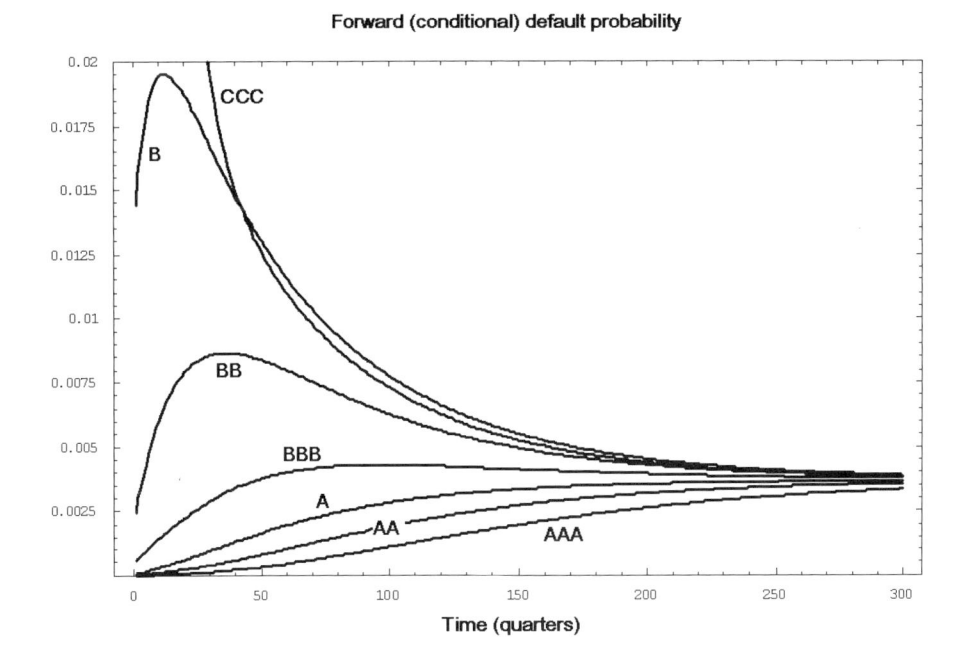

FIGURE 2.7: Forward PDs corresponding to Figure 2.6

applies to AA and A ratings. Therefore, we are not surprised to see more or less systematically higher model-based credit curves.

- For BBB-rated clients, the situation is different. Here, the data situation typically is fine and should yield reasonable multi-year default frequencies. We see in Figure 2.8 that up to the 8-year horizon the model-based and the empirical credit curves are not too far from each other. After the 8-year horizon the curves follow completely different paths. Because we do not have good reasons for blaming the data situation in the BBB case, we have to assume that the deviation comes from a suboptimal model fit. In Section 2.3.2, we will see that an improved model yields better matching credit curves.

- The model weakness also applies to BB and B ratings. For CCC we again have, besides a certain model weakness, to take data problems into account.

- In the lower right corner we see a comparison of average credit

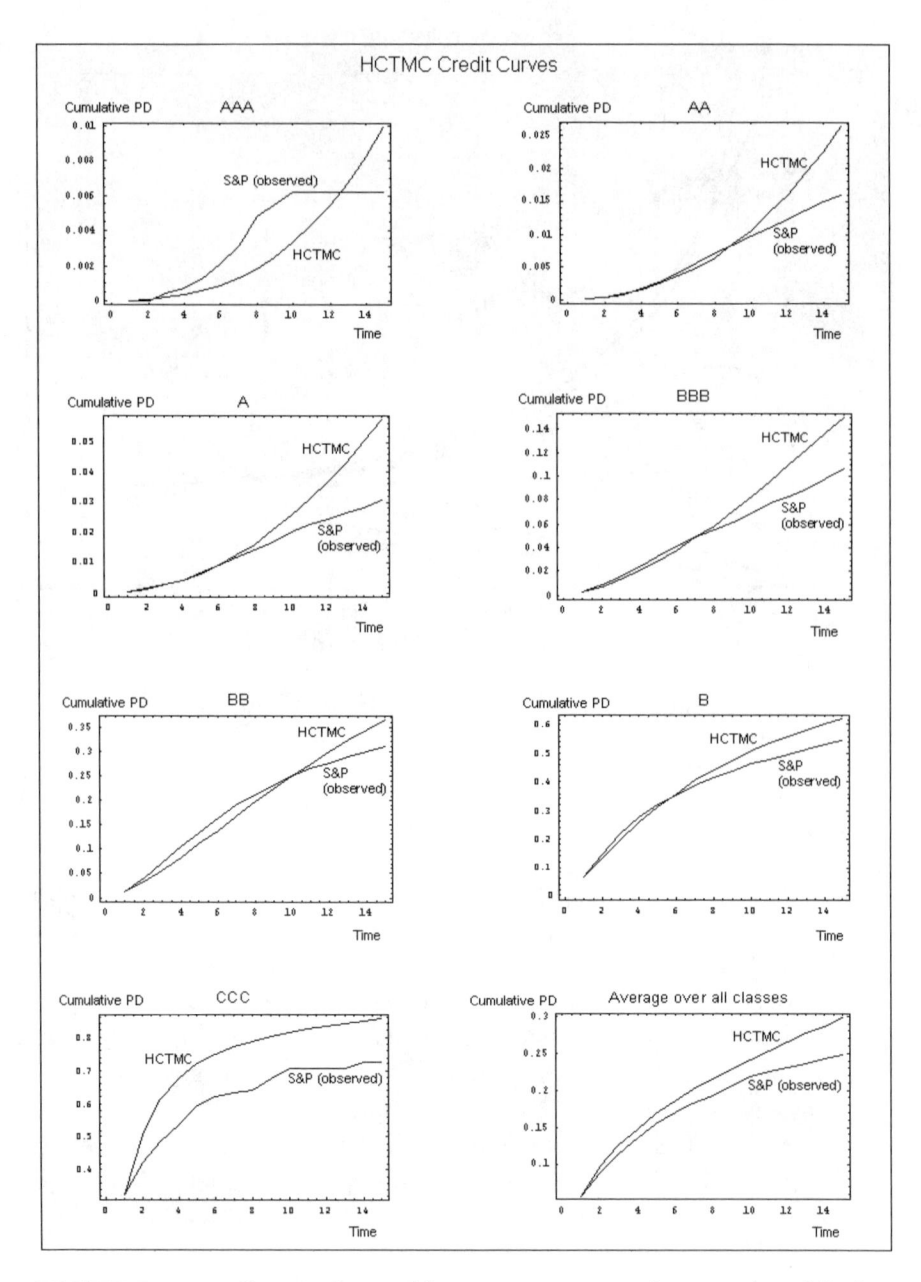

50   *Structured Credit Portfolio Analysis, Baskets & CDOs*

FIGURE 2.8: Comparison of homogeneous continuous-time Markov chain model-based PD term structures with purely empirical (observed) PD term structures from S&P

curves. One can think about averages in different ways. For reasons of simplicity, we worked with equally weighted arithmetic means. If one dives deeper into the raw data of default frequencies, e.g., in the bank-internal data warehouse, one can follow more sophisticated approaches. However, the comparison of average credit curves again reveals a systematically higher HCTMC-based PD term structure compared to empirical/observed S&P default frequencies.

Let us briefly recapitulate what we did in this section so that we have some motivation for going forward to the next section where we will modify the HCTMC approach to a non-homogeneous continuous-time Markov chain (NHCTMC) approach.

We started this section with a one-year migration matrix M from S&P. In addition to M, we also find in the S&P-report (see [108]) time series of empirical/observed multi-year default frequencies

$$(\hat{p}_R^{(t)})_{t=1,2,\ldots,15;\ R=AAA,AA,\ldots,CCC}$$

which theoretically can serve as estimates for multi-year PDs or credit curves. Based on M, we can calculate homogeneous Markov chain model-based multi-year PDs as in (2.16),

$$\left((M^t)_{row(R),8}\right)_{t=1,2,\ldots,15;\ R=AAA,AA,\ldots,CCC}.$$

Then, if multi-year default frequencies from S&P would be driven by a homogeneous Markov migration process represented by the one-year migration matrix M, we obtained

$$\text{distance}\left[(\hat{p}_R^{(t)})_{t;R},\ \left((M^t)_{row(R),8}\right)_{t;R}\right] \stackrel{!}{=} \text{small} \qquad (2.18)$$

where distance[·] is a suitable function for measuring the distance between credit curves. We wanted to be even more sophisticated and found a generator Q such that

$$M^t \approx \exp(tQ) \qquad (t=1,2,\ldots).$$

This enabled us to consider the migration process as a HCTMC such that multi-year default probabilities are estimated via (2.16). For discrete times $t = 1, 2, \ldots, 15$, Equation (2.18) should still hold, but according to Figure 2.8 that is not what we find and see.

So the question arises, *which of our assumptions in the chain of arguments discussed above is questionable?*

We will see in the next section that the approach itself can yield a good approximation of model-based multi-year PDs to observed default frequencies in line with Equation (2.18) if we drop the assumption of time-homogeneity. Homogeneity in the time dimension addresses the independence of migration rates from current time. More explicitly, migration rates for a HCTMC for a time interval $[s,t]$, $s < t$, depend on the length $(t-s)$ of the time period only and not on the starting time s. Homogeneity is reflected by multi-year migration matrices as powers of the one-year migration matrix, or, in more technical terms and in continuous time, by the *time-independent generator matrix* Q.

2.3.2 A Non-Homogeneous Markov Chain Approach

In this section, we modify the PD term structure model from the previous section by *dropping time-homogeneity*. In addition, we use a flexible functional approach with sufficient flexibility in order to fit our non-homogeneous continuous time Markov chain (NHCTMC) approach to observed multi-year default frequencies from S&P.

Starting point for our construction is the generator $Q = (q_{ij})_{1 \leq i,j \leq 8}$ from Table 2.6. But now, as already indicated, we do no longer assume that the transition rates q_{ij} are constant over time, leading to a HCTMC. Instead, we replace the time-homogeneous generator Q leading to migration matrices $\exp(tQ)$ for the time interval $[0,t]$ by some time-dependent generator

$$Q_t = \Phi_t * Q \tag{2.19}$$

where '$*$' denotes matrix multiplication and $\Phi_t = (\varphi_{ij}(t))_{1 \leq i,j \leq 8}$ is a diagonal matrix in $\mathbb{R}^{8 \times 8}$ with

$$\varphi_{ij}(t) = \begin{cases} 0 & \text{if } i \neq j \\ \varphi_{\alpha_i, \beta_i}(t) & \text{if } i = j \end{cases} \tag{2.20}$$

Because Φ_t is a diagonal matrix, Q_t satisfies Conditions 1, 2, and 3 in Remark 2.3.2. To see this, just note the obvious fact that scaling the rows of a Q-matrix leads to a Q-matrix again. The functions $\varphi_{\alpha,\beta}$ with respect to parameters α and β are defined as follows. Set

$$\varphi_{\alpha,\beta} : [0,\infty) \to [0,\infty), \quad t \mapsto \varphi_{\alpha,\beta}(t) = \frac{(1-e^{-\alpha t})t^{\beta-1}}{1-e^{-\alpha}}$$

Default Baskets 53

for *non-negative* constants α and β. Figure 2.9 illustrates the functions $t \mapsto t\varphi_{\alpha,\beta}(t)$ graphically for some given α's and β's.

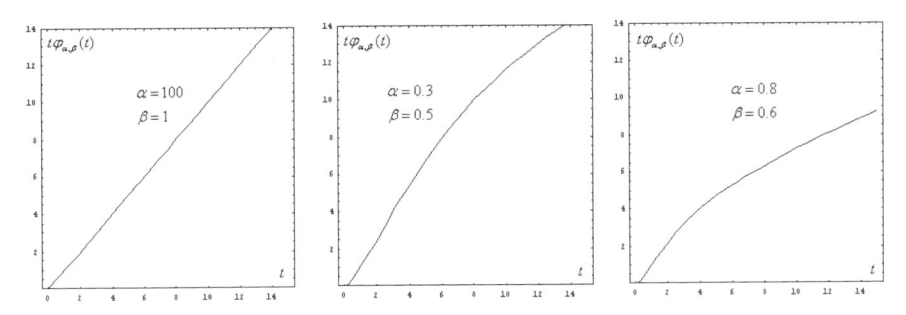

FIGURE 2.9: Illustration of the functions $t \mapsto t\varphi_{\alpha,\beta}$

Let us summarize some basic properties of the functions $\varphi_{\alpha,\beta}(t)$.

1. $\varphi_{\alpha,\beta}(1) = 1$ ('normalized' functions).

2. $t \mapsto t\varphi_{\alpha,\beta}(t)$ is increasing in the time parameter $t \geq 0$.

3. The first part of $t\varphi_{\alpha,\beta}$, namely $(1 - e^{-\alpha t})$, is the distribution function of an exponentially distributed random variable with intensity α. The second part of $t\varphi_{\alpha,\beta}$, namely t^β, can be considered as a time-slowing-down ($\beta < 1$) or time-accelerating ($\beta > 1$) adjustment term. Note that altogether $\varphi_{\alpha,\beta}$ exhibits certain similarities to the gamma distribution, typically applied in the context of queuing theory and reliability analysis. The scaling factor $(1 - e^{-\alpha})^{-1}$ is the normalizing multiplier forcing $\varphi_{\alpha,\beta}(1) = 1$.

Applying Proposition 2.3.3, we can now define migration matrices for given time periods $[0,t]$ via

$$M_t = \exp(tQ_t) \qquad (t \geq 0). \qquad (2.21)$$

The corresponding PD term structures are defined as usual by

$$\left((M_t)_{row(R),8}\right)_{t \geq 0;\ R=AAA,AA,...,CCC}.$$

Since the functional form of $(Q_t)_{t \geq 0}$ is given by Equation (2.20), generators Q_t are solely determined by two vectors $(\alpha_1,...,\alpha_8)$ and $(\beta_1,...,\beta_8)$ in $[0,\infty)^8$. We can now try to find α- and β-vectors such that

$$\text{distance}\left[(\hat{p}_R^{(t)})_{t;R},\ \left((M_t)_{row(R),8}\right)_{t;R}\right] \stackrel{!}{=} \text{small} \qquad (2.22)$$

TABLE 2.7: Optimized α- and β-vectors for time bending functions $\varphi_{\alpha,\beta}$

	α	β
AAA	0.34	0.89
AA	0.11	0.26
A	0.81	0.65
BBB	0.23	0.30
BB	0.32	0.56
B	0.23	0.40
CCC	2.15	0.46

analogous to Equation (2.18). This actually is a clear optimization problem: Find α- and β-vectors such that the distance or approximation error (2.22) attains its minimum value. Note that α_8 and β_8 have no meaning in the approach and can be fixed at some arbitrary value, e.g., $\alpha = \beta = 1$. As distance measure for the optimization problem (2.22) we use the mean-squared distance. As a result we obtain Table 2.7. Figure 2.10 illustrates the intuition underlying our approach and Figure 2.11 shows how well the NHCTMC model-based PD term structures fit the empirical/observed multi-year default frequencies from S&P.

A major difference between approaches underlying Figures 2.8 and 2.11 is that the HCTMC approach relies on observed migration rates only, whereas the NHCTMC approach relies on observed migration rates for the calibration of the generator Q as well as on observed multi-year default frequencies for finding α- and β-vectors such that the NHCTMC approach best possible approximates observed multi-year default frequencies.

At the end of this section, we want to briefly comment on the stochastic rationale of our approach. For the sake of a more convenient notation, let us denote by Ψ_t the diagonal matrix with diagonal elements

$$\psi_{ii}(t) = t\varphi_{\alpha_i,\beta_i}(t) \qquad (i = 1, ..., 8;\ t \geq 0).$$

The transition matrix M_t from Equation (2.21) for the time period $[0, t]$ can then be written as

$$M_t = \exp(\Psi_t * Q) \qquad (t \geq 0). \qquad (2.23)$$

First of all, we need the following proposition.

Default Baskets 55

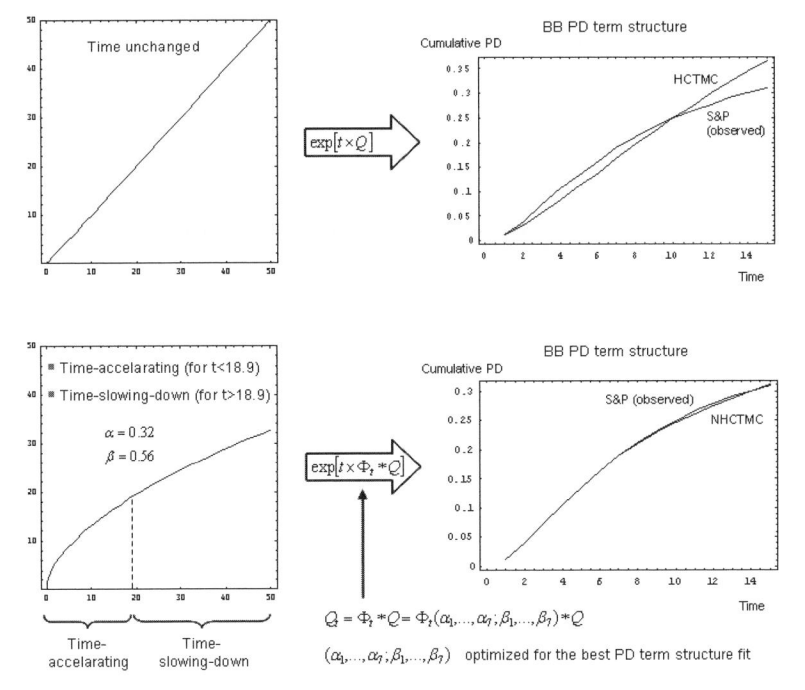

FIGURE 2.10: Illustration of the before-discussed (time-bending) NHCTMC approach to PD term structures

2.3.5 Proposition *Writing the migration matrix M_t in Markov kernel notation as*

$$P_{0,t} = M_t = \exp(Q_t) = \exp(\Psi_t * Q),$$

the following KOLMOGOROV *forward equation holds:*

$$\frac{\partial}{\partial t} P_{0,t} = \left(\frac{\partial}{\partial t} \Psi_t * Q\right) * P_{0,t}.$$

Proof. Term-by-term differentiation yields

$$\frac{\partial}{\partial t} P_{0,t} = \sum_{k=0}^{\infty} \frac{\partial}{\partial t} \frac{(\Psi_t * Q)^k}{k!} \quad (2.24)$$

$$= \sum_{k=1}^{\infty} \left(\frac{\partial}{\partial t} \Psi_t * Q\right) * \frac{(\Psi_t * Q)^{(k-1)}}{(k-1)!}$$

$$= \left(\frac{\partial}{\partial t} \Psi_t * Q\right) * P_{0,t}. \quad (2.25)$$

This proves the proposition. □

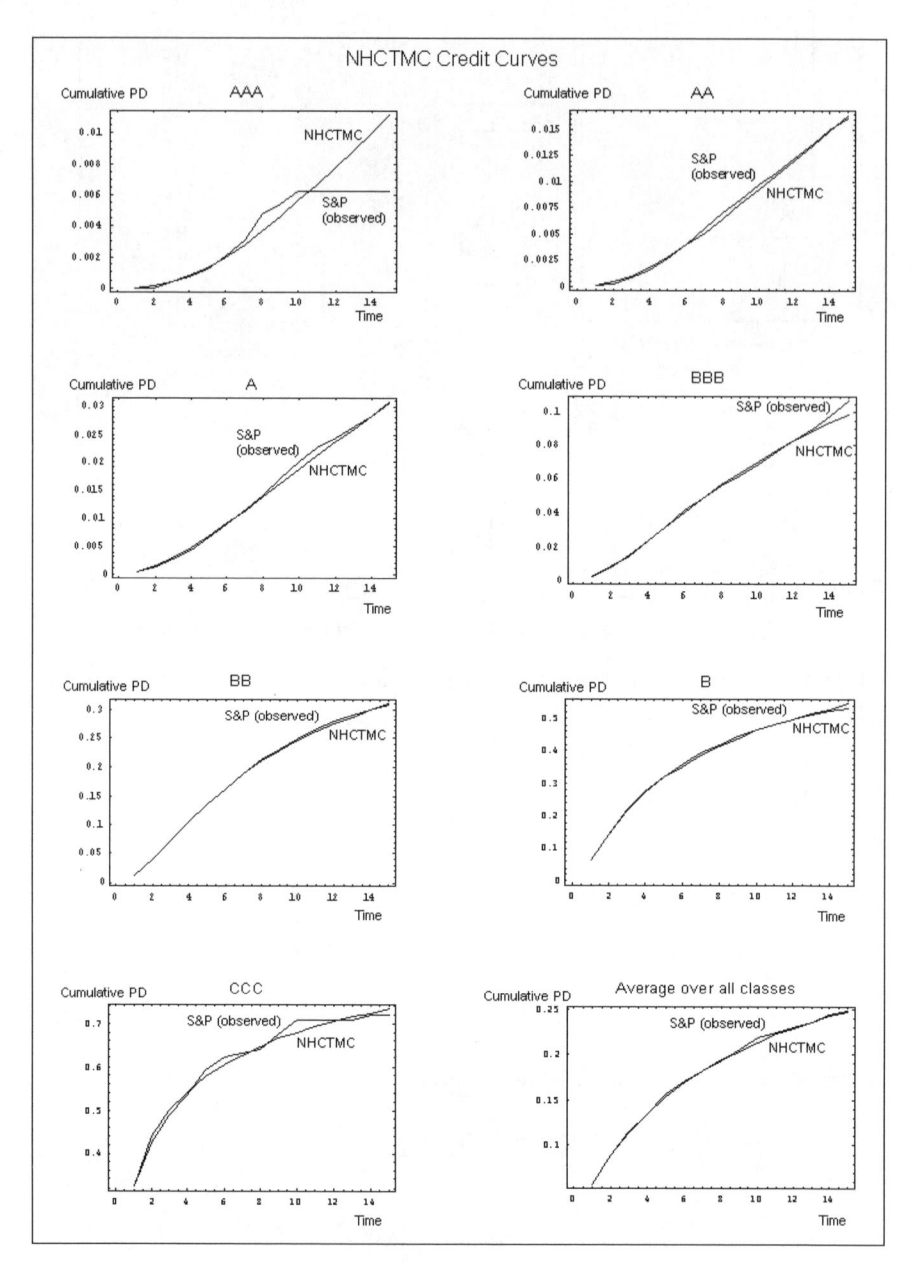

FIGURE 2.11: Comparison of non-homogeneous continuous-time Markov chain PD term structures with purely empirical (observed) PD term structures from S&P

Default Baskets 57

2.3.6 Proposition *The time-dependent generator Q_t defines an NHCTMC generating the PD term structures plotted in Figure 2.11.*

Proof. Because Ψ_t is a diagonal matrix, $(\partial/\partial t)\Psi_t$ is the diagonal matrix with entries $\psi'_{ii}(t)$. Therefore, the matrix $(\partial/\partial t)\Psi_t * Q$ is a Q-matrix, arguing in the same way as above when we said that $\Psi_t * Q$ is a Q-matrix and taking into account that $\psi'_{ii}(t) \geq 0$ at all times[5] t. As a consequence of general Markov theory (see ETHIER and KURTZ [44], Theorem 7.3 in Chapter 4, and LANDO and SKODEBERG [72]), Equation (2.24) is part of the *forward equation* (see Proposition 2.3.5) of a time-inhomogenous Markov chain $(X_t)_{t\geq 0}$ with state space $\{1, 2, ..., 8\}$ corresponding to a *semigroup* $\{P_{s,t} \mid 0 \leq s \leq t\}$ satisfying the *Kolmogorov backward* and *forward equations* associated with the family $\{(\partial/\partial t)\Psi_t * Q \mid t \geq 0\}$ defining the *infinitesimal generator* of the Markov process. Equation (2.24) shows that the NHCTMC $(X_t)_{t\geq 0}$ generates the PD term structures illustrated in Figure 2.11 via the default column of kernel-based transition matrices $P_{0,t} = M_t = \exp(\Psi_t * Q)$. □

Proposition 2.3.6 shows that our approach can be embedded in well understood general Markov process theory. The special functional form (2.19) we used in the definition of the time-dependent generator Q_t seems to work quite well when we look at Figure 2.11. However, there are various ways to construct migration processes and the approach elaborated here is just one out of several meaningful ways to calculate multi-year PDs. In Section 5, we list some references where the reader will find a wide range of other valuable methodologies.

Figure 2.11 contains a clear message: *observed multi-year default frequencies can be fitted seemingly well by Markovian PD term structures when one drops the homogeneity assumption.*

2.3.3 Extrapolation Problems for PD Term Structures

Figures 2.8 and 2.11 illustrate model-based PDs (HCTMC as well as NHCTMC approaches, respectively) in comparison to empirical (observed) multi-year default frequencies from S&P; see Table 2.8.[6]

The nice fit in Figure 2.11 based on our NHCTMC approach relies on two data sources, namely

---

[5] We have $(1 - \exp(-\alpha))\,\psi'_{ii}(t) = \alpha\exp(-\alpha t)t^\beta + (1-\exp(-\alpha t))\beta t^{\beta-1} \geq 0 \;\forall\; t \geq 0$.
[6] The one-year PD column coincided with Table 1.1.

TABLE 2.8: Empirical/observed multi-year default frequencies from S&P; see [108], Table 11

S&P hist.	1	2	3	4	5	6	7	8
AAA	0.00%	0.00%	0.04%	0.07%	0.12%	0.21%	0.31%	0.48%
AA	0.01%	0.03%	0.08%	0.16%	0.26%	0.40%	0.56%	0.71%
A	0.04%	0.13%	0.26%	0.43%	0.66%	0.90%	1.16%	1.41%
BBB	0.29%	0.86%	1.48%	2.37%	3.25%	4.15%	4.88%	5.60%
BB	1.28%	3.96%	7.32%	10.51%	13.36%	16.32%	18.84%	21.11%
B	6.24%	14.33%	21.57%	27.47%	31.87%	35.47%	38.71%	41.69%
CCC	32.35%	42.35%	48.66%	53.65%	59.49%	62.19%	63.37%	64.10%

S&P hist.	9	10	11	12	13	14	15
AAA	0.54%	0.62%	0.62%	0.62%	0.62%	0.62%	0.62%
AA	0.83%	0.97%	1.09%	1.23%	1.36%	1.50%	1.61%
A	1.71%	2.01%	2.24%	2.44%	2.64%	2.81%	3.08%
BBB	6.21%	6.95%	7.69%	8.32%	9.01%	9.81%	10.67%
BB	23.22%	24.84%	26.50%	27.84%	29.08%	29.93%	30.94%
B	43.92%	46.27%	48.19%	49.87%	51.41%	53.24%	54.73%
CCC	67.78%	70.80%	70.80%	70.80%	70.80%	72.26%	72.26%

- A one-year migration matrix

- Historical observations of cumulative default frequencies as reported in Table 2.8

The approach is quite sensitive to the latter-mentioned model input. This is not really surprising, given the flexibility of the model leading to the almost perfect fit shown in Figure 2.11. In other words, for a different time series of default frequencies we would get other α- and β-vectors parameterizing the NHCTMC. Therefore, the *data time horizon* (15 years in our case) constitutes a natural restriction of the model: beyond the data time horizon we cannot make any statements, any extrapolation beyond the data time horizon is kind of speculative. *We strongly recommend not to use the approach for extrapolations.*

The restriction to a given data time horizon is a serious model weakness of the NHCTMC approach. On the other side, as a rule of thumb it is wise anyway to apply models only in the range of time horizons where sufficient empirical evidence is given that the calibration of the model reflects 'true world' phenomena. In this book, we can pretty well live with the data time horizon restriction of the NHCTMC approach because baskets and CDO-type credit instruments typically have terms much shorter than 15 years.[7]

---

[7] A typical term for baskets and synthetic CDOs is 5 years; 3-/7-/10-year maturities are also common. Only for mortgage-backed securities (MBS), longer terms are theoretically thinkable.

Default Baskets

Note that for the HCTMC model, the situation is rather different because the HCTMC approach only needs a Q-matrix Q with

$$\exp(Q) \approx M$$

in order to generate the credit curve for a rating R via

$$\big((\exp(tQ))_{row(R),8}\big)_{t\geq 0}.$$

In Section 2.3.1 (see also Proposition 2.3.4) we found such an approximative generator Q for the one-year migration matrix M from S&P. Whereas for our basket and CDO examples we apply the NHCTMC-based credit curves throughout this book, we will, in addition, apply the HCTMC-based credit curves *in a few illustrative examples* referring to time horizons exceeding our data time horizon of 15 years. These exceptions to the rule will be clearly marked so that any confusion regarding a blend of credit curves can be excluded.

Here, we stop our discussion on credit curves. Because they constitute an important ingredient of time-dependent credit risk models, it is justified that we spent quite some time with discussions on this topic. Moreover, we want to encourage readers of this book to spend even more time with credit curves for investigating alternative approaches to the calibration of PD term structures at single-name level and their application at portfolio level, which is even more difficult based on challenges arising in the context of dependence modeling.

2.4 Duo Basket Evaluation for Multi-Year Horizons

Before we turn our attention to dependent default times, let us briefly extend some of our considerations on duo baskets to multi-year time horizons. For this purpose, we focus again on the duo basket consisting of assets A and B from Section 2.1.

In Equation (2.5), we calculated the default correlation $r = r_{A,B}$ as a function of the asset's PDs p_A and p_B as well as of the asset's CWI correlation $\varrho = \varrho_{A,B}$. Now, a typical assumption in credit risk models is to *keep the CWI correlation ϱ constant over different time horizons*. What does this mean for the default correlation r? It is a wrong conclusion to think that just because the correlation ϱ of underlying latent variables is fixed over time, the corresponding default correlation will also remain constant w.r.t. different evaluation horizons. Instead, we find that default correlations, for a given assumed level ϱ of CWI correlation, change over time, driven by the PD term structures[8]

$$(p_A^{(t)})_{t\geq 0} \quad \text{and} \quad (p_B^{(t)})_{t\geq 0}$$

of the underlying assets; see Equation (2.5). Figure 2.12 illustrates the change of default correlations for a given fixed CWI correlation when considered w.r.t. different time horizons.

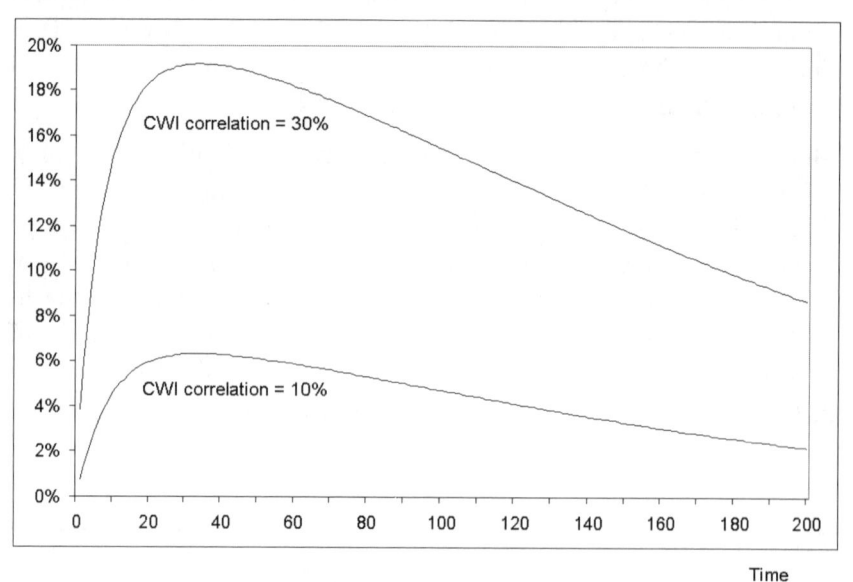

FIGURE 2.12: Changes of default correlation w.r.t. different time horizons between assets A and B for CWI correlations of 10% and 30% (credit curves generated by the HCTMC approach); note that only time horizons up to 20 or 30 years are relevant; nevertheless, the long-term time dependence is interesting to see and, therefore, illustrated

It is interesting to notice the sharp increase of the default correlation

---
[8]Note that $p_A^{(1)} = p_A$ and $p_B^{(1)} = p_B$.

in the first, e.g., 20 years and the slow decay or 'fading-out' of the default correlation over time. Because most[9] credit instruments have a maturity before the 20-year time horizon, the sharp increase of default correlation coincides on the time axis with the term of typical credit instruments.

Before we continue in the text, we want to briefly comment on the following 'subproblem' one has to solve if a chart like the plot in Figure 2.12 has to be generated.

For our duo basket, we assumed PDs for assets A and B of

$$p_A = 0.01 \quad \text{and} \quad p_B = 0.005$$

not contained in our assumed rating scale as shown in Table 1.1. Typically, banks will try to avoid the appearance of PDs not exactly matching a certain PD grid point in the bank's rating scale. On the other hand, PDs on continuous scale can play a certain role, e.g., if for some asset or client an external rating with an externally assigned PD is used instead of the bank-internal rating. For instance, if an asset carries a Moody's rating of Ba1 and historic default experience in Moody's reports suggests a PD of 50 bps for Ba1-rated clients, then, if we are willing to accept the 50 bps as PD assigned to the asset, we have a PD falling between PDs assigned to ratings BBB and BB in Table 1.1. On a rating scale with higher granularity, we would then probably say that in S&P-terminology this asset carries a BB+ rating. Regarding a PD term structure applied to this asset, our term structure monotony criteria mentioned at the beginning of Section 2.3.1 says that the PD term structure of the asset at all times must yield PDs lower than that of the BB term structure and higher than that of the BBB term structure.

Since we have BBB and BB term structures at our disposal, we may want to think about ways to derive term structures for assets A and B by means of term structures of BBB and BB. A common practitioner's *proxy* for the unknown PD evolution of p_A and p_B over time is the *log-linear interpolation* approach, where 'log-linear' means that we interpolate PDs at any time linear on logarithmic scale. More explicitly, we use the inequalities

$$p_{BBB}^{(1)} < p_A^{(1)} = p_A < p_{BB}^{(1)}$$

---
[9]Mortgage-backed loans could serve as exceptional counterexamples.

$$p^{(1)}_{BBB} < p^{(1)}_B = p_B < p^{(1)}_{BB}$$

to find weights $w_A \in (0,1)$ and $w_B \in (0,1)$ such that

$$p_A = p^{(1)}_A = \exp(w_A \ln p^{(1)}_{BBB} + (1-w_A) \ln p^{(1)}_{BB}) \qquad (2.26)$$

$$p_B = p^{(1)}_B = \exp(w_B \ln p^{(1)}_{BBB} + (1-w_B) \ln p^{(1)}_{BB}). \qquad (2.27)$$

In our concrete example, we get

$$w_A = 0.17 \quad \text{and} \quad w_B = 0.63$$

solving Equations (2.26) and (2.27); see Table 1.1 for the values of $p_{BBB} = p^{(1)}_{BBB}$ and $p_{BB} = p^{(1)}_{BB}$.

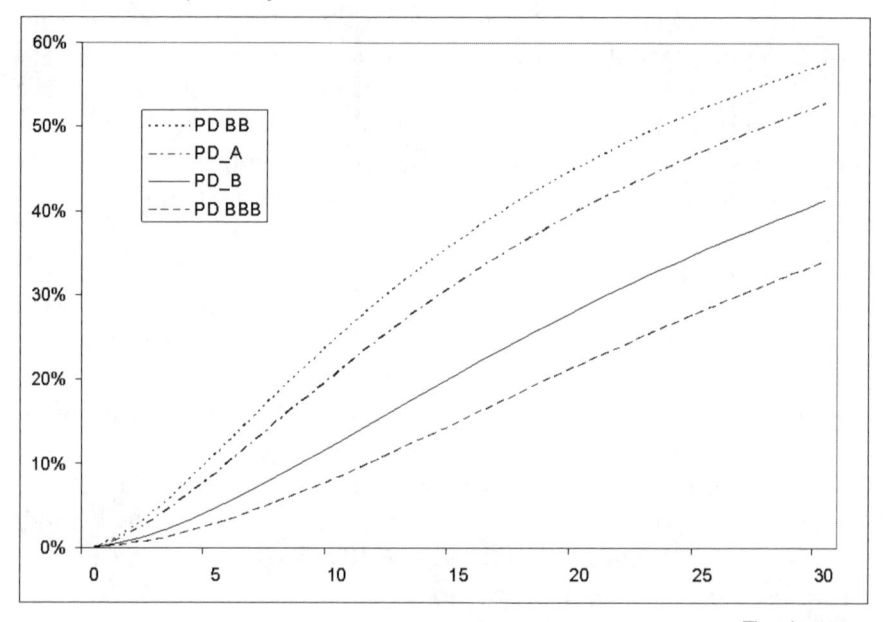

FIGURE 2.13: PD term structure interpolation for assets A and B (credit curves generated by the HCTMC approach)

The assumption in the log-linear interpolation approach is that the weights w_A and w_B can be used at any time $t > 0$ to interpolate between BBB and BB PD term structures in order to obtain PD term structures for assets A and B. That means that for any $t > 0$ we set

$$p_A^{(t)} = \exp(w_A \ln p_{BBB}^{(t)} + (1 - w_A) \ln p_{BB}^{(t)}) \qquad (2.28)$$

$$p_B^{(t)} = \exp(w_B \ln p_{BBB}^{(t)} + (1 - w_B) \ln p_{BB}^{(t)}) \qquad (2.29)$$

as illustrated in Figure 2.13.

Note again that PD term structures based on the the log-linear interpolation approach have to be considered as a *proxy* in cases where we do not have the chance to explicitly calibrate required PD term structures based on empirical data as we did in Section 2.3.

Another question we want to answer in this context is how joint default probabilities (JDPs) evolve over time. Figure 2.14 shows what we get in our particular example, calculated as in Remark 2.2.3.

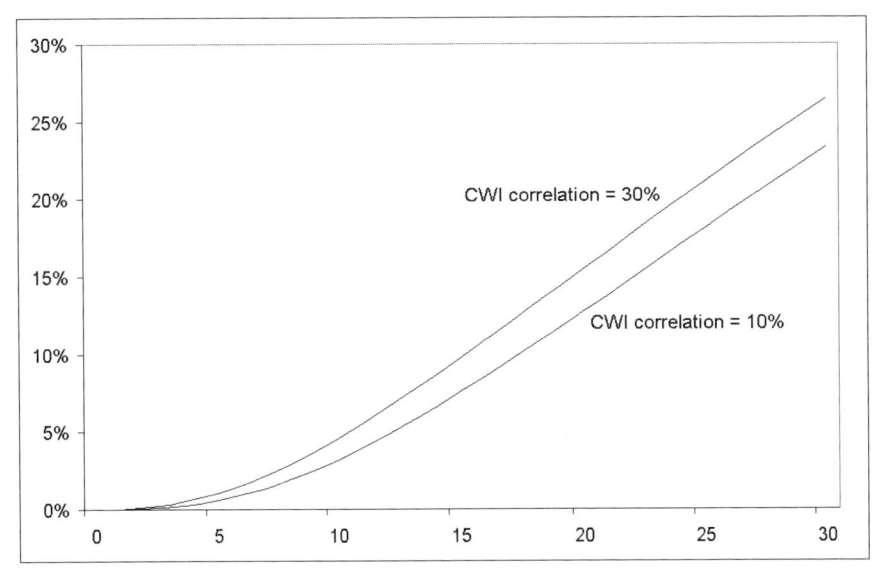

FIGURE 2.14: Time dependence of the joint default probability (JDP) for assets A and B with fixed CWI correlations of 10% and 30% (credit curves generated by the HCTMC approach)

At first sight, the difference between JDPs for CWI correlations at the 10% and 30% level does not seem to be that impressive. However, the appearance of Figure 2.14 is kind of misleading. If we look at short-term time horizons, we find that the percentage increase in the JDP when increasing the CWI correlation from 10% to 30% can be huge. This is illustrated in Figure 2.15.

For example, at the one-year horizon the JDP increases by 214% relative to the JDP at a CWI correlation of 10% when we increase the CWI correlation to 30%.

At the 5-year horizon, the relative percentage increase still is at the high level of almost 80%.

Therefore, it would be wrong to conclude from Figure 2.14 that JDPs moderately depend on the level of CWI correlations. Moreover, we should remark that *w.r.t. realistic time horizons JDPs are strongly driven by the assumed level of CWI correlation.*

At the same time, we conclude from Figure 2.15 that *for higher PDs the influence of the PDs on JDPs more and more dominates the influence of CWI correlations.* This conclusion can be made because JDPs at later time horizons are calculated in the same way as JDPs at earlier time horizons but with higher PDs.

This concludes the more general part of our discussion in this section and we now come back to our duo basket.

Let us now investigate how the first-to-default probability evolves over time. In line with Proposition 2.2.1, we can write

$$p_{1\text{st}}^{(t)} = \int_{-\infty}^{\infty} \left(g_{p_A^{(t)},\varrho}(y) + g_{p_B^{(t)},\varrho}(y)[1 - g_{p_A^{(t)},\varrho}(y)] \right) dN(y) \qquad (2.30)$$

for the first-to-default probability w.r.t. the t-year horizon. The conditional t-year PD of asset A in Equation (2.30) naturally equals

$$g_{p_A^{(t)},\varrho}(y) = N\left[\frac{c_A^{(t)} - \sqrt{\varrho}\, y}{\sqrt{1-\varrho}} \right] \qquad (2.31)$$

where the critical threshold is driven by the term structure $(p_A^{(t)})_{t \geq 0}$ via

$$c_A^{(t)} = N^{-1}[p_A^{(t)}] \qquad (t \geq 0). \qquad (2.32)$$

Analogous equations hold for asset B.

Default Baskets

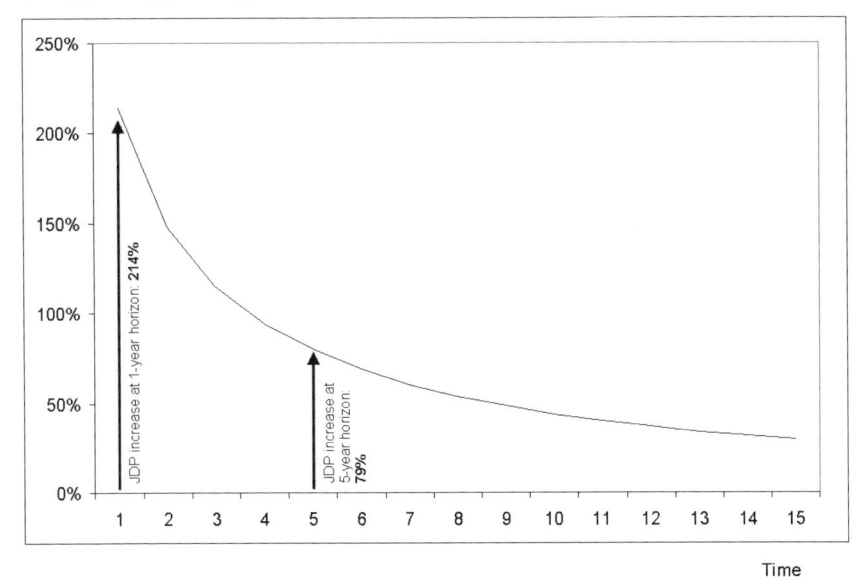

FIGURE 2.15: Time dependence of the relative percentage JDP increase arising from an increase of CWI correlations for assets A and B (credit curves generated by the HCTMC approach)

For the second-to-default probability of assets A and B we obtain

$$p_{2\text{nd}}^{(1)} = \int_{-\infty}^{\infty} g_{p_A^{(t)},\varrho}(y) g_{p_B^{(t)},\varrho}(y) \, dN(y). \tag{2.33}$$

in line with Proposition 2.2.2.

Based on Remark 2.2.3, showing that JDPs and second-to-default probabilities w.r.t. a fixed time horizon coincide, we already generated a plot of the second-to-default likelihood over time in Figure 2.14. For the first-to-default probability over time we calculated Figure 2.16. It shows the term structures of assets A and B as well as the term structure of the first-to-default likelihood w.r.t. three different CWI correlations (0%, 40%, and 70%) in order to illustrate the effect of correlations on the time evolution of the first-to-default probability.

In the discussion following Remark 2.2.4, we found that the perfect

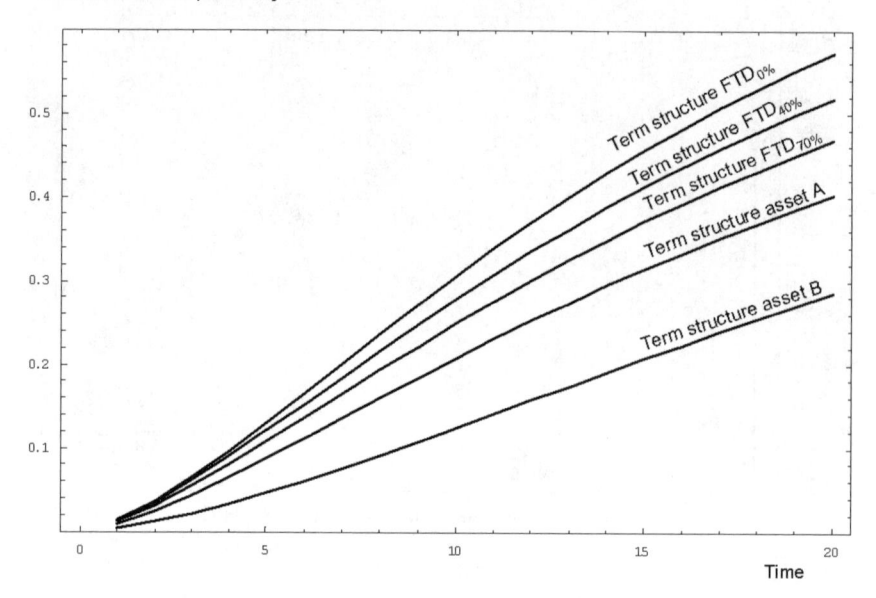

FIGURE 2.16: Time dependence of the first-to-default (FTD) probability of a duo basket consisting of assets A and B (credit curves generated by the HCTMC approach)

correlation case $\varrho = \varrho_{A,B} = 1$ yields

$$p^{(1)}_{1st} = \max[p_A, p_B].$$

This fact can be graphically guessed from Figure 2.16.

The more we increase the CWI correlation between assets A and B the more moves the term structure of the first-to-default toward the credit curve of asset A.

Remark 2.2.4 can also be graphically confirmed by Figure 2.16.

The lower the CWI correlation between assets A and B the higher the credit curve of the corresponding first-to-default.

2.5 Dependent Default Times

We now turn our attention to another core topic of this book by starting a discussion on *dependent default times*. In the literature, these objects are called *correlated default times* but we prefer the attribute 'dependent' (in the sense of 'not independent') instead of 'correlated', expressing the fact that dependencies imposed on our default time models are not exclusively based on linear correlation but also on other dependence measures; see Section 2.5.7. We begin with single-name default times in a first step and move toward dependent default times and applications in a second step.

2.5.1 Default Times and PD Term Structures

In the last section we extensively discussed PD term structures. A good justification for the time spent on this topic is that PD term structures and single-name default times are basically the same objects from a probability point of view. To see this and to start the discussion, let us denote by τ_R a (continuous) random variable defined on a suitable probability space $(\Omega, \mathcal{F}, \mathbb{P})$ representing the *time until default* (or, in short: *default time*) of a credit-risky instrument with a credit rating of R. Because τ_R represents a time, it ranges in $[0, \infty)$. For any given time horizon $t \geq 0$, the equation

$$\mathbb{P}[\tau_R \leq t] \stackrel{!}{=} p_R^{(t)} \qquad (2.34)$$

must hold, given the interpretation of the term structure $(p_R^{(t)})_{t \geq 0}$ of R-ratings as the *cumulative default probability* of R-rated instruments w.r.t. to any time horizon $t \geq 0$. It is worthwhile to state this simple but important relationship between PD term structures and default times in a separate remark.

2.5.1 Remark *The distribution of the default time of an R-rated instrument is uniquely determined by the PD term structure assigned to rating R. Single-name default times and PD term structures describe the same object from different angles.*

In other words, whenever people are convinced they have the right curve of cumulative default probabilities for a rating R, there is one and only one way, namely via Equation (2.34), to define the distribution of the default time for R-rated credit instruments.

2.5.2 Survival Function and Hazard Rate

Both of our term structure approaches (HCTMC and NHCTMC) generate smooth functions

$$t \mapsto p_R^{(t)}$$

for any rating R. Therefore, denoting the term structure functions in the form of a *default time distribution function*

$$\mathbb{F}_R(t) = \mathbb{P}[\tau_R \leq t] = p_R^{(t)}, \qquad (2.35)$$

we can derive the *default time density*

$$f_R : [0, \infty) \to [0, \infty)$$

of an R-rated credit instruments as the derivative of the default time distribution function as

$$f_R(t) = \frac{\partial}{\partial t} \mathbb{F}_R(t). \qquad (2.36)$$

Integrating the default time density gives us back the distribution function of τ_R via

$$\mathbb{F}_R(t) = \mathbb{P}[\tau_R \leq t] = \int_0^t f_R(s)\, ds.$$

In classical *survival analysis*, people also consider the *survival function*

$$S_R(t) = \mathbb{P}[\tau_R > t] = 1 - \mathbb{F}_R(t)$$

associated with rating R. Closely related to the survival function is the *hazard rate* function h_R of R-rated credit instruments defined by

$$h_R(t) = \frac{f_R(t)}{S_R(t)}. \qquad (2.37)$$

The relation between survival and hazard rate function is expressed in the following proposition.

2.5.2 Proposition Let S_R be the survival function and h_R be the hazard rate function of a given rating R. Then,

$$S_R(t) = \exp\left(-\int_0^t h_R(s)\,ds\right).$$

Proof. First of all, note that we can get back the default time density f_R from the survival function S_R via

$$f_R(t) = \frac{\partial}{\partial t}\mathbb{F}_R(t) = \frac{\partial}{\partial t}[1 - S_R(t)] = -\frac{\partial}{\partial t}S_R(t). \quad (2.38)$$

Therefore, we can write the hazard rate h_R as

$$h_R(t) = \frac{f_R(t)}{S_R(t)} = -\frac{1}{S_R(t)}\frac{\partial}{\partial t}S_R(t) = -\frac{\partial}{\partial t}\ln S_R(t).$$

Integration and taking exponentials proves the proposition. □

The hazard rate h_R has the following probabilistic interpretation:

$$\mathbb{P}[t \leq \tau_R \leq t + \delta \mid \tau_R \geq t] \approx \delta h_R(t) \quad \text{(for small } \delta\text{)}. \quad (2.39)$$

The following remark summarizes our findings above.

2.5.3 Remark *The PD term structure, the default time density, the survival function, and the hazard rate function all describe the same object, namely the distribution of the rating-dependent default time.*

Therefore, depending on the actual problem, credit risk modelers can work with the best suitable notion for the particular considered problem.

2.5.3 Calculation of Default Time Densities and Hazard Rate Functions

For illustration purposes, we now calculate default time densities in order to show the principal shapes we can expect for such probability densities. Hereby, we have to rely on our fallback credit curve model, namely the HCTMC approach, because the shape of default time densities becomes visible w.r.t. longer time horizons (beyond our data time horizon of 15 years) only.

There are basically two ways to calculate default time densities. First, we can use difference quotients as a proxy for derivatives and discretize the problem by calculating

$$f_R(t) \approx \frac{\mathbb{F}_R[t] - \mathbb{F}_R[t-\delta]}{\delta} = \frac{p_R^{(t)} - p_R^{(t-\delta)}}{\delta} \qquad (2.40)$$

for a specified 'mesh size' δ, which should be chosen sufficiently small in order to improve the approximation quality. Second, we can try to explicitly calculate the derivative (2.38). In case of the HCTMC approach, this can be done in fractions of a second via mathematical software like Mathematica or Matlab. We obtain the following:

$$\begin{aligned}
f_{AAA}(t) =\ & 2.311557 \times 10^{-6} \times e^{-0.648562 \times t} - 0.001020 \times e^{-0.276098 \times t} \\
& + 0.047183 \times e^{-0.172370 \times t} - 0.177495 \times e^{-0.131480 \times t} \\
& + 0.326801 \times e^{-0.092246 \times t} - 0.222215 \times e^{-0.067786 \times t} \\
& + 0.026762 \times e^{-0.014742 \times t}
\end{aligned}$$

$$\begin{aligned}
f_{AA}(t) =\ & -0.000034 \times e^{-0.648562 \times t} + 0.002697 \times e^{-0.276098 \times t} \\
& - 0.047946 \times e^{-0.172370 \times t} + 0.093266 \times e^{-0.131480 \times t} \\
& - 0.018913 \times e^{-0.092246 \times t} - 0.051534 \times e^{-0.067786 \times t} \\
& + 0.022472 \times e^{-0.014742 \times t}
\end{aligned}$$

$$\begin{aligned}
f_A(t) =\ & 0.000110 \times e^{-0.648562 \times t} - 0.006457 \times e^{-0.276098 \times t} \\
& + 0.035583 \times e^{-0.172370 \times t} - 0.017451 \times e^{-0.131480 \times t} \\
& - 0.027229 \times e^{-0.092246 \times t} - 0.003428 \times e^{-0.067786 \times t} \\
& + 0.019094 \times e^{-0.014742 \times t}
\end{aligned}$$

$$\begin{aligned}
f_{BBB}(t) =\ & -0.001178 \times e^{-0.648562 \times t} + 0.019624 \times e^{-0.276098 \times t} \\
& - 0.027258 \times e^{-0.172370 \times t} - 0.024451 \times e^{-0.131480 \times t} \\
& + 0.002838 \times e^{-0.092246 \times t} + 0.018179 \times e^{-0.067786 \times t} \\
& + 0.014387 \times e^{-0.014742 \times t}
\end{aligned}$$

$$\begin{aligned}
f_{BB}(t) =\ & 0.003384 \times e^{-0.648562 \times t} - 0.060121 \times e^{-0.276098 \times t} \\
& + 0.000187 \times e^{-0.172370 \times t} + 0.017316 \times e^{-0.131480 \times t} \\
& + 0.020854 \times e^{-0.092246 \times t} + 0.018687 \times e^{-0.067786 \times t} \\
& + 0.008521 \times e^{-0.014742 \times t}
\end{aligned}$$

$$\begin{aligned}f_B(t) = {} & -0.061491 \times e^{-0.648562 \times t} + 0.040093 \times e^{-0.276098 \times t} \\ & + 0.017396 \times e^{-0.172370 \times t} + 0.026446 \times e^{-0.131480 \times t} \\ & + 0.017124 \times e^{-0.092246 \times t} + 0.011624 \times e^{-0.067786 \times t} \\ & + 0.004281 \times e^{-0.014742 \times t}\end{aligned}$$

$$\begin{aligned}f_{CCC}(t) = {} & 0.385015 \times e^{-0.648562 \times t} + 0.016625 \times e^{-0.276098 \times t} \\ & + 0.006598 \times e^{-0.172370 \times t} + 0.008613 \times e^{-0.131480 \times t} \\ & + 0.006561 \times e^{-0.092246 \times t} + 0.003664 \times e^{-0.067786 \times t} \\ & + 0.001728 \times e^{-0.014742 \times t}\end{aligned}$$

Note that the simple form of the derivatives is due to the simple parameteric form of the HCTMC approach. To see this, denote by D the diagonal matrix of eigenvalues of Q and by L and R the matrix with orthonormal left and right eigenvectors respectively. The eigenvalues of Q from Table 2.6 are -0.6486, -0.2761, -0.1724, -0.1315, -0.0922, -0.0678, -0.0147, and 0. Then, Q admits the following representation:

$$Q = L * D * R$$

(here, '$*$' denotes matrix multiplication), which immediately leads to

$$\exp(tQ) = \sum_{k=0}^{\infty} \frac{t^k}{k!} L * D^k * R = L * \exp(tD) * R$$

where $\exp(tD)$ is a diagonal matrix with exponential functions on the diagonal. Differentiating the diagonalized matrix exponential leads to the form of default time densities as calculated above.

For the NHCTMC approach, default time densities gain in complexity due to the time-dependent scaling functions $\varphi_{\alpha,\beta}$ from (2.20) used in the definition of the time-dependent generator in (2.19).

Figure 2.17 shows the shape of default time densities generated by the HCTMC approach. In principal, we always expect default time densities to look like this. General comments are the following:

- The better the credit quality of an asset, the wider the default time density. The worse the credit quality of an asset, the more narrow the default time density.

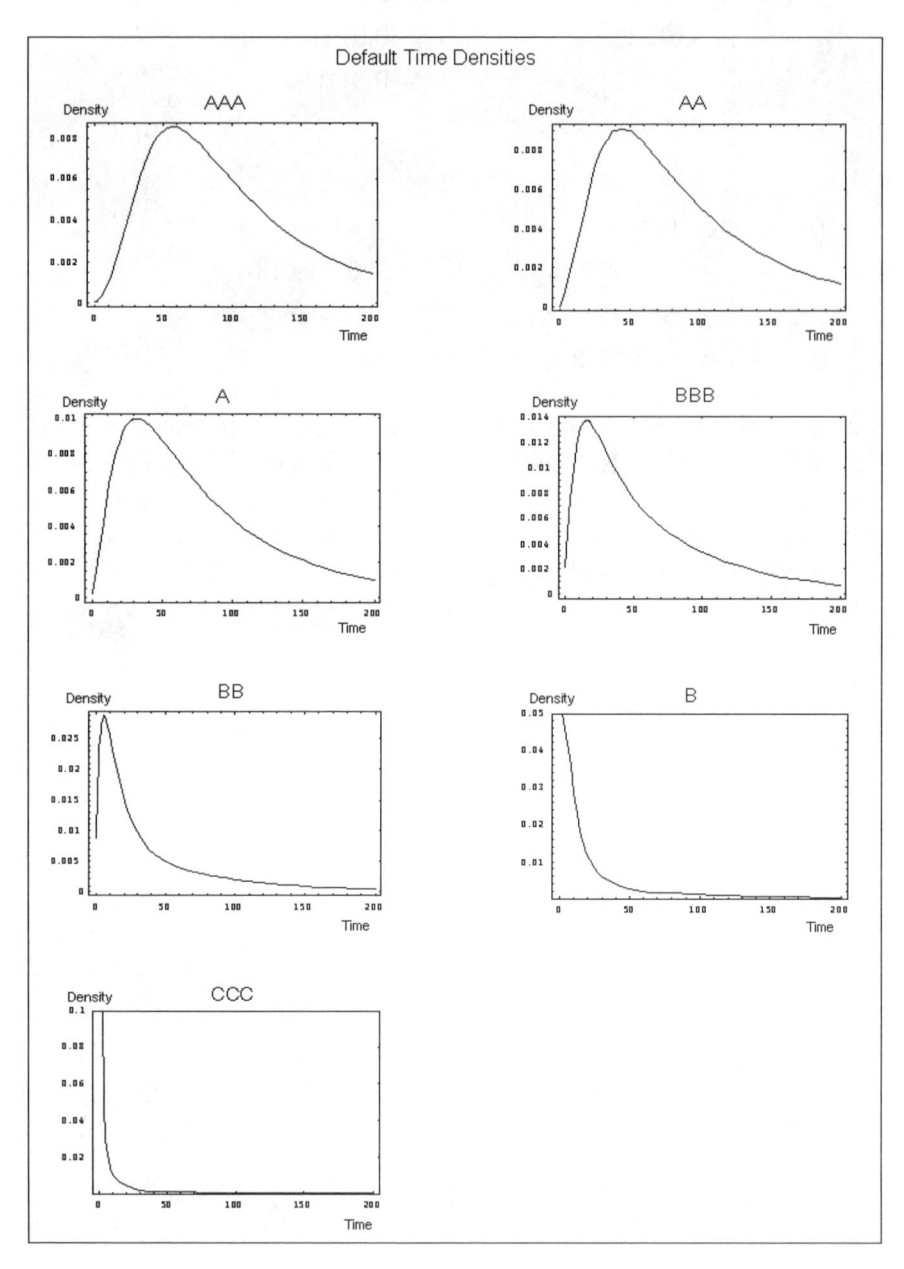

FIGURE 2.17: Illustration of the shape of default time densities (based on the HCTMC approach); the time axis counts in years

- Closely related, we find that naturally the mean default time of assets with high credit quality is far away in the future, whereas for assets with low credit quality the expected time until default is much shorter; compare, e.g., default time densities for AAA and CCC as extreme cases.

It is straightforward to calculate summary statistics for default time distributions. For instance, the *expected default time* of an *R*-rated credit instrument is given by

$$\mathbb{E}[\tau_R] = \int_0^\infty t f_R(t)\,dt$$

and the corresponding *default time standard deviation* equals

$$\sqrt{\mathbb{V}[\tau_R]} = \sqrt{\int_0^\infty (t - \mathbb{E}[\tau_R])^2 f_R(t)\,dt}.$$

Figure 2.18 graphically illustrates mean/expected default times and corresponding standard deviations as a function of the credit quality of the underlying asset. It is interesting to observe that the ratio

$$\frac{\sqrt{\mathbb{V}[\tau_R]}}{\mathbb{E}[\tau_R]} \tag{2.41}$$

increases stronger than linearly with decreasing credit quality. In fact, the ratio is more than four times larger for CCC than for AAA. This expresses the fact that low credit quality assets have a doubling effect at the credit risk side: *not only is their expected default time much nearer in the presence but also the deviation of default times from the expected default time can become higher than the mean default time itself.*

It is worthwhile to mention that *the ratio (2.41), increasing with decreasing credit quality, crosses the 100% level right at the borderline between investment and subinvestment grade*; see Figure 2.19.

Let us briefly comment on the numbers attached to Figure 2.18. How can we interpret the expected default time and its volatility? As a working example, we look at BB and, according to Figure 2.18, find that BB-rated credit instruments have an expected default time at the

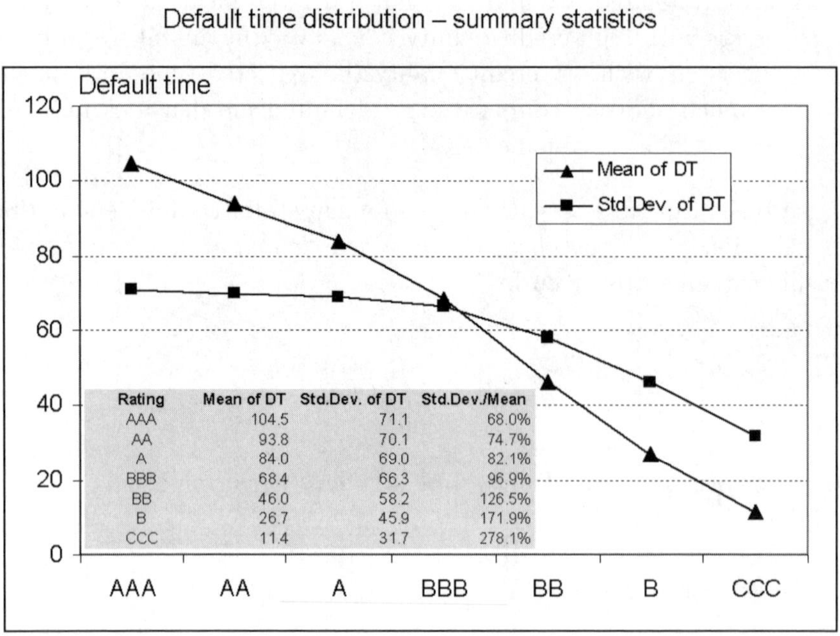

74   *Structured Credit Portfolio Analysis, Baskets & CDOs*

{graph showing Default time distribution – summary statistics, with Mean of DT and Std.Dev. of DT plotted across ratings AAA through CCC}

Rating	Mean of DT	Std.Dev. of DT	Std.Dev./Mean
AAA	104.5	71.1	68.0%
AA	93.8	70.1	74.7%
A	84.0	69.0	82.1%
BBB	68.4	66.3	96.9%
BB	46.0	58.2	126.5%
B	26.7	45.9	171.9%
CCC	11.4	31.7	278.1%

FIGURE 2.18: Summary statistics of default time densities (based on the HCTMC approach); time is measured in years

46-year time horizon. How does this compare to the one-year PD of BB? According to Table 1.1, we have

$$p_{BB} = p_{BB}^{(1)} = 1.28\%. \qquad (2.42)$$

Assuming a linear term structure[10], we find that p_{BB} fits slightly more often than 78 times into 100% such that from the 78-year horizon on the linear BB-credit curve must evolve flat at the 100% level. Figure 2.20 compares the two credit curves and the corresponding default time densities. It turns out that the expected default time for a linear BB-term structure roundabout equals 39 years. The HCTMC-based expected default time

$$\mathbb{E}[\tau_{BB}] = 46 \qquad (2.43)$$

is slightly later in time due to the shape of the BB-credit curve as shown in Figure 2.20. The linear BB-term structure puts BB-rated assets in

---
[10] A term structure is 'linear' if the cumulative n-year PD equals $n \times PD^{(1)}$ where $PD^{(1)}$ represents the one-year default probability.

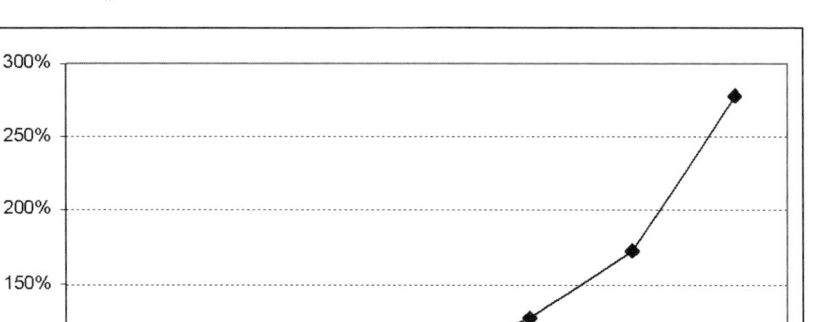

Ratio „standard deviation over mean" for default time distributions

FIGURE 2.19: The ratio (2.41) as a function of credit quality (based on the HCTMC approach)

a slightly worse light than necessary: In case of the HCTMC-based credit curve, we have for the corresponding default time density

$$f_{BB}(24) \approx 1.29\% \quad \text{and} \quad f_{BB}(25) \approx 1.24\%$$

such that after the 25th year the forward PDs in the HCTMC case are lower than the corresponding forward PDs (which are flat and equal to 1.28%) in the linear credit curve case. In addition, as already indicated in the discussion after Proposition 2.3.4, forward PDs for BB-rated credit instruments are decreasing in the HCTMC case and the default time distribution puts mass beyond the 78-year time horizon. Therefore, it is only natural that on average the linear BB-credit curve indicates an expected default time at some earlier time than we would obtain in case of HCTMC-based BB-credit curves.

As a last aspect in this context, we refer back to Remark 2.5.3 where we learned that default times can also be expressed in terms of hazard rates. If we calculate hazard rates for our HCTMC-based PD term structures, we obtain Figure 2.21. Proposition 2.5.2 constitutes

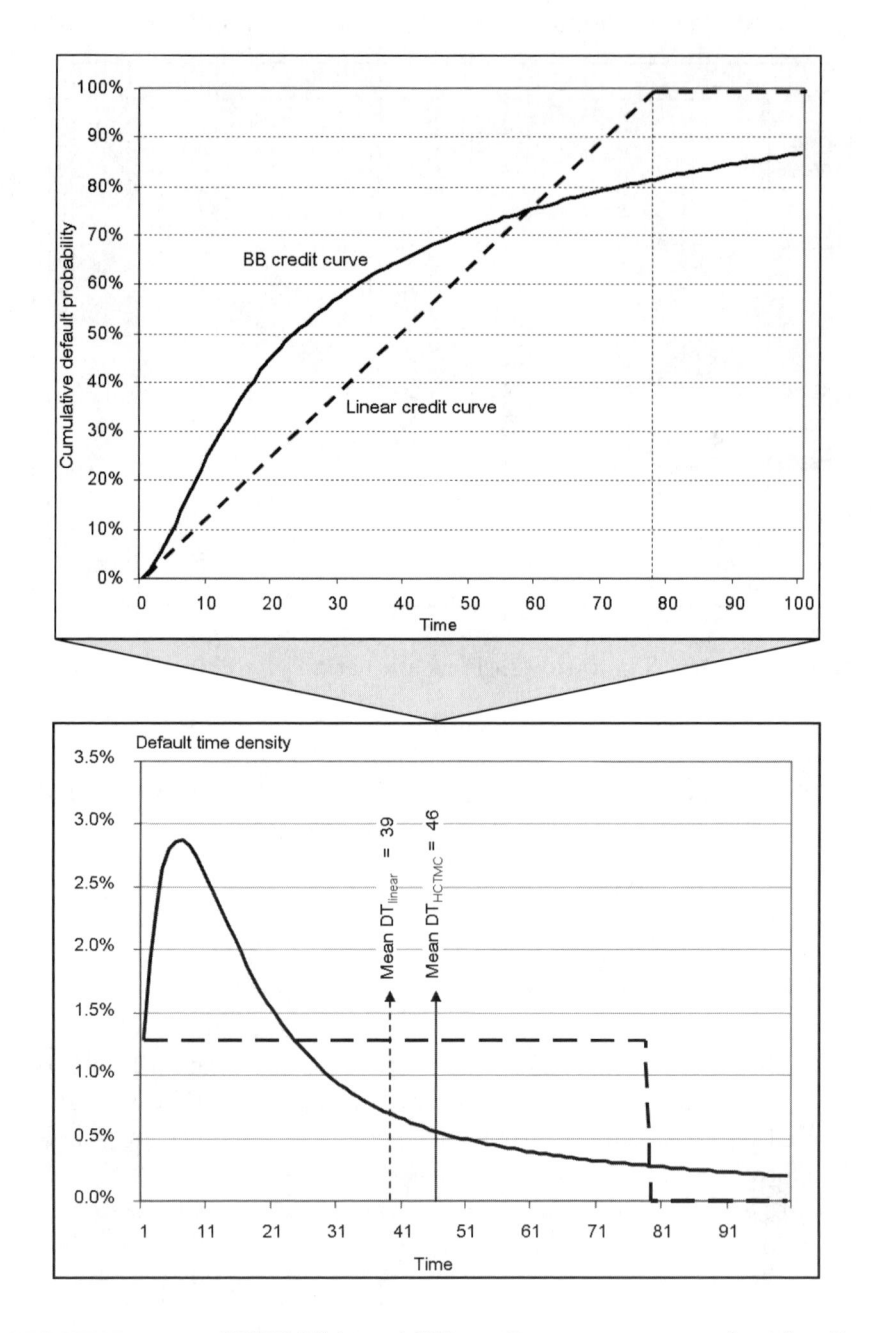

FIGURE 2.20: HCTMC-based BB-credit curve compared with a linear BB-credit curve; time is measured in years

Default Baskets

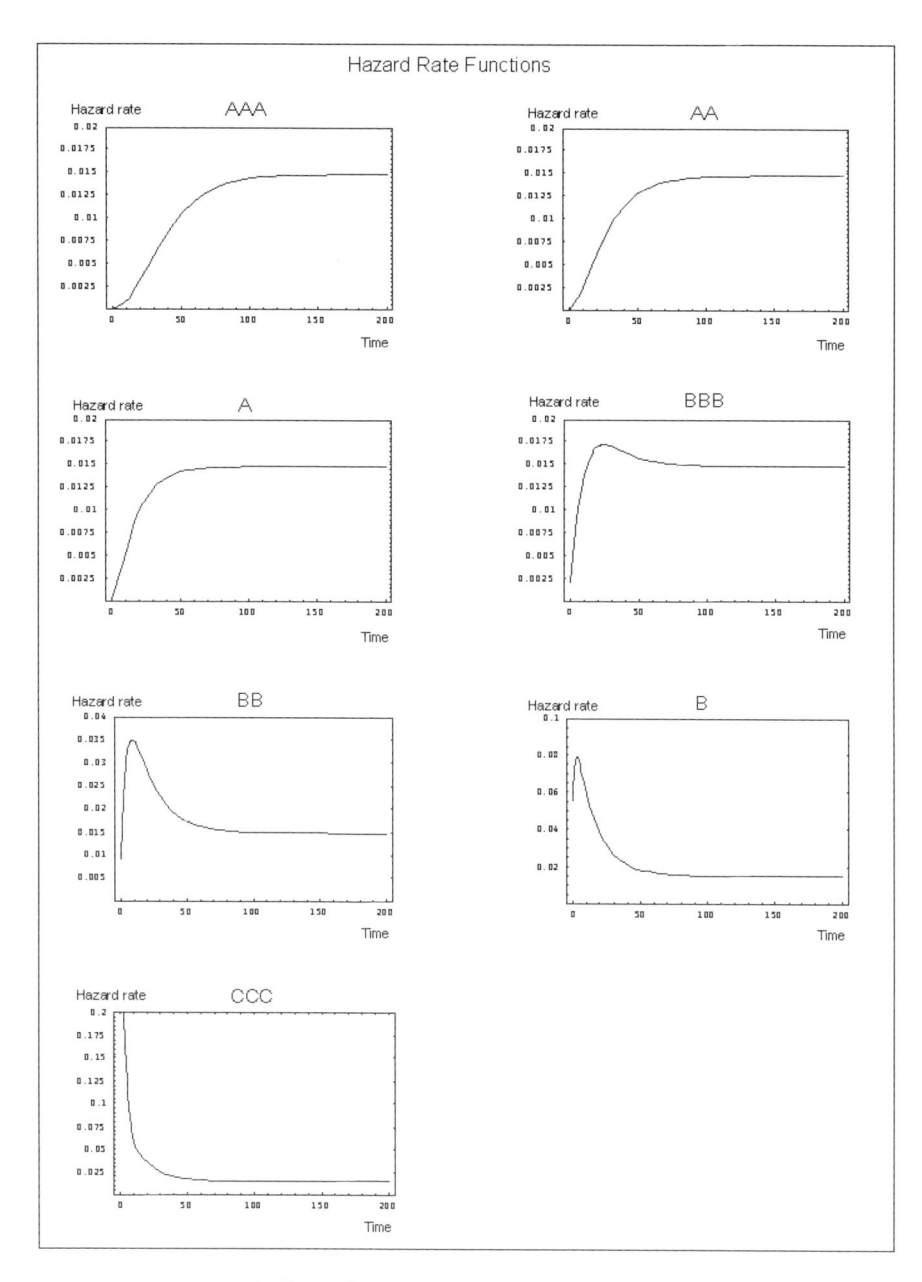

FIGURE 2.21: HCTMC-based hazard rates illustrating the discussion in Section 2.5.2; time is measured in years

a straightforward way to test by means of a simple calculation if the derived hazard rates are correct.

Based on the probabilistic interpretation of hazard rates, see (2.39), one could say that hazard rates are continuous-time forward default rates at an 'infinitesimal small' time scale. Therefore, it is not surprising that, in the same way as we discovered it for forward PDs, the hazard rate functions are monotonically increasing for good credit qualities and in the long run decreasing for bad credit qualities. The borderline between good and bad credit qualities again coincides with the borderline between investment and subinvestment grade: as can be seen in Figure 2.21, BBB is the first rating grade, moving from investment grade downward to subinvestment grade, where the hazard rate function is no longer monotonically increasing over time but almost flat for longer time horizons.

This concludes our discussion on single-name default times. In the next section, we turn our attention to dependence modeling. As a first step, we review a simple MERTON-style model approach, which leads us directly to the heart of our topic. At the end, we will have a nice concept for the construction of *dependent default times* at our disposal.

2.5.4 From Latent Variables to Default Times

In this section we briefly review a simplified version of a well-known asset value model framework for the CWIs of assets A and B in our duo basket and illustrate how one can obtain default times in such a framework. For this purpose, we assume a simple standard distribution for the time evolution of the CWI processes of assets A and B in our duo basket. Note that here we speak of CWI *processes* in contrast to our introductory remarks on CWIs in Chapter 1 where we pointed out that our CWI concept is a 'fixed evaluation horizon' approach. However, later in the discussion we will again drop the process view and consider CWIs w.r.t. fixed time horizons. The most commonly applied process in this context is *geometric Brownian motion* (gBm); see the seminal papers by BLACK and SCHOLES [23] and MERTON [86]. Under a gBm assumption for CWIs, we have

$$\mathrm{CWI}_A^{(t)} = \mathrm{CWI}_A^{(0)} \exp\bigl[(\mu_A - \tfrac{1}{2}\sigma_A^2)t + \sigma_A B_A^{(t)}\bigr] \quad (2.44)$$

$$\mathrm{CWI}_B^{(t)} = \mathrm{CWI}_B^{(0)} \exp\bigl[(\mu_B - \tfrac{1}{2}\sigma_B^2)t + \sigma_B B_B^{(t)}\bigr]$$

where the processes $(B_A^{(t)})_{t\geq 0}$ and $(B_B^{(t)})_{t\geq 0}$ are correlated *Brownian motions* with a correlation equal to the CWI correlation $\varrho = \varrho_{A,B}$ arising in Equation (2.3). The parameters μ_A and σ_A (as well as μ_B and σ_B) refer to the usual interpretation of CWIs in terms of *asset value processes* such that default is triggered by a function depending on assets and liabilities of the considered firm over time. The parameters μ_A and σ_A can then be seen as *mean rate of return* and *volatility* of asset A. Note that in this model framework the expectation and volatility functions of CWI_A are given by

$$\mathbb{E}[\text{CWI}_A^{(t)}] = \text{CWI}_A^{(0)} \exp(\mu_A t) \tag{2.45}$$

$$\mathbb{V}[\text{CWI}_A^{(t)}] = [\text{CWI}_A^{(0)}]^2 \exp(2\mu_A t)\left(\exp(\sigma_A^2 t) - 1\right). \tag{2.46}$$

Analogous equations hold for asset B.

Now let us fix some time horizon $T > 0$. We introduce indicator variables

$$L_A^{(T)} = \mathbf{1}_{\{\text{CWI}_A^{(T)} < \tilde{c}_A^{(T)}\}} \tag{2.47}$$

$$L_B^{(T)} = \mathbf{1}_{\{\text{CWI}_B^{(T)} < \tilde{c}_B^{(T)}\}} \tag{2.48}$$

where $\tilde{c}_A^{(T)}$ and $\tilde{c}_B^{(T)}$ denotes the default-critical thresholds w.r.t. the time horizon T for obligors A and B, respectively, in line with our CWI-based model triggering default of credit-risky instruments (as discussed in previous sections of this book). As a next step, we recalculate the default likelihood as

$$\begin{aligned}
\mathbb{P}[L_A^{(T)} = 1] &= \mathbb{P}[\text{CWI}_A^{(T)} < \tilde{c}_A^{(T)}] \\
&= \mathbb{P}\left[\text{CWI}_A^{(0)} \exp\left[(\mu_A - \tfrac{1}{2}\sigma_A^2)T + \sigma_A B_A^{(T)}\right] < \tilde{c}_A^{(T)}\right] \\
&= \mathbb{P}\left[\text{CWI}_A^{(0)} \exp\left[(\mu_A - \tfrac{1}{2}\sigma_A^2)T + \sigma_A B_A^{(T)}\right] < \tilde{c}_A^{(T)}\right] \\
&= \mathbb{P}\left[B_A^{(T)} < \frac{\ln(\tilde{c}_A^{(T)}/\text{CWI}_A^{(0)}) - (\mu_A - \tfrac{1}{2}\sigma_A^2)T}{\sigma_A}\right]
\end{aligned} \tag{2.49}$$

The variable $B_A^{(T)}$ is normally distributed, $B_A^{(T)} \sim N(0,T)$, where $N(\mu, \sigma^2)$ denotes the normal distribution with mean μ and variance σ^2. Therefore, we can define a *standardized default point* $c_A^{(T)}$ by

$$c_A^{(T)} = \frac{\ln(\tilde{c}_A^{(T)}/\text{CWI}_A^{(0)}) - (\mu_A - \tfrac{1}{2}\sigma_A^2)T}{\sigma_A \sqrt{T}} \tag{2.50}$$

such that Equation (2.49) turns into

$$\mathbb{P}[L_A^{(T)} = 1] = \mathbb{P}[\text{CWI}_A < c_A^{(T)}] \quad (2.51)$$

with a standard normal random variable $\text{CWI}_A \sim B_A^{(1)} \sim N(0,1)$. Note that the calculation just shown holds *in distribution only*; the time dynamic of the Brownian motion process is no longer reflected when switching from $B_A^{(T)}$ to $\text{CWI}_A \sim B_A^{(1)}$ in (2.51) via standardization. Again, analogous equations hold for asset B. CWI_A and CWI_B can now be considered like a general *timeless* credit worthiness index such that the impact of bad or good CWI realizations is that default happens earlier or later, respectively.

We can link Equation (2.51) to the credit curve $(p_A^{(t)})_{t\geq 0}$ of asset A by writing (for any $t \geq 0$)

$$\mathbb{P}[\text{CWI}_A < c_A^{(t)}] = \mathbb{P}[L_A^{(t)} = 1] \overset{!}{=} p_A^{(t)} \quad (2.52)$$

which holds by definition of the probabilities $p_A^{(t)}$. From here we can conclude that

$$c_A^{(t)} = N^{-1}[p_A^{(t)}] = N^{-1}[\mathbb{F}_A(t)] \quad (2.53)$$

where $\mathbb{F}_A[\cdot]$ again denotes the term structure function, or, as we called it previously, the default time distribution function

$$\mathbb{F}_A : t \mapsto p_A^{(t)}.$$

Inserting Equation (2.53) into Equation (2.51) yields

$$\mathbb{P}[L_A^{(t)} = 1] = \mathbb{P}[\mathbb{F}_A^{-1}(N[\text{CWI}_A]) < t] \quad (2.54)$$

$$\mathbb{P}[L_B^{(t)} = 1] = \mathbb{P}[\mathbb{F}_B^{-1}(N[\text{CWI}_B]) < t] \quad (2.55)$$

for all $t \geq 0$. Following the suggestion in LI [74, 75], we can set

$$\tau_A = \mathbb{F}_A^{-1}(N[\text{CWI}_A]) \quad \text{and} \quad \tau_B = \mathbb{F}_B^{-1}(N[\text{CWI}_B]) \quad (2.56)$$

and interpret these random variables as the *default times* of assets A and B, respectively. We can then re-discover the functions $\mathbb{F}_A$ and $\mathbb{F}_B$ as the cumulative distribution functions of the default times τ_A and τ_B because by construction we have

$$\begin{aligned}\mathbb{P}[\tau_A < t] &= \mathbb{P}[\mathbb{F}_A^{-1}(N[\text{CWI}_A]) < t] \\ &= \mathbb{P}\left[\text{CWI}_A < N^{-1}[\mathbb{F}_A(t)]\right] \\ &= N\left[N^{-1}[\mathbb{F}_A(t)]\right] \\ &= \mathbb{F}_A(t)\end{aligned}$$

based on $\text{CWI}_A \sim N(0,1)$ due to our distributional assumptions.

Equation (2.56) includes an approach for the *simulation of default times*. We will come back to this feature later in this section; see Figure 2.22.

For dependence considerations, the t-year joint default probability $\text{JDP}_{A,B}^{(t)}$ of assets A and is a good starting point. In line with Equation (2.14) where the case $t=1$ is described and based on Equation (2.51), the t-year JDP of assets A and B is given by

$$\begin{aligned}\text{JDP}_{A,B}^{(t)} &= \mathbb{P}[\mathbf{1}_{\{\text{CWI}_A^{(t)} < \tilde{c}_A^{(t)}\}} = 1, \mathbf{1}_{\{\text{CWI}_B^{(t)} < \tilde{c}_B^{(t)}\}} = 1] \\ &= \mathbb{P}[\text{CWI}_A < c_A^{(t)}, \text{CWI}_B < c_B^{(t)}] \end{aligned} \quad (2.57)$$

where $c_A^{(t)}$ and $c_B^{(t)}$ denote the standardized default points of assets A and B, respectively. In the definition of gBm (right after Equation (2.44)) we assumed the Brownian motions to be correlated with a CWI correlation of $\varrho = \varrho_{A,B}$. This implies that $(\text{CWI}_A, \text{CWI}_B)$ follows a bi-variate normal distribution with correlation ϱ. Based on Equations (2.53) and (2.57), we have

$$\begin{aligned}\text{JDP}_{A,B}^{(t)} &= \mathbb{P}[\text{CWI}_A < c_A^{(t)}, \text{CWI}_B < c_B^{(t)}] \\ &= N_2[c_A^{(t)}, c_B^{(t)}; \varrho] \\ &= N_2[N^{-1}[p_A^{(t)}], N^{-1}[p_B^{(t)}]; \varrho] \\ &= N_2[N^{-1}[\mathbb{F}_A(t)], N^{-1}[\mathbb{F}_B(t)]; \varrho] \end{aligned} \quad (2.58)$$

where $N_2[\cdot, \cdot\,; \varrho]$ denotes the standard bi-variate normal distribution function with a correlation of ϱ; recall that we applied the distribution function N_2 already in Equation (2.5).

The formula for the JDP in Equation (2.58) can be written in the following form,

$$\text{JDP}_{A,B}^{(t)} = C(\mathbb{F}_A(t), \mathbb{F}_B(t)) \quad (2.59)$$

with a function $C : [0,1]^2 \to [0,1]$ defined by

$$C(u_A, u_B) = N_2[N^{-1}[u_A], N^{-1}[u_B]; \varrho]. \quad (2.60)$$

C is a *bi-variate distribution function with uniform marginal distributions* C_A and C_B. To see this, choose a Gaussian pair (X, Y) with zero mean and correlation ϱ. Then,

$$\begin{aligned}
C_A(u_A) &= C(u_A, 1) \\
&= \mathbb{P}[X \leq N^{-1}[u_A], Y \leq N^{-1}[1]] \\
&= \mathbb{P}[N[X] \leq u_A] \\
&= u_A
\end{aligned} \tag{2.61}$$

due to $N[X] \sim U([0,1])$ where $U([0,1])$ denotes the uniform distribution on the unit square.

Functions like C in Equation (2.60) are called *copula functions*. Our particular example in Equation (2.60) is the *Gaussian* or *normal copula function*, where for abbreviation reasons the word 'function' most often is suppressed such that people briefly say 'Gaussian copula' or 'normal copula'. In the next section, we will study copula functions and provide, in addition to the Gaussian copula, three other copulas we found useful in the context of basket and CDO modeling and evaluation.

Before we turn our attention to copulas as dependence modeling tools, we want to write Equations (2.58) and (2.59) in a slightly more general form allowing for individual time horizons for assets A and B,

$$\begin{aligned}
\mathbb{P}[\tau_A < t_A, \tau_B < t_B] &= \mathbb{P}[\mathbb{F}_A^{-1}(N[\mathrm{CWI}_A]) < t_A, \mathbb{F}_B^{-1}(N[\mathrm{CWI}_B]) < t_B] \\
&= \mathbb{P}[\mathrm{CWI}_A < c_A^{(t_A)}, \mathrm{CWI}_B < c_B^{(t_B)}] \\
&= N_2[N^{-1}[\mathbb{F}_A(t_A)], N^{-1}[\mathbb{F}_B(t_B)]; \varrho] \\
&= C(\mathbb{F}_A(t_A), \mathbb{F}_B(t_B)).
\end{aligned} \tag{2.62}$$

Equation (2.62) means that the joint distribution of default times τ_A and τ_B has a *Gaussian copula representation* in a way that the joint distribution function of the two default times can be written as the Gaussian copula evaluated at the marginal default time distribution functions evaluated at the considered time horizons t_A and t_B.

2.5.4 Remark *Equation (2.62) indicates that we can combine single-name default time distributions (in our example $\mathbb{F}_A$ and $\mathbb{F}_B$) to a joint or correlated/dependent multi-variate default times distribution via any specified copula function C.*

In the next section, we will elaborate this idea and provide some examples and test calculations.

Default Baskets

At the end of this section we show how (2.56) can be used to simulate dependent default times in a straightforward manner. The following algorithm is a widely used application of the theory just discussed in order to simulate correlated default times.

Starting point for the simulation scheme is Equation (2.56),

$$\tau_A = \mathbb{F}_A^{-1}(N[\text{CWI}_A]) \quad \text{and} \quad \tau_B = \mathbb{F}_B^{-1}(N[\text{CWI}_B])$$

The ingredients in the default time definition are

- The term structure or default time distribution functions $\mathbb{F}_A$ and $\mathbb{F}_B$ of the assets in the considered portfolio (duo basket in our example)

- The latent variables CWI_A and CWI_B originally used to trigger default at the one-year time horizon

The latter-mentioned incorporate the dependence structure of the normal copula function due to their joint distribution being bi-variate normal with correlation ϱ. In Equation (2.3) we explicitly generated a normal copula dependency via factor equations

$$\text{CWI}_A = \sqrt{\varrho}\, Y + \sqrt{1-\varrho}\, \varepsilon_A$$

$$\text{CWI}_B = \sqrt{\varrho}\, Y + \sqrt{1-\varrho}\, \varepsilon_B$$

with independent standard normal random variables Y, ε_A, and ε_B. The simulation scheme for dependent default times τ_A and τ_B then relies on *scenario generation* in a Monte Carlo simulation as follows:

1. Step: randomly draw a realization $\hat{Y}$, $\hat{\varepsilon}_A$, and $\hat{\varepsilon}_B$ from independent random variables $Y, \varepsilon_A, \varepsilon_B \sim N(0,1)$.

2. Step: calculate CWI realizations

$$\hat{\text{CWI}}_A = \sqrt{\varrho}\, \hat{Y} + \sqrt{1-\varrho}\, \hat{\varepsilon}_A$$

$$\hat{\text{CWI}}_B = \sqrt{\varrho}\, \hat{Y} + \sqrt{1-\varrho}\, \hat{\varepsilon}_B$$

based on realizations of random variables from the 1. Step.

3. Step: apply the standard normal distribution function to CWI realizations in order to obtain $N[\hat{\text{CWI}}_A]$ and $N[\hat{\text{CWI}}_B]$.

4. **Step:** Invert the term structure (default time distribution) functions in order to obtain $\mathbb{F}_A^{-1}$ and $\mathbb{F}_B^{-1}$ for calculating

$$\hat{\tau}_A = \mathbb{F}_A^{-1}(N[\hat{\mathrm{CWI}}_A]), \qquad \hat{\tau}_B = \mathbb{F}_B^{-1}(N[\hat{\mathrm{CWI}}_B]).$$

The pair $(\hat{\tau}_A, \hat{\tau}_B)$ is called a *default time scenario* for the duo basket simulation.

5. **Step:** Go back to the 1. Step as many times as default time scenarios are required for the calculation of stable summary statistics.

Figure 2.22 illustrates the simulation scheme. Note that in case of more than two assets in the underlying portfolio the simulation will be slightly more complicated because some underlying factor model structure (see Figure 1.7) has to be incorporated in the simulation scheme. In more technical terms, this means that the systematic factor Y has to be replaced by some systematic (composite) factor of the considered asset and the uniform correlation parameter ϱ will be replaced by the asset's R-squared or beta factor.

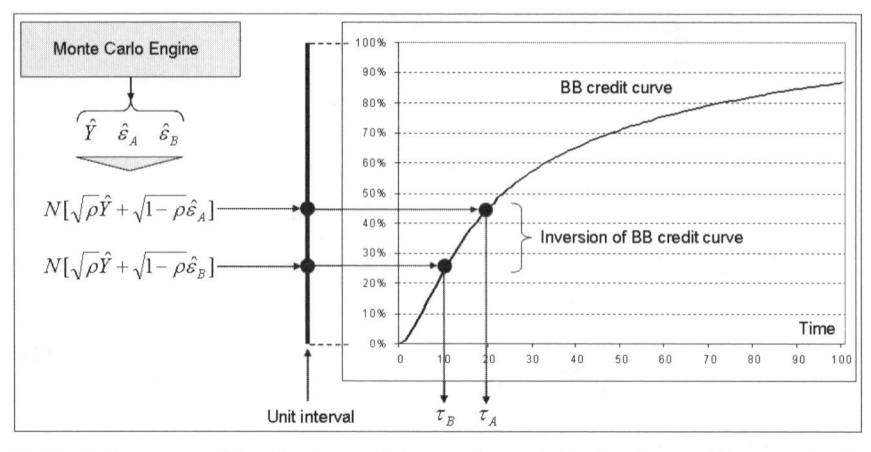

FIGURE 2.22: Simulation of dependent default times (illustrative)

Following this simulation scheme, we obtain a set

$$\left(\hat{\tau}_A^{(i)}, \hat{\tau}_B^{(i)}\right)_{i=1,\ldots,n}$$

of *default time scenarios* where n denotes the number of scenarios in the Monte Carlo simulation. Given n is sufficiently large, we can hope to get stable estimates of relevant credit risk quantities calculated by means of the default time scenarios. For example, the first-to-default probability w.r.t. a time horizon $T > 0$ can be estimated from default time scenarios by

$$\hat{p}_{1st}^{(T)} = \frac{1}{n} \sum_{i=1}^{n} \mathbf{1}_{\{\min(\hat{\tau}_A^{(i)}, \hat{\tau}_B^{(i)}) < T\}} \qquad (2.63)$$

and the corresponding second-to-default probability is given by

$$\hat{p}_{2nd}^{(T)} = \frac{1}{n} \sum_{i=1}^{n} \mathbf{1}_{\{\max(\hat{\tau}_A^{(i)}, \hat{\tau}_B^{(i)}) < T\}}. \qquad (2.64)$$

In the next section we will refine the simulation scheme by considering different dependence structures induced by different copulas, hereby following Remark 2.5.4.

2.5.5 Dependence Modeling via Copula Functions

In the previous section, we introduced the copula functions as a concept naturally arising in a simplified Gaussian asset value framework. We have seen in Equation (2.61) that the Gaussian copula is a multi-variate distribution function with uniform marginal distributions. This motivates the following more general definition.

2.5.5 Definition *A copula (function) is a multi-variate distribution (function) such that its marginal distributions are uniform in the unit interval. As in the previous section, we denote copulas as*

$$C(u_1, ..., u_m) : [0,1]^m \to [0,1]$$

if considered in R^m, as it will be the case for portfolios with m assets.

The following two theorems show that copulas offer a universal tool for constructing and studying multi-variate distributions. We start with the fact that multi-variate distributions, in general, admit a so-called *copula representation*; see also our discussion right after Equation (2.62) where we found the Gaussian copula representation of our CWI-based default times τ_A and τ_B.

2.5.6 Theorem (SKLAR [106], [107]) *For any m-dimensional distribution function $\boldsymbol{F}$ with marginal distributions $F_1, ..., F_m$, there exists a copula function C such that*

$$\boldsymbol{F}(x_1, ..., x_m) = C(F_1(x_1), ..., F_m(x_m)) \qquad (x_1, ..., x_m \in \mathbb{R}).$$

Moreover, if the marginal distributions $F_1, ..., F_m$ are continuous, then C is unique.

Sketch of Proof. Define the following function on $[0,1]^m$,

$$C(u_1, ..., u_m) = \boldsymbol{F}(F_1^{-1}(u_1), ..., F_m^{-1}(u_m)). \qquad (2.65)$$

Then one only has to verify that C is a copula representing $\boldsymbol{F}$; see also NELSON [90]. □

In Equation (2.62) and Remark 2.5.4 we already found that the converse direction also holds. Not only can any multi-variate distribution be written by means of a copula representation. Any given marginal distributions can also be combined to a multi-variate distribution by 'binding together' the given marginals via any chosen copula function. This is summarized in the following proposition.

2.5.7 Proposition *Given a copula C and (marginal) distribution functions $F_1, ..., F_m$ on $\mathbb{R}$, the function*

$$\boldsymbol{F}(x_1, ..., x_m) = C(F_1(x_1), ..., F_m(x_m)) \qquad (x_1, ..., x_m \in \mathbb{R})$$

defines a multi-variate distribution function with marginal distribution functions $F_1, ..., F_m$.

Proof. The proof is straightforward. □

Next, we want to bring the copula concept alive by discussing several copula functions later in the book applied to structured credit products.

2.5.5.1 The Independence Copula

An especially straightforward example of a copula function is the so-called *independence copula*. It is defined by

$$C_\perp(u_1, ..., u_m) = \prod_{k=1}^{m} u_k \qquad (u_1, ..., u_m \in [0,1]). \qquad (2.66)$$

A family of continuous random variables $X_1, ..., X_m$ with marginal distribution functions $F_1, ..., F_m$ is independent if and only if their joint distribution function can be written as

$$\mathbb{P}[X_1 \leq x_1, ..., X_m \leq x_m] = C_\perp(F_1(x_1), ..., F_m(x_m))$$

for all $x_1, ..., x_m$ in the range of $X_1, ..., X_m$. The proof is obvious due to the product formula for independent random variables.

2.5.5.2 The Gaussian Copula

In Equation (2.62) we found that the Gaussian copula naturally arises in the context of CWIs if the CWI processes are assumed to follow a geometric Brownian motion leading to Gaussian log-returns w.r.t. any given time horizon. The general definition of the *Gaussian copula* is

$$C_{m,\Gamma}(u_1, ..., u_m) = N_m[N^{-1}[u_1], ..., N^{-1}[u_m]; \Gamma] \qquad (2.67)$$

where $N[\cdot]$ denotes the standard normal distribution function as usual, $N^{-1}[\cdot]$ denotes its inverse and $N_m[...; \Gamma]$ refers to the multi-variate Gaussian distribution function on $\mathbb{R}^m$ with correlation matrix

$$\Gamma = (\varrho_{ij})_{1 \leq i,j \leq m}$$

and zero mean. In the applications we have in mind, the parameter m typically refers to the number of obligors or assets in some considered reference portfolio. Gaussian copulas and their application to baskets and CDOs have been studied in various papers; see, e.g., [48], [74], [102], just to mention a few examples. One can safely say that whenever people are not explicitly addressing the problem of 'how to find the appropriate copula' they most often rely on a Gaussian copula. A possible mathematical justification for choosing the Gaussian copula as kind of a standard case is explained in Appendix 6.5, where we outline that the Gaussian copula is an *entropy maximizing distribution* in a certain context, where 'maximum entropy' means the same as 'most randomly behaving'; see Appendix 6.5.

A major advantage of the Gaussian or normal copula is that only the linear correlations between assets, represented by the correlation matrix Γ, have to be calibrated. Practitioners know that for some asset classes like, e.g., loans to private clients, already the calibration of Γ (typically done via a factor model) is quite some challenge. Switching to a more complex copula with a need to calibrate even more parameters (e.g., as it is the case for the Student-t copula, where in addition to Γ also the degrees of freedom have to be calibrated) will certainly make the situation more problematic.

In the sequel, we always denote the Gaussian copula by $C_{m,\Gamma}$.

2.5.5.3 The Student-t Copula

If one would make a ranking of the most widely used copulas, the *Student-t copula* would occupy rank 2, after the Gaussian copula at rank 1. It is defined by

$$C_{m,\Gamma,d}(u_1,...,u_m) = \Theta_{m,\Gamma,d}[\Theta_d^{-1}[u_1],...,\Theta_d^{-1}[u_m]] \qquad (2.68)$$

where $\Theta_{m,\Gamma,d}$ denotes the multi-variate t-distribution function with d degrees of freedom and (linear) correlation matrix $\Gamma \in \mathbb{R}^{m \times m}$, Θ_d denotes the t-distribution function with d degrees of freedom and Θ_d^{-1} refers to the corresponding inverse.

In Appendix 6.3, we provide a brief 'wrap-up' on the t-distribution and introduce the notation used later on. We also mention that with increasing degrees of freedom d the t-dependence gets closer and closer to a Gaussian dependence. This is a nice feature of the t-distribution, which one can use in simulation models. One can start with a t-copula with large d and then successively decrease the degrees of freedom d in order to generate fatter and fatter tails with decreasing d.

The Student-t copula will always be denoted by $C_{m,\Gamma,d}$ such that the degrees of freedom in the subscript line distinguish t-copulas from Gaussian copulas in our notation.

2.5.5.4 Archimedean Copulas (e.g., Clayton Copulas)

In Equation (2.66) we introduced the independence copula $C_\perp$. This copula can also be written as

$$C_\perp(u_1,...,u_m) = \exp\big[-[(-\ln u_1) + \cdots + (-\ln u_m)]\big].$$

Therefore, the function

$$\varphi_\perp : [0,1] \to [0,\infty], \quad u \mapsto \varphi_\perp(u) = -\ln u$$

can be used to write the independence copula in the form

$$C_\perp(u_1,...,u_m) = \varphi_\perp^{-1}(\varphi_\perp(u_1) + \cdots + \varphi_\perp(u_m)). \qquad (2.69)$$

Copulas, which have a representation like the independence copula in (2.69), are called *Archimedean copulas*. We denote such copulas by C_φ and call the function φ the *generator* of C_φ. Example (2.69) shows that

$$C_\perp = C_{\varphi_\perp}.$$

The observation that the independence copula can we written as an Archimedean copula naturally raises the question for which φ

$$C_\varphi(u_1, ..., u_m) = \varphi^{-1}(\varphi(u_1) + \cdots + \varphi(u_m)) \quad (2.70)$$

satisfies the necessary conditions for being a copula function. It turns out (see, e.g., [90]) that, in order to be the generator of an Archimedean copula, functions φ have to fulfill certain criteria. It is beyond the scope of our little excursion to dive too deeply into details but we should at least present a collection of criteria guaranteeing that φ generates an Archimedean copula. The list of criteria is the following. Let

$$\varphi : [0, 1] \to [0, \infty]$$

be a function satisfying the following conditions:

1. φ is strictly decreasing;

2. $\varphi(0) = \infty$;

3. $\varphi(1) = 0$;

4. φ^{-1} is 'completely monotonic', meaning that φ^{-1} has derivatives of all orders and satisfies

$$(-1)^q (\varphi^{-1})^{(q)}(y) \geq 0 \quad (y \in [0, \infty); q = 0, 1, 2, ...)$$

where $f^{(q)}$ denotes the qth-order derivative of a function f.

If these conditions are fulfilled, φ is the generator of an Archimedean copula C_φ defined as in (2.70).

Let us briefly comment on the generator criteria. Conditions 2 and 3, combined with Condition 4, which yields the continuity of φ, imply that the range of φ is all of $[0, \infty]$. This guarantees that the inverse φ^{-1} is defined on the whole interval $[0, \infty]$. In the literature, Condition 2 carries the attribute *'strict'* such that one could say that we only consider strict generators.

What is Condition 4 good for? It looks strange at first sight but there are good reasons from analysis and probability theory why it shows up in this context. As said, it is beyond the scope of this book to provide a survey on complete monotonicity but a few remarks may help the reader to at least become kind of 'familiar' with that new notion.

The hour of birth of complete monotonicity goes back to the mathematician FELIX HAUSDORFF [56] who introduced the notion to the mathematical community more than 80 years ago.

Since then, analysts and probabilists paid quite some attention to the question which well-known functions are completely monotonic (see, e.g., MILLER and SAMKO [84] for a newer investigation in this direction) and what implication and applications can be generated from completely monotonic functions. Here is an important result:

2.5.8 Theorem (KIMBERLING [69], Theorem 2) *Let $F_1, ..., F_m$ be a family of one-dimensional distribution functions. Let*

$$\varphi : [0,1] \to [0,\infty]$$

be a completely monotonic function. Then there exists a probability space $(\Omega, \mathcal{F}, \mathbb{P})$ and random variables $X_1, ..., X_m$ defined on that space such that

- F_i *is the distribution function of X_i for $i = 1, 2, ..., m$;*

- *the joint distribution function $\boldsymbol{F}_{i_1,...,i_q}$ of any subfamily of random variables $X_{i_1}, ..., X_{i_q}$ with $i_1, ..., i_q \in \{1, 2, ..., m\}$ satisfies*

$$\begin{aligned}\boldsymbol{F}_{i_1,...,i_q}(x_{i_1}, ..., x_{i_q}) &= \varphi^{-1}(\varphi(F_{i_1}(x_{i_1})) + \cdots + \varphi(F_{i_q}(x_{i_q}))) \\ &= C_\varphi(F_{i_1}(x_{i_1}) + \cdots + F_{i_q}(x_{i_q})).\end{aligned}$$

In other words, given marginal distribution functions and a completely monotonic function φ with suitable domain and range one can combine the marginal distribution functions to multi-variate distributions with a dependence structure determined by the Archimedean copula function generated by φ.

A side-result of KIMBERLING's theorem is the following: if we choose $F_1, ..., F_m$ to equal the uniform distribution function in the unit interval, we have $F_k(x) = x$ for all k and x such that Theorem 2.5.8 implies that C_φ indeed is a copula function if φ is completely monotonic. This justifies Condition 4 in the generator criteria list.

A nice way to construct Archimedean copulas can be found in [113] (see 'BERSTEIN's Theorem'). There it is shown that a function φ is completely monotonic if and only if φ^{-1} has a representation as the one-sided Laplace transform of a distribution function $\mathbb{F}$ on $[0, \infty)$,

$$\varphi^{-1}(y) = \int_0^\infty e^{-yt} d\mathbb{F}(t), \qquad (2.71)$$

such that the integral is convergent for every positive y. This opens a rich source for the construction of Archimedean copulas just by taking the Laplace transform of suitable measures or distribution functions. We will use (2.71) in Section 2.5.6.5.

We already discussed the independence copula as a natural example of an Archimedean copula. Another Archimedean copula becoming more and more popular in finance applications is the *Clayton copula* generated by

$$\varphi_\eta(u) = u^{-\eta} - 1 \qquad (u \in [0,1])$$

where it is assumed that $\eta > 0$. The inverse of φ_η is given by

$$\varphi_\eta^{-1}(v) = (1+v)^{-1/\eta}.$$

The Clayton copula, therefore, writes as

$$\begin{aligned} C_{\varphi_\eta}(u_1, ..., u_m) &= C_\eta(u_1, ..., u_m) \\ &= (u_1^{-\eta} + \cdots + u_m^{-\eta} - m + 1)^{-1/\eta}. \end{aligned} \qquad (2.72)$$

The Clayton copula C_η only depends on one single parameter η such that the flexibility to calibrate it to empirical data is somewhat limited. However, as we will see later, the Clayton copula can be used for stress testing by means of extreme tail events.

Clayton copulas seem exotic at first sight but naturally arise in statistical problem solving; see Appendix 6.4. Moreover, based on the special form of the Laplace transform of the gamma distribution (see Appendix 6.1), it is not overly difficult to show that CreditRisk[+], which is a widely known industry portfolio model (see [35]), has a representation via the Clayton copula; see [41]. The link between properties of the gamma distribution and CreditRisk[+] arises from the construction of CreditRisk[+] based on Poisson mixture distributions with gamma-distributed (random) intensities; see [25], Chapter 2.

There are various other interesting and applicable Archimedean copulas in the literature, which can be applied in finance; see [32]. However, we restrict ourselves to the Clayton copula to have at least one representative of the Archimedean copula family in our 'repertoire'.

2.5.5.5 The Comonotonic Copula

The *comonotonic copula* is the complete opposite to the independence copula. It is defined as

$$C_\diamond(u_1, ..., u_m) = \min\{u_1, ..., u_m\} \qquad (u_1, ..., u_m \in [0, 1]).$$

Note that in the literature the comonotonic copula sometimes is called the *upper Frechet copula* due to the upper bound it represents: for any copula function C we have

$$C(u_1, ..., u_m) \leq C_\diamond(u_1, ..., u_m) \qquad (u_1, ..., u_m \in [0, 1]).$$

To see this, just note that

$$C(u_1, ..., u_m) \leq \min\{C(1, ..., u_k, ..., 1) \mid k = 1, 2, ..., m\}$$
$$= C_\diamond(u_1, ..., u_m).$$

Let us make an easy illustrative example for better understanding the comonotonic copula. Assume we are given two normal random variables $X, Y \sim N(0, 1)$ and let us define a bi-variate distribution on R^2 via Proposition 2.5.7,

$$\boldsymbol{F}_{(X,Y)}(x, y) = C_\diamond(N[x], N[y]) \qquad (x, y \in \mathbb{R}).$$

Then, we can write

$$\mathbb{P}[X \leq x, Y \leq y] = C_\diamond(N[x], N[y])$$
$$= \min\{N[x], N[y]\}$$
$$= N[\min\{x, y\}]$$
$$= \mathbb{P}[Z \leq \min\{x, y\}]$$

where $Z \sim N(0, 1)$ is a standard normal random variable. Our little calculation shows that the dependence induced by the comonotonic copula is so strong that the bi-variate joint distribution of X and Y can be reduced to the distribution of one single random variable Z. In fact, the *comonotonic copula induces the strongest possible dependence structure among random variables*.

In Section 3.3.6.3, we develop a very efficient simulation scheme for baskets and CDOs, based on a comonotonic copula approach.

2.5.6 Copulas in Practice

In the previous section we introduced the copula concept and discussed several common examples of copula functions. If we want to use our findings for applications in practice, we need to come up with ways to *sample scenarios* out of copula-induced multi-variate distributions.

Based on $\mathbb{F}_X[X] \sim U([0,1])$ for any random variable X with distribution function $\mathbb{F}_X$ one can obtain random draws from X by simulation of a uniform random variable $U \sim U([0,1])$ and then setting

$$X = \mathbb{F}_X^{-1}[U] \qquad (2.73)$$

where $\mathbb{F}_X^{-1}[\cdot]$ refers to the generalized inverse defined in Chapter 1. Random sampling from $U([0,1])$ can be done via pseudo random numbers or quasi random random numbers like, e.g., the simulation method based on *Faure sequences* as proposed in [87]. However, (2.73) generates realizations of a single random variable only, it does not provide a way to sample from a multi-variate distribution. It turns out that different simulation techniques are appropriate for different multi-variate distributions. In the sequel, we comment on the normal, Student-t, and Clayton copula.

2.5.6.1 Gaussian Copula with Gaussian Marginals

In the context of Figure 1.7 we discussed the factor decomposition of a client's credit risk into systematic risk factors plus some firm-specific effect. We briefly commented on the *separation* as well as on the *combined* approaches to build up a factor model. Such factor models lay the foundation not only for the interpretation of underlying credit risk drivers but serve also as a starting point for corresponding Monte Carlo simulations. In the sequel, we demonstrate the application of the Gaussian copula in such a factor model approach.

Let us assume we have m assets in the portfolio. For each name, we want to derive a factor decomposition as in Figure 1.7 for the name's CWI. The one-year time horizon representation we have in mind for the CWI of name i is given by

$$\text{CWI}_i = \beta_i \sum_{n=1}^{N} w_{i,n} \Psi_n + \varepsilon_i \qquad (2.74)$$

being in line with Equations (1.6) and (1.7). Here, we suppress the time-superscript index in (2.74) for the sake of a slightly simplified notation. The weights $w_{i,n}$ have to be positive numbers[11] weights and the indices Ψ_n are chosen to be multi-variate normally distributed with a certain correlation matrix and normalized variances equal to 1 (see Table 6.3 for an example) such that the CWI vector of the portfolio at the one-year horizon is distributed according to a Gaussian copula with Gaussian marginal distributions,

$$(\text{CWI}_1, ..., \text{CWI}_m) \;\sim\; C_{m,\Gamma}(N[\cdot], ..., N[\cdot]).$$

In order to achieve this, the residual variables ε_i have to be distributed as an independent family

$$\varepsilon_i \sim N(0, 1 - R_i^2), \tag{2.75}$$

independent of systematic indices $\Psi_1, ..., \Psi_N$, where R_i^2 is defined as

$$\mathbb{V}\left[\beta_i \sum_{n=1}^{N} w_{i,n} \Psi_n\right] = \beta_i^2 \sum_{k,l=1}^{N} w_{i,k} w_{i,l} \text{Cov}[\Psi_k, \Psi_l] =: R_i^2. \tag{2.76}$$

Given Equations (2.75) and (2.76), we can summarize that

- CWI_i is standard normally distributed for every $i = 1, ..., m$ and
- the multi-variate distribution of the CWIs is (by construction) a Gaussian distribution with correlation matrix Γ such that

$$\mathbb{P}[\text{CWI}_1 \leq c_1, ..., \text{CWI}_m \leq c_m] \;=\; C_{m,\Gamma}(N[c_1], ..., N[c_m])$$

as it should be the case for a Gaussian copula with standard normal marginals.

Note that the discussion above is typical for a *combined factor approach*. In a *separated factor approach*, the covariance matrix Σ can be chosen to be a diagonal matrix if the factor decomposition is aggregated to the most aggregated level of independent regional and industrial factors; see, e.g., [25], Chapter 1.

The discussion above can without facing difficulties or challenges be translated into a Monte Carlo simulation scheme in which CWI realizations are sampled from a multi-variate distribution consisting of a Gaussian copula with standard normal marginals.

---
[11]Later in the text (see Section 2.6.1) we will encounter an explicit example.

2.5.6.2 Gaussian Copula with Student-t Marginals

Now, let us assume that internal data suggests that the multi-variate CWI dependency is correctly described by a Gaussian copula but that CWI marginals have fatter tails than a Gaussian copula such that their randomness better can be described by Student-t distribution. Then, we can modify the simulation approach above and simulate in addition to the random variables in (2.74) m i.i.d. $\chi^2(d)$-distributed random variables $X_1, ..., X_m$. Equation (2.74) then reads as

$$\text{CWI}_i = \Theta_d^{-1}\left[N\left[\beta_i \sum_{n=1}^{N} w_{i,n}\Psi_n + \varepsilon_i\right]\right] \qquad (2.77)$$

where d denotes the degrees of freedom arising from our data analysis at marginal distribution level and Θ_d denotes the distribution function of a one-dimensional Student-t distribution with d degrees of freedom; see Appendix 6.3. Then,

- CWI_i is t-distributed with d degrees of freedom because in (2.77) the inverse t-distribution function is applied to a uniform random variable (namely, the standard normal distribution function applied to a standard normal CWI); see also (2.73);

- the multi-variate CWI distribution is given by

$$\mathbb{P}[\text{CWI}_i \leq c_i \; \forall \; i] = \mathbb{P}\left[\beta_i \Phi_i + \varepsilon_i \leq N^{-1}[\Theta_d(c_i)] \; \forall \; i\right]$$
$$= N_m[N^{-1}[\Theta_d(c_1)], ..., N^{-1}[\Theta_d(c_m)]; \Gamma]$$
$$= C_{m,\Gamma}(\Theta_d[c_1], ..., \Theta_d[c_m])$$

so that indeed the CWIs follow a multi-variate distribution, which is a combination of a normal copula and t-distributed marginals.

2.5.6.3 Student-t Copula with Student-t Marginals

For the Student-t copula we can in principal follow the same lines as for the Gaussian copula. Equation (2.74) has to be modified toward

$$\text{CWI}_i = \sqrt{d}\left(\beta_i \sum_{n=1}^{N} w_{i,n}\Psi_n + \varepsilon_i\right)/\sqrt{X} \qquad (2.78)$$

with a $\chi^2(d)$-distributed random variable (independent of the Ψ_n's and ε_i's; see Appendix 6.3) to transform the Gaussian copula from (2.74) into a Student-t copula with d degrees of freedom and (linear) correlation matrix Γ. Hereby, all other distribution assumptions on the Ψ_n's and ε_i's can be kept.

2.5.6.4 Student-t Copula with Gaussian Marginals

To combine the t-copula from the previous section with Gaussian instead of t-distributed marginals we have to further transform the CWIs from (2.78) via

$$\text{CWI}_i = N^{-1}\left[\Theta_d\left[\sqrt{d}\left(\beta_i \sum_{n=1}^{N} w_{i,n}\Psi_n + \varepsilon_i\right)/\sqrt{X}\right]\right] \qquad (2.79)$$

such that the CWI's have standard normally distributed marginals because the inverse of the standard normal distribution function is applied to uniform random variables. In addition, the joint CWI distribution satisfies

$$\mathbb{P}[\text{CWI}_i \leq c_i \ \forall \ i] = \mathbb{P}\left[Z_i \leq \Theta_d^{-1}(N[c_i]) \ \forall \ i\right]$$
$$= \Theta_{m,\Gamma,d}[\Theta_d^{-1}(N[c_1]), ..., \Theta_d^{-1}(N[c_m])]$$
$$= C_{m,\Gamma,d}(N[c_1], ..., N[c_m])$$

such that the t-copula with Gaussian marginals is confirmed.

2.5.6.5 Simulations Based on Archimedean Copulas

For the simulation of CWIs based on Archimedean copulas we need the following well-known theorem.

2.5.9 Theorem (MARSHALL and OLKIN [80]) *Let $\varphi : [0,1] \to [0,\infty]$ be a generator of an Archimedean copula as described in Section 2.5.5.4. Based on (2.71) we know that the inverse φ^{-1} of the generator φ admits a representation*

$$\varphi^{-1}(y) = \int_0^\infty e^{-yt} d\mathbb{F}_X(t)$$

as the Laplace transform of the distribution function $\mathbb{F}_X$ of a random variable X in $[0,\infty)$. Then, if $U_1, ..., U_m \sim U([0,1])$ is a sequence of i.i.d. uniformly distributed random variables, the variables

$$Y_i = \varphi^{-1}\left(-\frac{\ln U_i}{X}\right)$$

have a joint distribution
$$\mathbb{P}[Y_1 \leq y_1, ..., Y_m \leq y_m] = \varphi^{-1}(\varphi(y_1) + \cdots + \varphi(y_m))$$
determined by the φ-generated Archimedean copula as in (2.70).

Proof. The dependence between variables Y_i is exclusively generated by the 'mixing variable' X. We obtain
$$\begin{aligned}\mathbb{P}[Y_i \leq y_i \mid X = x] &= \mathbb{P}[-\ln U_i \leq \varphi(y_i)X \mid X = x] \\ &= \mathbb{P}[U_i \leq \exp(-\varphi(y_i)X) \mid X = x] \\ &= \exp(-\varphi(y_i)x)\end{aligned}$$
due to the uniform distribution of U_i. The joint distribution of the Y_i's can be written in a typical form using that conditioning on the mixing variable X will make the Y_i's independent,
$$\begin{aligned}\mathbb{P}[Y_1 \leq y_1, ..., Y_m \leq y_m] &= \int_0^\infty \mathbb{P}[Y_1 \leq y_1, ..., Y_m \leq y_m \mid X = x]\, d\mathbb{F}_X(x) \\ &= \int_0^\infty \prod_{i=1}^m \mathbb{P}[Y_i \leq y_i \mid X = x]\, d\mathbb{F}_X(x) \\ &= \int_0^\infty \prod_{i=1}^m \exp(-\varphi(y_i)x)\, d\mathbb{F}_X(x) \\ &= \int_0^\infty \exp\Big(-x\sum_{i=1}^m \varphi(y_i)\Big) d\mathbb{F}_X(x) \\ &= \mathcal{L}_X\Big[\sum_{i=1}^m \varphi(y_i)\Big] \\ &= \varphi^{-1}(\varphi(y_1) + \cdots + \varphi(y_m))\end{aligned}$$
where $\mathcal{L}_X$ denotes the Laplace transform of $\mathbb{F}_X$. □

Let us apply Theorem 2.5.9 to the Clayton copula defined in (2.72),
$$C_\eta(u_1, ..., u_m) = (u_1^{-\eta} + \cdots + u_m^{-\eta} - m + 1)^{-1/\eta}$$
for some $\eta > 0$. The generator of the Clayton copula is
$$\varphi_\eta(u) = u^{-\eta} - 1 \qquad (u \in [0,1])$$

and the inverse of φ_η is given by

$$\varphi_\eta^{-1}(v) = (1+v)^{-1/\eta}.$$

We have to find the mixing variable X in order to enable the application of Theorem 2.5.9. In other words, we are looking for a distribution function $\mathbb{F}_X$ such that

$$\varphi_\eta^{-1}(v) = (1+v)^{-1/\eta} = \int_0^\infty e^{-vt} d\mathbb{F}_X(t).$$

It turns out that if we choose

$$X \sim \Gamma(1/\eta, 1)$$

where $\Gamma(\alpha, \beta)$ denotes the gamma distribution with parameters α and β we obtain the desired Laplace transform; see Appendix 6.1. Based on that, Theorem 2.5.9 provides clear guidance on how to simulate CWIs with normal or t-distributed marginals combined to a multi-variate distribution via the Clayton copula C_η. More explicitly, we choose i.i.d. random variables $U_1, ..., U_m \sim U([0,1])$ and a random variable $X \sim \Gamma(\alpha, \beta)$ independent of all other involved random variables. Based on Theorem 2.5.9,

$$\text{CWI}_i = N^{-1}\left[\varphi^{-1}\left(-\frac{\ln U_i}{X}\right)\right]$$

generates a multi-variate CWI distribution with

$$\mathbb{P}[\text{CWI}_1 \leq c_1, ..., \text{CWI}_m \leq c_m] = \varphi^{-1}(\varphi(N[y_1]) + \cdots + \varphi(N[y_m])).$$

Note that φ^{-1} maps from $[0, \infty]$ to $[0, 1]$ such that application of the inverse normal distribution function is justified. Student-t or other alternative marginals in a Clayton copula can be generated analogously.

We stop our general discussion on copula functions here. There are many more aspects worthwhile to be discussed and presented but for our purposes we are fully equipped now with everything we need in the sequel. Note that Section 2.5.9 provides a word of caution regarding 'wild' applications of copula-related concepts. In general, everything said on data-based calibrations in previous sections in other contexts applies also to copula functions: *empirical data should be the foundation for any decision regarding copula choice and calibration.* If data

Default Baskets

is insufficient for coming to a conclusion, expert judgement is the only way out. Here, the Gaussian copula or maybe the Student-t copula are natural choices; see also Appendix 6.5. A realistic scenario, e.g., in non-public debt, is that it is even difficult to estimate linear correlations from historical data. Parameterizing more complicated copula functions is not feasible in various relevant cases, unfortunately. However, copula variations can be applied in the context of stress and scenario analysis as we will see later in Chapter 3.

2.5.7 Visualization of Copula Differences and Mathematical Description by Dependence Measures

This section is purely dedicated to generically illustrate the difference arising from application of different copula functions. For this purpose, we simulate samples from a pair of random variables

$$(X,Y) \sim C(\mathbb{F}_X[\cdot], \mathbb{F}_Y[\cdot])$$

where C equals the

- Independence copula
- Comonotonic copula
- Gaussian copula with linear correlation of 50%
- Student-t copula with linear correlation of 50% and 5 degrees of freedom
- Clayton copula with $\eta = 1$
- Clayton copula with $\eta = 5$

and the marginals $\mathbb{F}_X$ and $\mathbb{F}_Y$ are standard normally distributed (Figure 2.23) and t-distributed with 5 degrees of freedom (Figure 2.24).

Let us briefly comment on the copula-induced differences we see in Figures 2.23 and 2.24. First of all, we find quite some differences in the plot of random points (X,Y). In general, from a purely eye-catching perspective the plots with Student-t marginals appear much 'wilder' than the plots with Gaussian marginals. A second observation is that the Student-t copula seems to generate a greater dependency in the lower and upper tail than the Gaussian copula. The Clayton copula

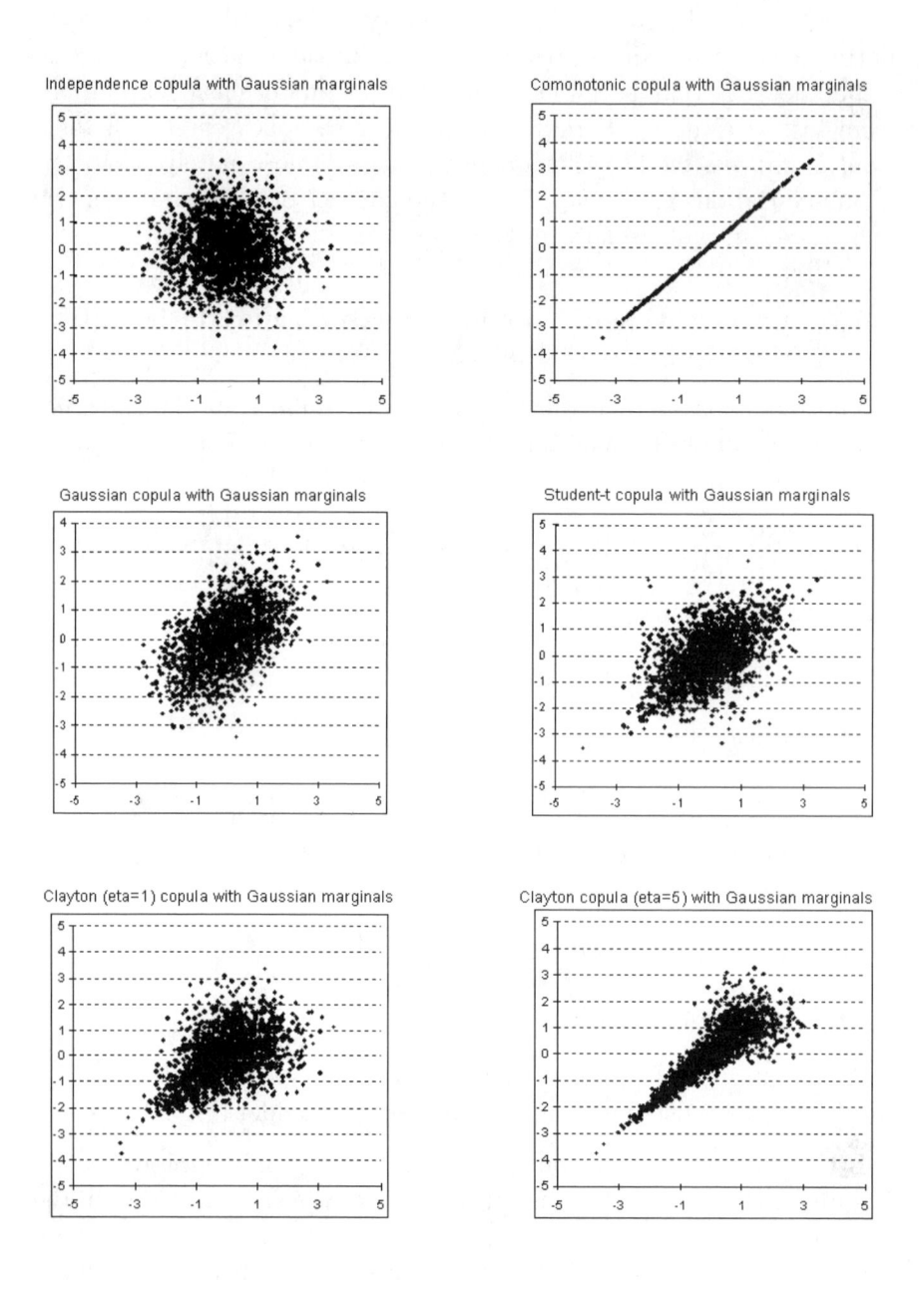

FIGURE 2.23: Different copulas combined with Gaussian marginals

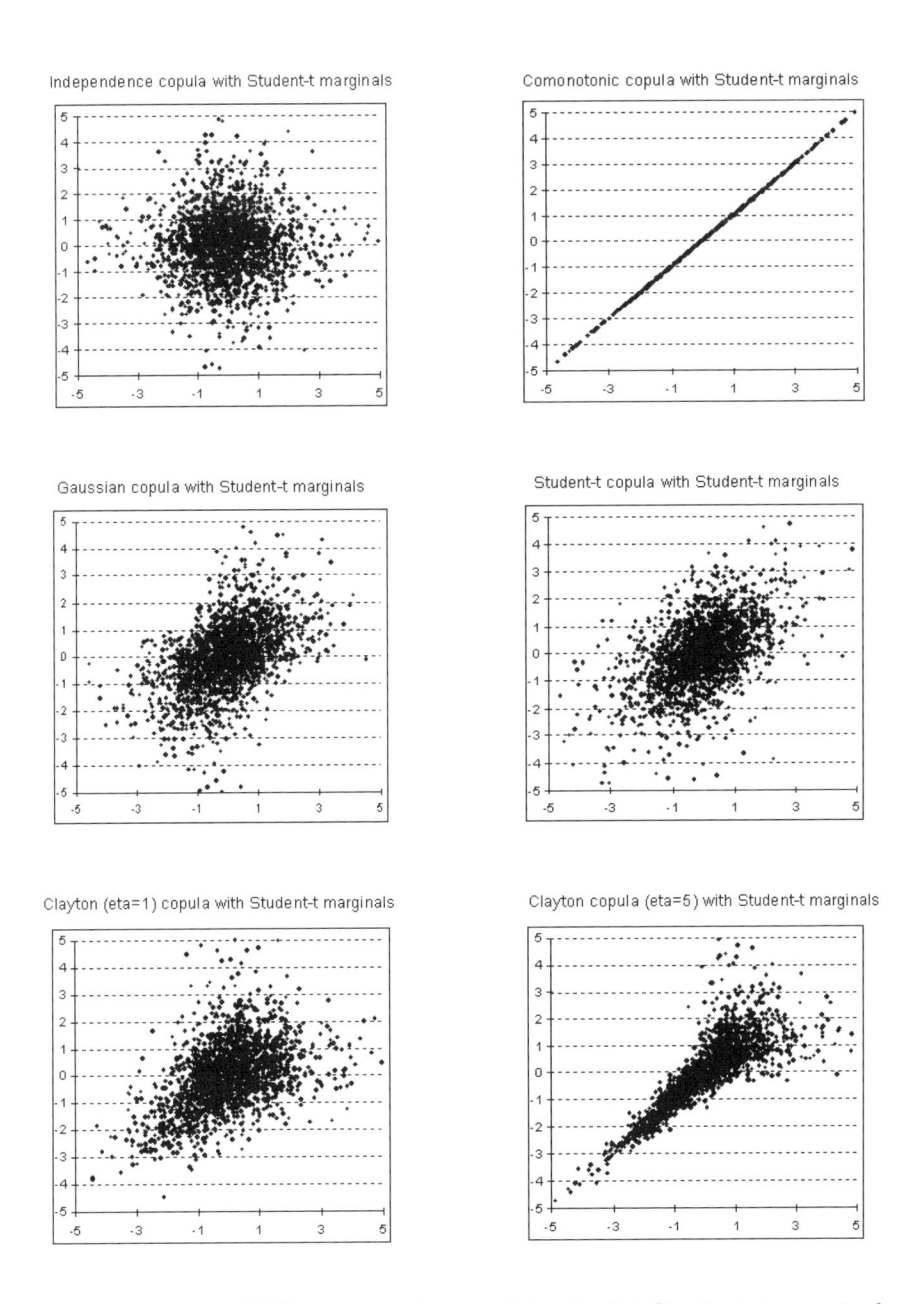

FIGURE 2.24: Different copulas combined with Student-t marginals

seems to generate a strong dependency in the lower tail, which seems to increase with increasing parameter η.

Qualitative judgements like the statements above generally do not satisfy mathematicians. Instead, we need some mathematical description for the different effects on the dependence structure of (X, Y) generated by the choice of a particular copula function. And indeed, there is a whole theory at our disposal, which we can use to describe dependencies in a more formal way. It is beyond the scope of this book to attach a fully-fledged survey on dependence measures, but a brief excursion may be helpful for a slightly better understanding of different dependence concepts. Interested readers find an excellent survey on dependence measures and copula function in EMBRECHTS et al. [42]. Other sources for further studies are [32] and [58].

First of all, let us remark that there are two key properties a reasonable dependence measure $\Delta[\cdot, \cdot]$ at least should have:

- Δ should be symmetric: $\Delta[X, Y] = \Delta[Y, X]$

- Δ should be standardized/normalized: $-1 \leq \Delta[X, Y] \leq 1$

Another desirable property of a dependence measure Δ is the following invariance property:

- $\Delta(T(X), Y) = \Delta(X, Y)$ for any strictly increasing function T on range(X) and

- $\Delta(T(X), Y) = -\Delta(X, Y)$ for any strictly decreasing function T on range(X)

We will later remark that classical correlation fails regarding this property for general non-linear T's, whereas rank-transformed correlations (see, e.g., Section 2.5.7.2) satisfy the invariance condition due to strictly monotonic functions preserving or inverting rankings, respectively.

In the sequel, we assume X and Y to be continuous.

2.5.7.1 Dependence Measure No. 1: Correlation

Correlation is the best known as well as a, regarding its meaning, sometimes 'misinterpreted' notion of dependence between random variables. It is the classical dependence measure in the world of Gaussian and, more general, so-called elliptical and spherical distributions. The definition of correlation can be found in any text on probability theory,

$$\text{Corr}[X,Y] = \frac{\text{Cov}[X,Y]}{\sqrt{\mathbb{V}[X]}\sqrt{\mathbb{V}[Y]}} = \frac{\mathbb{E}[XY] - \mathbb{E}[X]\mathbb{E}[Y]}{\sqrt{\mathbb{V}[X]}\sqrt{\mathbb{V}[Y]}}.$$

To give an example, let us assume that X and Y are standard normally distributed. Then, zero correlation is synonymous to independence. Moreover, let us formally write a *linear regression* equation

$$Y = \rho X + \varepsilon$$

which attempts to explain Y by X plus an independent standard normally distributed residual effect ε. Obviously we find

$$\text{Corr}[X,Y] = \text{Corr}[X, \rho X + \varepsilon] = \rho$$

such that correlation in our example but also in general is a *measure for linear dependence*, nothing more and nothing less.

Correlation as a dependence measure is symmetric and normalized. Furthermore, it is invariant under shifts and scalings,

$$\text{Corr}[aX + b, cY + d] = \frac{a \times c}{|a| \times |c|} \text{Corr}[X,Y],$$

for any non-zero numbers a and c and numbers b and d. Thus, correlation is a dependence measure *invariant under strictly increasing linear transformations*. However, a disadvantage of correlation is that it is not invariant under general non-linear strictly increasing transformations. Other dependence measures (see Sections 2.5.7.2 and 2.5.7.3) do better from this perspective. We, therefore, now briefly turn our attention to measures of *monotonic dependence* between random variables. As a reference for the sequel we refer to [58], 2.1.9. Correlation and the following two dependence measures are *bi-variate* notions. To apply them to families of random variables with more than two variables the bi-variate dependence figures are collected into an $m \times m$-matrix if m denotes the number of involved random variables. The most prominent representative of such matrices is the *correlation matrix*.

Note that correlation as well as other bi-variate dependence notions like the one in the next two sections only provide information about certain *pairwise* dependencies. Multiple dependencies for more than two variables (see, e.g., the black intersection in Figure 2.41 for an illustration of 'triple overlapping') cannot be captured by a matrix and require a more complex description.

2.5.7.2 Dependence Measure No. 2: Spearman's Rank Correlation

The first kind of rank correlation we look at is *Spearman's rank correlation*, or, as it is often called in the literature, *Spearman's rho*. It follows a very simple idea: instead of measuring correlation between two random variables X and Y we transform the variables uniformly into the unit interval and measure the correlation of the transformed random variables. More explicitly, if $\mathbb{F}_X$ and $\mathbb{F}_Y$ denote the distribution function of X and Y, respectively, then

$$\mathbb{F}_X(X), \ \mathbb{F}_Y(Y) \ \sim \ U([0,1])$$

are uniformly distributed in the unit interval. Potential outliers in samples of X and Y are now captured between the boundaries 0 and 1 of the unit interval. Spearman rank correlation or rho is defined as

$$\rho_S(X,Y) \ = \ \text{Corr}[\mathbb{F}_X(X), \mathbb{F}_Y(Y)].$$

We already mentioned that correlation is symmetric and normalized. Spearman's rank correlation inherits both properties. If X and Y are independent, the transformed variables are also independent and, therefore, Spearman's rank correlation of X and Y is zero. As already mentioned right before Section 2.5.7.1, rank correlations are invariant or inverted under strictly monotonic transformations because they preserve or invert rankings, respectively.

2.5.7.3 Dependence Measure No. 3: Kendall's Rank Correlation

Another common kind of rank correlation is *Kendall's rank correlation*, or, *Kendall's tau*. It is defined in the following way. Denote by $\mathbb{F}_{(X,Y)}$ the joint distribution function of X and Y. Choose two independent random draws (X_1, Y_1) and (X_2, Y_2) from $\mathbb{F}_{(X,Y)}$. Then, Kendall's rank correlation or tau is defined as

$$\tau(X,Y) \ = \ \mathbb{P}[(X_1 - X_2)(Y_1 - Y_2) > 0] - \mathbb{P}[(X_1 - X_2)(Y_1 - Y_2) < 0].$$

The condition
$$(X_1 - X_2)(Y_1 - Y_2) \ > \ 0$$

is fulfilled if one of the two pairs (X_1, Y_2) and (X_2, Y_2) dominates the other pair in both components (the pairs are *concordant*), whereas the second condition
$$(X_1 - X_2)(Y_1 - Y_2) \ < \ 0$$

refers to the case where each pair has one component dominating the corresponding component of the other pair (the pairs are *discordant*). Therefore, Kendall's tau measures the likelihood of concordance of two pairs of random variables minus (or corrected by) the likelihood of discordance of the pairs. Figure 2.25 graphically illustrates concordant and discordant pairs (X_1, Y_1) and (X_2, Y_2).

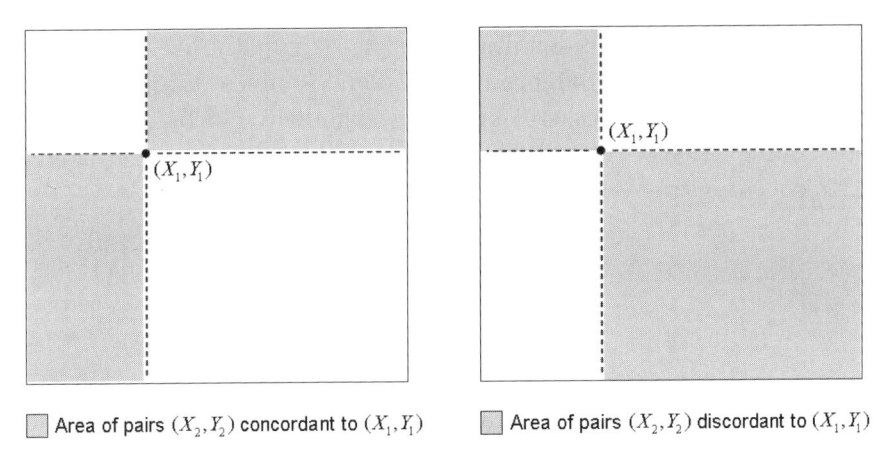

■ Area of pairs (X_2, Y_2) concordant to (X_1, Y_1) ■ Area of pairs (X_2, Y_2) discordant to (X_1, Y_1)

FIGURE 2.25: Illustration of concordant and discordant pairs

To give an example of extreme dependence w.r.t. Kendall's tau, assume $\mathbb{F}_{(X,Y)}$ admits a strictly increasing transform T such that

$$\mathbb{P}[Y = T(X)] = 1. \qquad (2.80)$$

This implies that whenever in the definition of $\tau(X, Y)$ we have $X_1 > X_2$ we also have $Y_1 > Y_2$. In other words, the concordance likelihood equals 1 and the discordance likelihood equals 0 such that $\tau(X, Y)$ attains the maximum possible value of 1. Analogously, if (2.80) holds with a strictly decreasing function T then the discordance likelihood is 1 and the concordance likelihood is 0 such $\tau(X, Y)$ attains its minimum value -1. This shows that $\tau(X, Y)$ is a normalized dependence measure.

It is obvious that Kendall's tau is symmetric in its arguments. Its invariance (inversion) under strictly increasing (decreasing) transformations also should be clear: transforming one component in a strictly increasing way keeps the ordering in this component, whereas strictly decreasing transformations exchange the likelihoods of concordance and discordance and, therefore, can be viewed as multiplying $\tau(X,Y)$ by -1.

A last case we want to look at is the case of independent components,

$$\mathbb{F}_{(X,Y)}[X \leq x, Y \leq y] = \mathbb{F}_X(x)\mathbb{F}_Y(y) \quad \forall\, x, y.$$

Based on intuition, we expect that independence makes concordance and discordance in the definition of $\tau(X,Y)$ equally likely such that $\tau(X,Y) = 0$. Indeed, independent random variables X and Y have a Kendall's tau equals to zero. A formal argument can be derived as a side-consequence of the following calculations. As suggested by Figure 2.25, we can write

$$\mathbb{P}[X_2 < X_1, Y_2 < Y_1] = \int \mathbb{P}[X_2 < x_1, Y_2 < y_1] d\mathbb{F}(x_1, y_1)$$
$$= \int \mathbb{F}_{(X,Y)}(x_1, y_1)\, d\mathbb{F}_{(X,Y)}(x_1, y_1)$$

by a standard conditioning argument. Based on a symmetry argument and the assumed continuity[12] of variables X and Y, we obtain

$$\tau(X,Y) = \mathbb{P}[(X_1 - X_2)(Y_1 - Y_2) > 0] - \mathbb{P}[(X_1 - X_2)(Y_1 - Y_2) < 0]$$
$$= 2\,\mathbb{P}[(X_1 - X_2)(Y_1 - Y_2) > 0] - 1$$
$$= 4 \int \mathbb{F}_{(X,Y)}(x_1, y_1)\, d\mathbb{F}_{(X,Y)}(x_1, y_1) - 1$$

such that in case of independent variables X and Y we get

$$\tau(X,Y) = 4 \int \mathbb{F}_X(x_1)\mathbb{F}_Y(y_1)\, d\mathbb{F}_{(X,Y)}(x_1, y_1) - 1$$
$$= 4\,\mathbb{E}[\mathbb{F}_X(X)]\,\mathbb{E}[\mathbb{F}_Y(Y)] - 1 = 0$$

based on the fact that $\mathbb{F}_X(X), \mathbb{F}_Y(Y) \sim U([0,1])$ such that the product of expectations $\mathbb{E}[\mathbb{F}_X(X)] = \mathbb{E}[\mathbb{F}_Y(Y)] = 0.5$ equals $1/4$.

Rank correlations like Spearman's rho and Kendall's tau quantify the level of *monotonic dependence* as indicated in the introduction to our 'mini excursion' on dependence measures. The meaning of this notion should be clear now due to our discussion in the last two sections.

---

[12]Implying $\mathbb{P}[(X_1 - X_2)(Y_1 - Y_2) = 0] = 0$.

2.5.7.4 Dependence Measure No. 4: Tail Dependence

Tail dependence is the last dependence measure we want to briefly introduce. A comprehensive discussion can be found in [58]. Tail dependence typically is denoted by λ. One distinguishes between *upper tail dependence* defined by

$$\lambda_U(X,Y) = \lim_{\gamma \to 1} \mathbb{P}[Y > \mathbb{F}_Y^{-1}(\gamma) \mid X > \mathbb{F}_X^{-1}(\gamma)] \qquad (2.81)$$

and *lower tail dependence* specified as

$$\lambda_L(X,Y) = \lim_{\gamma \to 0} \mathbb{P}[Y \leq \mathbb{F}_Y^{-1}(\gamma) \mid X \leq \mathbb{F}_X^{-1}(\gamma)]. \qquad (2.82)$$

Hereby, the assumption has to be made that the limits exist. Upper (lower) tail dependence quantifies the likelihood to observe a large (low) value of Y given a large (low) value of X. The notion of tail dependence is independent of marginal distributions and a function purely of the (unique - because X and Y are assumed to be continuous) copula underlying the joint distribution of X and Y. To see this, we follow [58], 2.1.10., and re-write the tail dependence measures in the form

$$\lambda_U(X,Y) = \lim_{\gamma \to 1} \frac{1 - 2\gamma + C(\gamma,\gamma)}{1-\gamma} \qquad (2.83)$$

$$\lambda_L(X,Y) = \lim_{\gamma \to 0} \frac{C(\gamma,\gamma)}{\gamma} \qquad (2.84)$$

where C denotes the copula underlying $\mathbb{F}_{(X,Y)}$,

$$\mathbb{F}_{(X,Y)}(x,y) = C(\mathbb{F}_X(x), \mathbb{F}_Y(y)) \qquad \forall\, x, y. \qquad (2.85)$$

To see that (2.81) and (2.83) are really defining the same object, re-write the conditional probability from (2.81) as

$$\frac{\mathbb{P}[Y > \mathbb{F}_Y^{-1}(\gamma), X > \mathbb{F}_X^{-1}(\gamma)]}{\mathbb{P}[X > \mathbb{F}_X^{-1}(\gamma)]} =$$

$$\frac{1 - \mathbb{P}[Y > \mathbb{F}_Y^{-1}(\gamma)] - \mathbb{P}[X > \mathbb{F}_X^{-1}(\gamma)] + \mathbb{P}[Y \leq \mathbb{F}_Y^{-1}(\gamma), X \leq \mathbb{F}_X^{-1}(\gamma)]}{\mathbb{P}[X > \mathbb{F}_X^{-1}(\gamma)]}$$

and take into account that by (2.85) we have

$$\mathbb{P}[Y \leq \mathbb{F}_Y^{-1}(\gamma), X \leq \mathbb{F}_X^{-1}(\gamma)] = C(\mathbb{F}_X(\mathbb{F}_X^{-1}(\gamma)), \mathbb{F}_Y(\mathbb{F}_Y^{-1}(\gamma))) = C(\gamma,\gamma)$$

as well as

$$\mathbb{P}[X > \mathbb{F}_X^{-1}(\gamma)] = 1 - \mathbb{P}[X \leq \mathbb{F}_X^{-1}(\gamma)] = 1 - \gamma.$$

Analogously, it can be verified that (2.82) and (2.84) are compatible.

Tail dependence is a tool widely applied in *extreme value theory* because it is a concept referring to the lower and upper tails representing extreme scenarios of considered multi-variate distributions. Equations (2.83) and (2.84) show that it defines symmetric (considered as in our example in the bi-variate case) dependence measures invariant under strictly increasing transformations. Note that copulas, in general, are invariant w.r.t. strictly increasing transformations in one or several variables. To see this, consider again Equation (2.85) and transform X by a strictly increasing transformation T. Then,

$$\begin{aligned}\mathbb{P}[T(X) \leq x, Y \leq y] &= \mathbb{P}[X \leq T^{-1}(x), Y \leq y] \\ &= C(\mathbb{F}_{T(X)}(x), \mathbb{F}_Y(y))\end{aligned}$$

such that the copula for the joint distribution, transformed in one component, does not change: the transformation concerns only marginal distributions. Altogether we see that tail dependence behaves as we want it to behave as a sound measure of dependence.

2.5.7.5 Application of Dependence Measures to Different Copulas

We conclude our little excursion on dependence measures by applying the different dependence notions to the copula functions playing the dominant role in this book, hereby demonstrating that Figures 2.23 and 2.24 can be explained in precise mathematical terms.

We start with the Gaussian and Student-t copulas. By definition, only the matrix of linear correlations Γ plays a role in the determination of dependencies in a multi-variate normal distribution with correlation matrix Γ. Next, let us look at rank correlations. The following relations between linear correlation and Spearman's rho and Kendall's tau

$$\text{Corr}[X, Y] = 2 \sin\left(\frac{\pi}{6} \rho_S\right) \tag{2.86}$$

$$\text{Corr}[X, Y] = \sin\left(\frac{\pi}{2} \tau\right) \tag{2.87}$$

hold for the Gaussian as well as for the Student-t copula. This is a result from a more general context; see LINDSKOG et al. [77]. As a

Default Baskets

consequence, rho and tau can be derived from linear correlations just by applying the arcsin-function and some scaling. Figure 2.26 shows plots of the respective functional relationships.

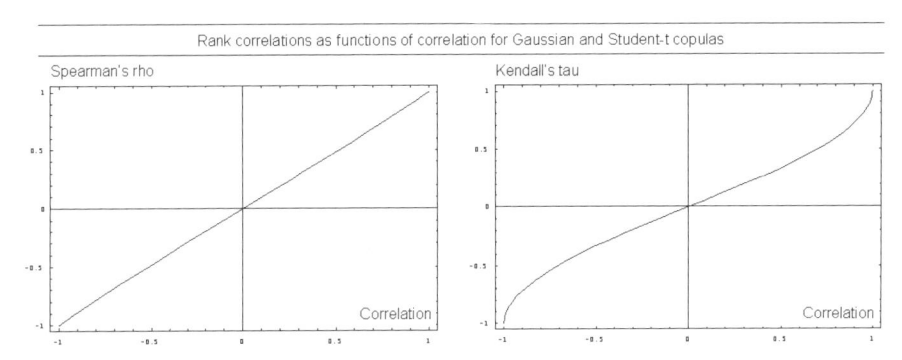

FIGURE 2.26: Rank correlations as functions of correlation for Gaussian and Student-t copula functions

Let us now look at tail dependence. For the Gaussian copula case we consider (X, Y) with a bi-variate normal distribution. It turns out that the Gaussian copula does not exhibit any tail dependence at all:

$$\lambda_U(X, Y) = \lambda_L(X, Y) = 2 \lim_{x \to \infty} \left(1 - N\left[x \, \frac{\sqrt{1 - \text{Corr}[X, Y]}}{\sqrt{1 + \text{Corr}[X, Y]}} \right] \right) = 0$$

(see [42]), whereas for the Student-t copula, now assuming that (X, Y) is bi-variate t-distributed, we obtain

$$\lambda_U(X, Y) = \lambda_L(X, Y) =$$

$$= 2 \left(1 - \Theta_{d+1}\left[\sqrt{d+1} \, \frac{\sqrt{1 - \text{Corr}[X, Y]}}{\sqrt{1 + \text{Corr}[X, Y]}} \right] \right) \quad (2.88)$$

where Θ_d denotes the t-distribution function with d degrees of freedom and $\text{Corr}[X, Y]$ denotes the linear correlation of the bi-variate t-copula inherited from the bi-variate Gaussian component (before multiplication with the χ^2-component; see Appendix 6.3). Note that equality of upper and lower tail dependence follows from the *radial symmetry of elliptical distributions*. For a proof of (2.88) we refer to LINDSKOG

[76]. In other words, the Gaussian copula leads to *asymptotically independent* extreme (tail) events, whereas the Student-t copula exhibits tail dependence, which fades-out[13] with increasing degrees of freedom. Figure 2.27 illustrates the tail dependence of the Student-t copula. It shows that tail dependence also increases with increasing linear correlation. However, even in case of zero or negative linear correlation we recognize positive tail dependence.

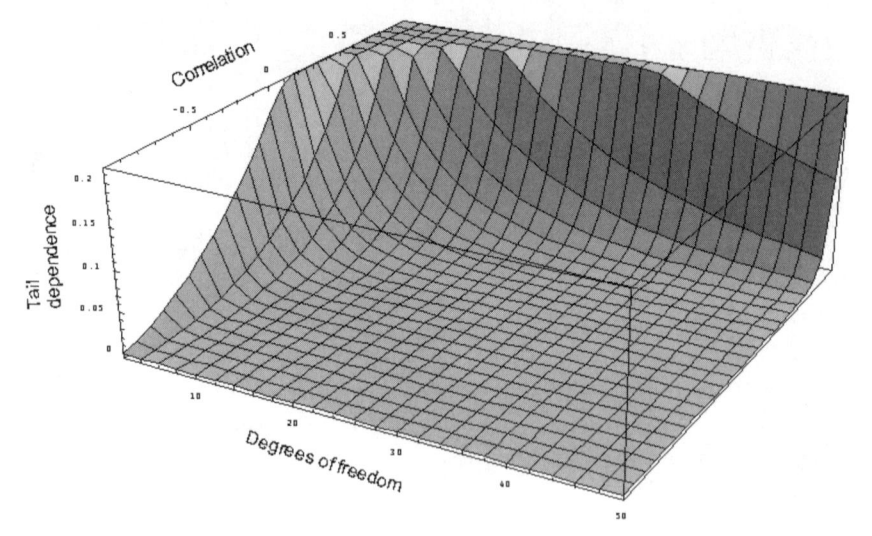

FIGURE 2.27: Tail dependence as a function of linear correlation $(-1 \leq \varrho \leq 1)$ and the degrees of freedom $(2 \leq d \leq 50)$ for Student-t copula functions

Next, let us focus on the Clayton copula. Proofs regarding the calculation of tail dependence typically can be based on Equations (2.83) and (2.84) in case of explicitly described copulas like the Clayton copula. However, we can refer to LINDSKOG [76] for proofs of the following statements.

As graphically indicated by Figures 2.23 and 2.24, the Clayton copula

---

[13]This is quite natural given the convergence of t-distributions to the normal distribution if $d \to \infty$.

exhibits positive lower tail dependence but zero upper tail dependence,

$$\lambda_L(X,Y) = 2^{-1/\eta} \quad \text{and} \quad \lambda_U(X,Y) = 0$$

(see also [90] for more general statements for Archimedean copulas). In the limits, the Clayton copula exhibits perfect dependence for $\eta \to \infty$ and independence for $\eta \to 0$. One can also observe this in Figures 2.23 and 2.24 where $\eta = 5$ generates seemingly more lower tail dependence than the case $\eta = 1$. As a last remark on Clayton copulas in this context, we remark that Kendall's tau[14] for Clayton copulas equals

$$\tau = \frac{\eta}{\eta + 2}$$

as can be found in most of the research papers already quoted.

2.5.7.6 Model Implications from Copula Choices

What are the model implications arising from our discussion on copulas and dependence measures? In this section we want to briefly indicate some consequences in the form of remarks as a 'wrap-up' of the last sections. Later in Section 2.5.8 we then apply our findings to our duo basket example in order to bring the concept alive.

Let us start with the Gaussian copula. It constitutes the most widely used multi-variate distribution (see also Appendix 6.5), not only in credit risk modeling but also in other areas of pricing and risk management, e.g., in the famous BLACK & SCHOLES option pricing formulas. In credit risk modeling problems, the Gaussian copula is a convenient choice because the modeler only needs to calibrate *linear correlations* based on empirical data. This exercise typically is reduced to the problem of finding a suitable *factor model* (see Figure 1.7) based on which correlations between credit-risky assets can be determined. A major disadvantage of the Gaussian copula is its *lack of tail dependence*. Market observations show that in times of extreme events *asymptotic independence* as it is the case for Gaussian copulas cannot be assumed. In fact, the opposite is the case: there is strong evidence that *in times of extreme events dependencies have a potential to increase rather than decrease*. A shocking example for this is the September 11 terror act,

---

[14]Theorem 5.4. in [76] presents an easy handable formula for the calculation of Kendall's tau for Archimedean copulas.

which not only constitutes the worst terror act ever in the history of human civilization but also provided insight in the 'dependence mechanics' in an industry (in this case the airline/aircraft industry) after a shock event: correlations within the aircraft/airline sector sharply increased due to the extreme event (terror act in this case). Altogether one can say that *a realistic model for risks should allow for tail dependence* and should not exhibit asymptotic independence like the Gaussian copula.

Somehow related to the Gaussian copula but *incorporating tail dependencies* is the Student-t copula. Therefore, if we want to rely on a model not too far from the Gaussian 'base case model' but no longer asymptotically independent we can work with the Student-t copula. In Figure 2.27 we plotted the tail dependence of such a model as a function of linear correlation and, more important, the degrees of freedom. However, the Student-t copula, as an elliptical distribution, exhibits *radial symmetry* such that both directions for extreme events are possible, extreme *downside risks* as well as extreme *upside chances*. Note that latent variable models as we apply them in this book (see, e.g., Equation (2.3)) typically *associate downside risk with the lower tail and upside chances with the upper tail; see Appendix 6.6*. Summarizing, we can state that the Student-t copula provides a more realistic model than the Gaussian copula but has limited applications for stressing a risk model w.r.t. extreme downside risk because tail dependencies are symmetric in the lower *and* upper tails.

For stressing a credit risk model toward extreme downside risk without allowing for upside chances one needs an *asymmetric* copula with lower tail dependence. An example complying with these conditions is the Clayton copula with parameter η. As we have seen in the last sections, the higher η the greater the lower tail dependence of the Clayton copula. As already mentioned, the Clayton copula lacks a certain calibration flexibility due to its one-parameter form. Nevertheless, it is a great tool for stress testing. For this purpose, we demonstrated in Section 2.5.6.5 how one can combine given marginal distribution with a Clayton copula in a Monte Carlo simulation.

As a last remark in this section, we want to mention the *Gumbel copula*, which, in contrast to the Clayton copula, which generates lower tail dependence, generates upper tail dependence. Analogously to the Clayton copula, the Gumbel copula is an Archimedean copula defined by the generator $\psi_\eta(t) = (-\ln t)^\eta$ such that

$$C_{\psi_\eta}(u,v) = \exp(-[(-\ln u)^\eta + (-\ln v)^\eta])^{1/\eta} \qquad (\eta \geq 1).$$

Kendall's tau for the Gumbel copula equals

$$\tau = 1 - \frac{1}{\eta}$$

and the upper tail dependence of the Gumbel copula is given by

$$\lambda_U = 2 - 2^{1/\eta}$$

(the proof again is based on (2.83)). It is an asymmetric copula but exhibiting tail dependence in the opposite direction than the Clayton copula. The dependence is perfect for $\eta \to \infty$. Independence corresponds to the case $\eta = 1$. The Gumbel copula can be applied in situations where we want to test the impact of extremely positive events (presumably based on great optimism), e.g., in a latent variable model according to (2.3) or in some other comparable threshold model, eventually with an underlying multi-factor model.

This concludes our general discussion on copulas and dependence measures and we return to the discussion of basket products. However, from here on our findings regarding copulas and their application constitute a fixed part of our modeling toolkit.

2.5.8 Impact of Copula Differences to the Duo Basket

In this section, we want to apply the copulas belonging to our modeling toolkit to our duo basket example in order to illustrate the impact a copula change can make. Starting point is Equation (2.62)

$$\mathbb{P}[\tau_A < t_A, \tau_B < t_B] = C(\mathbb{F}_A(t_A), \mathbb{F}_B(t_B)) \qquad (2.89)$$

specifying the joint distribution of default times of assets A and B via some copula combined with marginal default time distributions. Recall that the marginal distributions $\mathbb{F}_A$ and $\mathbb{F}_B$ are determined by the PD term structures of assets A and B via Equation (2.34) such that

$$\mathbb{F}_A(t) = p_A^{(t)} \quad \text{and} \quad \mathbb{F}_B(t) = p_B^{(t)} \qquad (t \geq 0).$$

As term structures for assets A and B we now work with the NHCTMC approach-based well-fitting credit curves from Figure 2.11. Based on

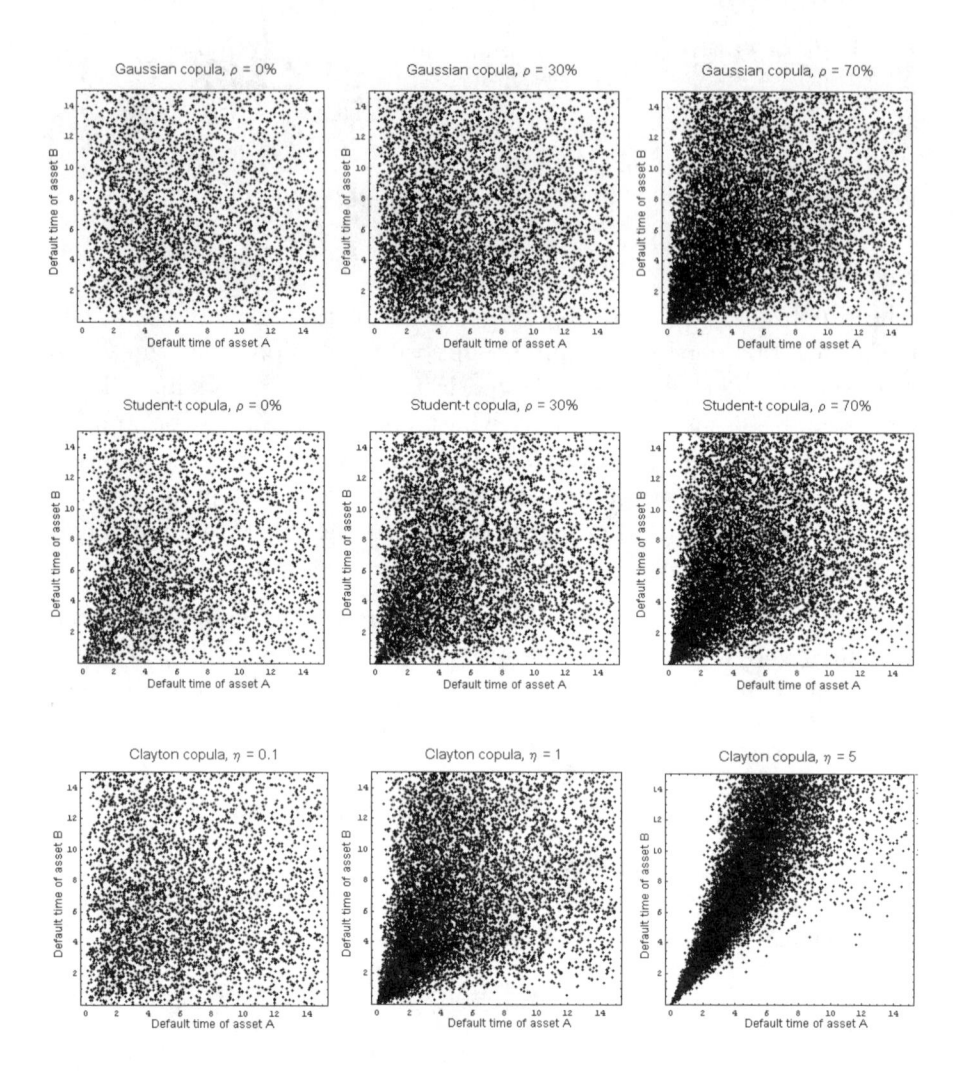

FIGURE 2.28: Scatterplot of dependent default times of assets A and B in the duo basket for different Gaussian and Student-t (with 5 degrees of freedom) copula functions; time is measured in years

the discussion in Section 2.3.3 we have to restrict ourselves in the sequel to time horizons $T \leq 15$ years, hereby respecting our data time horizon.

Figure 2.28 shows what we get if we implement Equation (2.89) for different copulas. We considered Gaussian copulas with 0, 30%, and 70% CWI correlation, Student-t-copulas with 5 degrees of freedom and the same levels of CWI correlation as in the Gaussian case, and Clayton copulas with $\eta = 0.1$, $\eta = 1$, and $\eta = 5$. Observations we make are:

- Zero CWI correlation case: for the Gaussian copula, the scatterplot of default times looks noisy, structural effects are almost negligible; the only effect, which is moderately visible in the cloud of points, is the slight tendency of points to cluster closer to the y-axis corresponding to earlier default times of asset A. This can be explained by the two times higher PD of asset A. For the Student-t copula we see a structure in the scatterplot already in the zero correlation case. This is explainable by the tail dependence of the t-copula according to Equation (2.88).

- Positive CWI correlation case: the Gaussian as well as the Student-t copula exhibit more and more structure in the default time scatterplot the more we increase the underlying CWI correlation. In case of the Student-t copula, the superposition of correlation as well as tail dependence effects leads to an even more coordinated default timing of assets A and B.

- For the Clayton copula, the strong lower tail dependence is visible in the case $\eta = 1$ and even stronger in the case $\eta = 5$. The case $\eta = 0.1$ is close to the independence case ($\eta = 0$) for Clayton copulas such that in this case hardly any coordination between default times of assets A and B can be observed.

Figure 2.28 translates the generic correlation plots from Figure 2.23 and 2.24 into the language of default times such that marginal distributions are naturally determined by the PD term structures of considered assets and the chosen copula function combines marginal default time distributions to a (joint) distribution of the default time vector (τ_A, τ_B). The illustration shows that the respective copula choice really makes a difference. In Section 2.6 we will see that analogous findings apply to larger baskets or credit portfolios.

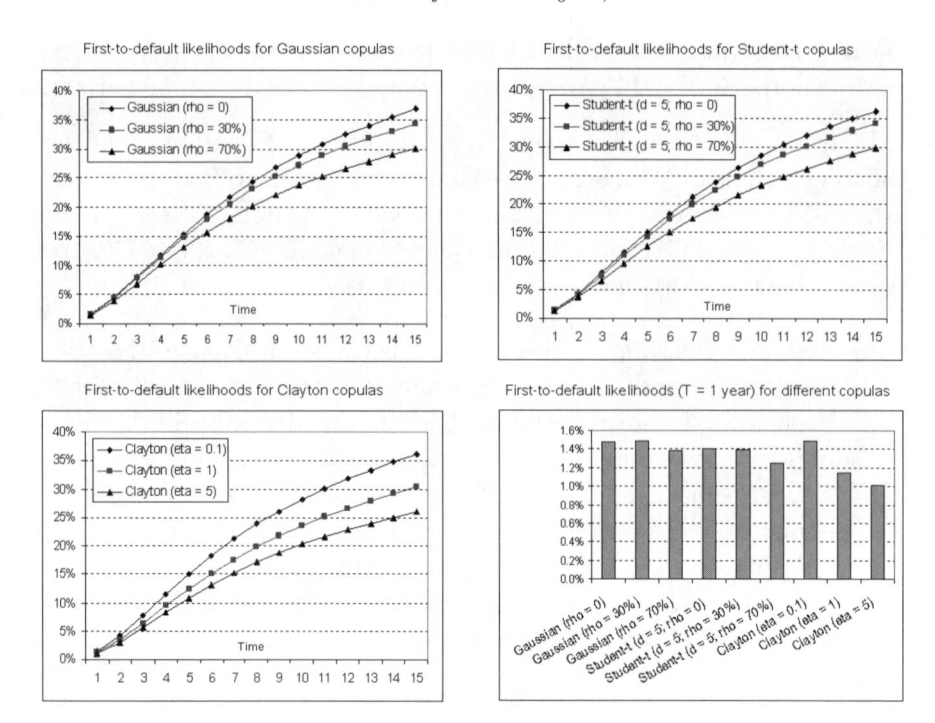

116   *Structured Credit Portfolio Analysis, Baskets & CDOs*

FIGURE 2.29: First-to-default likelihoods of assets A and B w.r.t. different time horizons and different copula functions; for the Student-t copulas 5 degrees of freedom are assumed

Let us now investigate the impact of copula changes to first-to-default and second-to-default likelihoods, hereby following up on our discussion in Sections 2.2 and 2.4 where we solely relied on the Gaussian copula with different CWI correlations. We can do this exercise with analytical formulas (e.g., via numerical integration over mixing variables by an appropriate software like Mathematica or Matlab) or by applying formulas (2.63) and (2.64),

$$\hat{p}_{1st}^{(T)} = \frac{1}{n}\sum_{i=1}^{n} \mathbf{1}_{\{\min(\hat{\tau}_A^{(i)}, \hat{\tau}_B^{(i)}) < T\}}$$

$$\hat{p}_{2nd}^{(T)} = \frac{1}{n}\sum_{i=1}^{n} \mathbf{1}_{\{\max(\hat{\tau}_A^{(i)}, \hat{\tau}_B^{(i)}) < T\}},$$

to the sample of 100,000 default time scenarios we already generated for Figure 2.28. In this case, we decide for the latter-mentioned. This will

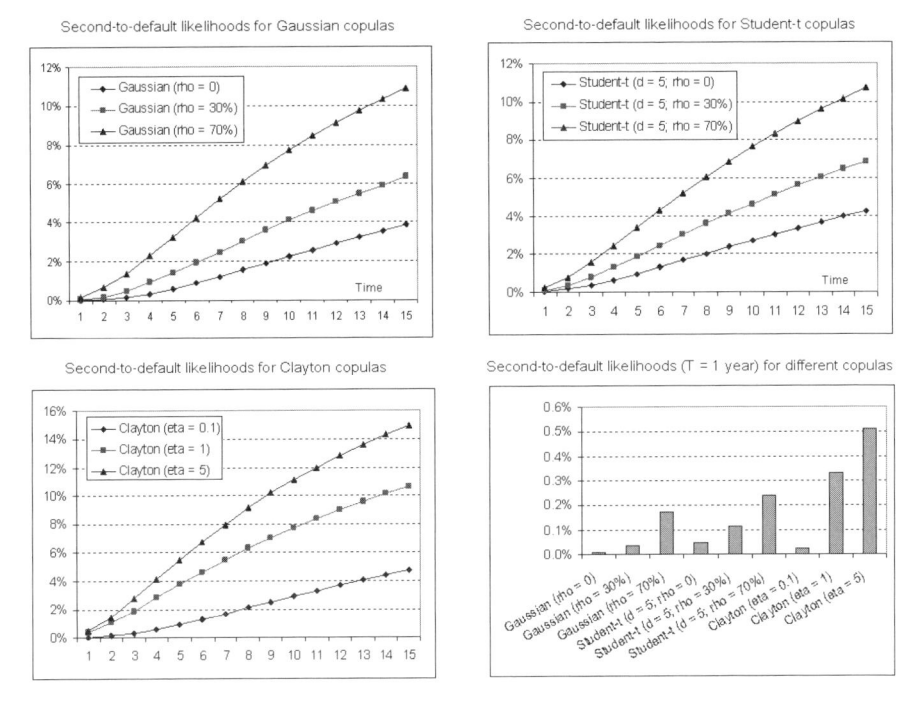

FIGURE 2.30: Second-to-default likelihoods of assets A and B w.r.t. different time horizons and different copula functions; for the Student-t copulas 5 degrees of freedom are assumed

involve some sample random fluctuations but nevertheless illustrate the effect of copula changes is a satisfactory manner.

We consider
$$\hat{p}_{1\text{st}}^{(T)} \quad \text{and} \quad \hat{p}_{2\text{nd}}^{(T)}$$
for time horizons $T = 1, 2, ..., 15$ years. Figures 2.29 and 2.30 show the results of our simulation study. We observe comparable results to what we already discovered in Section 2.4 for the Gaussian world.

- The underlying CWI correlation as well as the non-linear tail dependencies heavily drive the second-to-default likelihoods. The first-to-default likelihoods are also affected but at a more moderate level than in case of the second-to-default; see also Figure 2.3 for the one-year horizon. One can expect that *in practical applications first-to-default insurers are slightly more interested in the distribution of default probabilities of the underlying reference pool* in a basket or CDO transaction than in the underlying dependence structure, given that the reference pool is not an extremely exotic portfolio regarding its industry and country mix.

- In general, it remains correct what we already mentioned in the Gaussian copula case: *the more uncoordinated assets default, the worse the position of a first-to-default insurer.* This observation can be carried over from the linear correlation as a driver of 'coordinated moves' to the non-linear tail dependence as a driver of coordinated tail behavior. In the Student-t copula case we again observe an amplification of this effect due to a superposition of linear correlation and tail dependence, which even is positive in the zero correlation case.

- The strong effect of copulas on the second-to-default likelihood can be explained by the already-mentioned fact that the second-to-default likelihoods coincide with the joint default probability (JDP) w.r.t. the considered time horizon (see also Figure 2.15) because the JDP can be directly written as a function

$$\text{JDP}_{A,B}^{(T)} = \mathbb{P}[\tau_A < T, \tau_B < T] = C(\mathbb{F}_A(T), \mathbb{F}_B(T)),$$

of the applied copula function C.

Figures 2.29 and 2.30 nicely demonstrate how (extreme) copulas like the Clayton copula can be used to generate 'stressed' (artificial) tail behavior. We will come back to this point when we simulate baskets with more than two assets and CDO transactions.

Here, we stop our duo basket 'toy example' and turn our attention in the following examples to baskets with more than two assets. Before that, we want to give a 'word of caution'.

2.5.9 A Word of Caution

If one goes through the literature of the last, e.g., five years on risk modeling then one could easily get the impression that copulas are the current most popular 'toys' in the risk modeling community. We believe that the success of copulas as risk measurement tools is based on two major reasons:

- *The typical organizational separation of rating development and portfolio modeling in different quant teams in many banks.* This separation into single-name modeling and portfolio modeling finds a lot of support in the decompostion of multi-variate distributions into marginal distributions and a copula function.

- *The seeming flexibility arising from a separate treatment of single-name risks and their multi-variate dependence function.*

Let us comment on the first-mentioned reason. We are convinced that *it is conceptually wrong to separate rating and portfolio modeling teams* although one of the authors has to admit that his unit is organized in exactly this way, to some extent for historic reasons, to some extent for a better leverage of skill sets of quants. As an argument for our statement, consider CWI processes as *perfect scores* indicating default if and only if CWIs fall below a threshold, say, a score cutoff level. In this paradigm, CWI scores work like *crystal balls*; see also Figure 1.2. Now, the discriminatory power of a rating system as introduced in Chapter 1 (measured, e.g., in terms of AUROC), therefore, can be expressed in terms of, for instance, the correlation between the rating score and the perfect score given by CWIs. Given the dependence between the CWIs and the dependence between CWIs and rating scores, it is recommendable *to develop rating and portfolio models in an integrated approach*. In other words, modeling the rating or scoring system and the portfolio dependencies sequentially and separated is kind of comparable to a maximum likelihood estimation of a two-parametric distribution where one tries to estimate the two parameters sequentially one after another instead, as it is standard in statistics, trying to estimate the two parameters simultaneously in one step. Given the organizational constrains of separated rating and portfolio modeling teams, the separation into marginals and a copula is convenient. Conceptually more justified would be an integrated development of ratings and portfolio measures, bringing us back to the good old multi-variate distribution theory.

Regarding the second-mentioned reason, one should not forget that multi-variate distributions have to be fitted to empirical data before one can do something with it. Hereby, a separated view of marginals and copulas finds natural limitations. In MIKOSCH [83], the reader finds a survey on a whole bunch of limitations of the copula approach as well as some critical remarks regarding the current popularity of copula functions and the seeming enthusiasm regarding the application of copula techniques to problems in finance and risk measurement and management.

Despite all criticism and warnings, we find the copula concept useful and see various applications for it. Our basic understanding is that it is essential that in practical applications credit risk modelers keep a certain level of 'healthy suspect' against model findings. To be more explicit, *credit risk modeling problems do not very often find their answer in the form of a precise 'point estimate' but rather have to be answered by means of a potential solution space* taking model uncertainties and parameter fuzziness into account. This 'model principle' applies to parameters (e.g., attachment points in PD calibrations), model choices (e.g., threshold approach versus intensity-based approach), as well as to the choice of the copula for a certain model (e.g., Gaussian copula as a base case supplemented by a Student-t copula to test tails symmetrically as well as by a Clayton copula to test asymmetrically the lower tail, downside risk, without allowing for upside potential; see Section 2.5.7.6). Along these lines it makes a lot of sense to work with different copula functions in the sense of different *distributional scenarios*. The use of different copulas applied to the same modeling problem, supplemented by other scenario analyses, is exercised over and over in later sections. Transparency in model and parameter choices as well as transparency regarding possible ranges of model outcomes and implications should be considered as a natural 'codex of serious modeling' in any (credit) risk modeling project.

2.6 Nth-to-Default Modeling

In this section we turn our attention to baskets with more than two assets. For this purpose, we have to select a sample of reference names. Appendix 6.9 provides a universe of 100 credit-risky names from which we chose 10 names for inclusion in a reference portfolio underlying our basket example. Table 2.9 lists the 10 chosen names.

Next, we model the default time vector

$$(\tau_1, ..., \tau_{10})$$

of the reference portfolio. We consider three different copula functions,

Default Baskets

TABLE 2.9: Reference portfolio of names for our basket example; the 10 names constitute a subportfolio of the sample portfolio consisting of 100 names described in Appendix 6.9

Asset No.	Rating	PD	Region	Industry	Beta	Exposure	LGD
3	AAA	0.00%	1	8	65%	10,000,000	50%
4	AAA	0.00%	2	6	69%	10,000,000	50%
14	AA	0.01%	2	1	50%	10,000,000	50%
15	AA	0.01%	5	4	66%	10,000,000	50%
34	A	0.04%	3	2	45%	10,000,000	50%
35	A	0.04%	5	1	56%	10,000,000	50%
46	BBB	0.29%	1	7	39%	10,000,000	50%
63	BBB	0.29%	4	9	44%	10,000,000	50%
70	BBB	0.29%	5	4	36%	10,000,000	50%
79	BB	1.28%	3	4	44%	10,000,000	50%

namely, the Gaussian copula, the Student-t copula, and the Clayton copula. In the following three sections we briefly recall the equations necessary to get the Monte Carlo simulations implemented.

2.6.1 Nth-to-Default Basket with the Gaussian Copula

For our simulation study, we need to explicitly elaborate Formula (2.74) from Section 2.5.6,

$$\text{CWI}_i = \beta_i \sum_{n=1}^{N} w_{i,n} \Psi_n + \varepsilon_i \qquad (2.90)$$

$$(i \in I = \{3, 4, 14, 15, 34, 35, 46, 63, 70, 79\})$$

where CWI_i describes the CWI of client i at the one-year horizon. For any client i, the beta of the client can be read off from Tables 6.1, 6.2, or 2.9. The random variables Ψ_n describe systematic factors, which are represented by 5 regional and 10 industrial indices:

- $N = 15$
- $\Psi_1, ..., \Psi_5$ are regional indices
- $\Psi_6, ..., \Psi_{15}$ are industrial indices

In a 'real life' situation, region 1 could be the United States, region 2 could be Europe, region 3 could be Latin America, region 4 could be Africa, and region 5 could be Asia Pacific. However, for our illustrative example this does not play any role at all. The 10 industries could refer to typically applied broad industry categories like health care and financial institutions, and so on.

Table 6.3 in the appendix contains the correlation matrix of the 15 systematic indices. For example, regions 2 and 4 exhibit a low CWI or index correlation of 10%, whereas regions 1 and 2 are highly correlated at a level of 70%. The correlation matrix also reports on index correlations between regions and industries. For instance, region 5 and industry 8 are fairly high correlated, which is a typical example for a region where some special industry is dominating the overall economic performance of a region. The firms and systematic factors in our example are fictitious, but by throwing in some heterogeneity we attempt to imitate some 'real life' effects modelers face in their daily work. For reasons of simplicity we assume that

- every company has a 100% weight in one region and a 100% weight in one industry only.

This makes subsequent calculations a lot easier. The 5 *region* and 10 *industry indices* are assumed to be

- multi-variate standard normally distributed with a correlation matrix given by Table 6.3 in the appendix.

For instance, regions 2 and 4 exhibit a low CWI or index correlation of 10%, whereas regions 1 and 2 are highly correlated at a level of 70%. The correlation matrix also reports on index correlations between regions and industries. To give another example, region 5 and industry 8 are fairly high correlated, which is a typical example for a region where some special industry is dominating the overall economic performance of a region. The firms and systematic factors in our example are fictitious, but by throwing in some heterogeneity we attempt to imitate some 'real life' effects modelers face in their daily work.

CWI correlations in this model framework can be calculated as follows. Based on Equation (2.90), firms are correlated exclusively via their systematic *composite factors*, which, due to our factor model assumption (100% in one region, 100% in one industry), is given by

$$\sum_{n=1}^{N} w_{i,n} \Psi_n = \nu_i \times \left(\Psi_{n_{\text{region}}(i)} + \Psi_{5+n_{\text{industry}}(i)} \right) \qquad (2.91)$$

where n_{region} and n_{industry} denote the regional index and the industrial index, respectively, where the client takes on a 100% weight. The offset 5 in $\Psi_{5+n_{\text{industry}}(i)}$ refers to our notation of 10 industry indices numbered by $n = 6, ..., 15$. The scaling factor ν_i in Equation (2.91) is a *normalizing* constant determined for each name i in order to make the variance of the composite factor (2.91) equal to 1. To achieve this, we have to set ν_i equal to

$$\nu_i = \mathbb{V}[\Psi_{n_{\text{region}}(i)} + \Psi_{5+n_{\text{industry}}(i)}]^{-1/2}. \qquad (2.92)$$

Based on our distributional assumptions (see above) on region and industry indices and based on the correlation Table 6.3 we can easily calculate ν_i for each name by the formula

$$\nu_i = (2 + 2\text{Corr}[\Psi_{n_{\text{region}}(i)}, \Psi_{5+n_{\text{industry}}(i)}])^{-1/2}. \qquad (2.93)$$

The form of the composite factor (2.91) shows that two firms are positively correlated if and only if they either have the same region, or the same industry, or both. By construction, the composite factor for name i then boils down to

$$\beta_i \sum_{n=1}^{N} w_{i,n} \Psi_n = \beta_i \frac{\Psi_{n_{\text{region}}(i)} + \Psi_{5+n_{\text{industry}}(i)}}{\sqrt{2 + 2\text{Corr}[\Psi_{n_{\text{region}}(i)} + \Psi_{5+n_{\text{industry}}(i)}]}} \qquad (2.94)$$

which shows (due to Equation (2.92) the the R-squared of firm i equals

$$R_i^2 = \beta_i^2 \qquad (2.95)$$

such that the residual, firm-specific effects ε_i in Equation (2.90) have to be chosen w.r.t. normal distributions

$$\varepsilon_i \sim N(0, 1 - \beta_i^2) \qquad (2.96)$$

(independent and independent of the Ψ_n's) in order to achieve

$$\mathbb{V}[\text{CWI}_i] = 1 \quad (i = 1, ..., m \text{ resp. } i \in I)$$

for the CWIs of names in the reference portfolio as defined in Equation (2.90). According to Equation (2.95) we can calculate the distribution of R-squared parameters in the reference portfolio easily from given betas; see Figure 6.7 in Appendix 6.9.

124   *Structured Credit Portfolio Analysis, Baskets & CDOs*

TABLE 2.10: CWI correlation matrix of the basket reference portfolio described in Table 2.9

	3	4	14	15	34	35	46	63	70	79
3	100%	28%	19%	22%	14%	18%	17%	12%	12%	15%
4	28%	100%	23%	21%	10%	19%	17%	10%	12%	10%
14	19%	23%	100%	13%	9%	21%	11%	7%	7%	8%
15	22%	21%	13%	100%	12%	24%	12%	9%	24%	19%
34	14%	10%	9%	12%	100%	10%	8%	6%	7%	13%
35	18%	19%	21%	24%	10%	100%	10%	7%	13%	8%
46	17%	17%	11%	12%	8%	10%	100%	7%	7%	9%
63	12%	10%	7%	9%	6%	7%	7%	100%	5%	5%
70	12%	12%	7%	24%	7%	13%	7%	5%	100%	10%
79	15%	10%	8%	19%	13%	8%	9%	5%	10%	100%

The CWI correlation of names i and j with $i \neq j$ is given by

$$\text{Corr}[\text{CWI}_i, \text{CWI}_j] = \beta_i \beta_j \nu_i \nu_j (\rho_1 + \rho_2 + \rho_3 + \rho_4) \qquad (2.97)$$

where the correlations $\rho_1, \rho_2, \rho_3, \rho_4$ are given by

$$\rho_1 = \text{Corr}[\Psi_{n_{\text{region}}(i)}, \Psi_{n_{\text{region}}(j)}]$$
$$\rho_2 = \text{Corr}[\Psi_{n_{\text{region}}(i)}, \Psi_{5+n_{\text{industry}}(j)}]$$
$$\rho_3 = \text{Corr}[\Psi_{5+n_{\text{industry}}(i)}, \Psi_{n_{\text{region}}(j)}]$$
$$\rho_4 = \text{Corr}[\Psi_{5+n_{\text{industry}}(i)}, \Psi_{5+n_{\text{industry}}(j)}]$$

and ν_i, ν_j are the variance normalizing factors defined in Equations (2.92) and (2.93).

Let us calculate an example for illustration purposes. We want to calculate the CWI correlation between names 3 and 4. The betas of the two names can be found in Table 6.1:

$$\beta_3 = 65\%, \qquad \beta_4 = 69\%, \qquad \beta_3 \beta_4 = 44.9\%.$$

Name 3 is in region 1 and industry 8, whereas name 4 is in region 2 and industry 6. From Table 6.3 we read off

$$\text{Corr}[\Psi_{n_{\text{region}}(3)}, \Psi_{n_{\text{region}}(4)}] = 70\%$$
$$\text{Corr}[\Psi_{n_{\text{region}}(3)}, \Psi_{5+n_{\text{industry}}(4)}] = 48\%$$
$$\text{Corr}[\Psi_{5+n_{\text{industry}}(3)}, \Psi_{n_{\text{region}}(4)}] = 33\%$$
$$\text{Corr}[\Psi_{5+n_{\text{industry}}(3)}, \Psi_{5+n_{\text{industry}}(4)}] = 11\%.$$

In addition, we read off from Table 6.3 that

$$\text{Corr}[\Psi_{n_{\text{region}}(3)}, \Psi_{5+n_{\text{industry}}(3)}] = 35\%$$
$$\text{Corr}[\Psi_{n_{\text{region}}(4)}, \Psi_{5+n_{\text{industry}}(4)}] = 25\%$$

Formula (2.93) then helps us to calculate the normalizing constants as

$$\nu_3 = \frac{1}{\sqrt{2 + 2 \times 35\%}} = 60.9\%$$

$$\nu_4 = \frac{1}{\sqrt{2 + 2 \times 25\%}} = 63.3\%$$

such that altogether we obtain

$$\text{Corr}[\text{CWI}_3, \text{CWI}_4] = 44.9\% \times 60.9\% \times 63.3\% \times$$
$$\times (70\% + 48\% + 33\% + 11\%) = 28.0\%$$

as the pairwise CWI correlation between names 3 and 4. Table 2.10 shows the CWI correlation matrix of the 10 names from Table 2.9.

Based on Formula (2.97) we can calculate all pairwise CWI correlation in the reference portfolio of the 100 names described in Appendix 6.9, from which the 10 names in Table 2.9 underlying our basket example are taken. Hereby, we can restrict the calculation of CWI correlations to all entries located in the upper triangle of the portfolio's 100×100-correlation matrix. Altogether we have to calculate

$$(100^2 - 100)/2 = 100(100 - 1)/2 = 4,950$$

pairwise correlations. Figure 2.31 shows the result of this exercise.

The Monte Carlo simulation generating $(\tau_1, ..., \tau_{10})$ now proceeds as indicated in Figure 2.22:

- Simulate regional and industrial indices $\Psi_1, ..., \Psi_{15}$ w.r.t. a multivariate standard normal distribution with correlation matrix Γ given by Table 6.3.

- Simulate residual (idiosyncratic) effects $\varepsilon_1, ..., \varepsilon_m$ (with $m = 10$ for our basket example) w.r.t. normal distributions

$$\varepsilon_i \sim N(0, 1 - \beta_i^2)$$

as explained in the context of Equation (2.96).

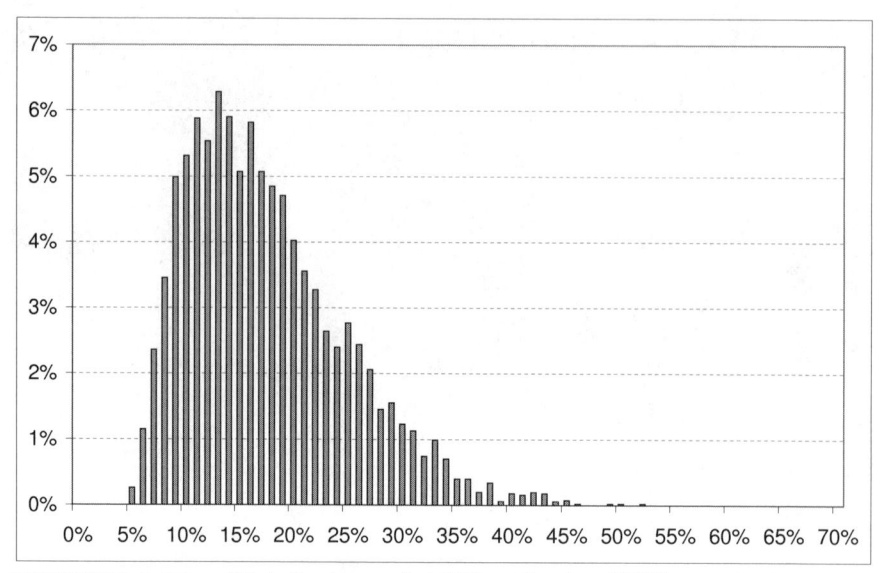

FIGURE 2.31: Distribution of 4,950 pairwise CWI correlations corresponding to our 100-names sample portfolio described in Appendix 6.9 (x-axis: CWI correlation; y-axis: relative frequency)

- Calculate normalizing constants ν_i as in Equation (2.93) and determine CWIs for the one-year horizon as

$$\mathrm{CWI}_i \;=\; \beta_i \, \frac{\Psi_{n_{\mathrm{region}}(i)} + \Psi_{5+n_{\mathrm{industry}}(i)}}{\sqrt{2 + 2\mathrm{Corr}[\Psi_{n_{\mathrm{region}}(i)} + \Psi_{5+n_{\mathrm{industry}}(i)}]}} \;+\; \varepsilon_i \quad (2.98)$$

in line with Equation (2.94) showing the form composite factors take on in our particular model.

- Simulate default times of the 10 names w.r.t. the Gaussian copula by inversion of the NHCTMC-based PD term structures

$$p_i^{(t)} \;=\; \mathbb{F}_i(t) \qquad (0 \leq t \leq 15)$$

and application of the inverses $\mathbb{F}_i^{-1}$ to the normal distribution function evaluated at CWI_i,

$$\tau_i \;=\; \mathbb{F}_i^{-1}(N[\mathrm{CWI}_i]) \qquad (i \in I)$$

following Equations (2.62) and (2.89).

Before we actually do the Monte Carlo simulation, we briefly supplement our discussion in this section by corresponding modifications to reflect the Student-t and the Clayton copulas.

2.6.2 Nth-to-Default Basket with the Student-t Copula

For modeling the basket with the Student-t copula we can pretty much stick to what we already developed for the Gaussian copula case. We only need to simulate in addition a sample of $\chi^2(d)$-distributed random variables and 'mix' the Gaussian CWI vector accordingly,

$$X \sim \chi^2(d)$$

$$\text{CWI} = \sqrt{d}\left(\beta_i \frac{\Psi_{n_{\text{region}}(i)} + \Psi_{5+n_{\text{industry}}(i)}}{\sqrt{2 + 2\text{Corr}[\Psi_{n_{\text{region}}(i)} + \Psi_{5+n_{\text{industry}}(i)}]}} + \varepsilon_i\right)/\sqrt{X},$$

being in line with Equation(2.98) and the discussion in Section 2.5.6. Everything else is analogous to the simulation in Section 2.6.1: we clearly proceed via

$$\tau_i = \mathbb{F}_i^{-1}(\Theta_d[\text{CWI}_i]).$$

Again, we set the degrees of freedom equal to $d = 5$ in our examples.

2.6.3 Nth-to-Default Basket with the Clayton Copula

For the Clayton copula, the situation is slightly different. First of all, we should mention that we set the parameter η in the definition of the Clayton copula to $\eta = 2$. In Section 2.5.7.5, we mentioned that the Clayton copula has lower tail dependence

$$\lambda_U = 2^{-1/\eta} \stackrel{\eta=2}{=} \frac{1}{\sqrt{2}}.$$

Therefore, we expect a moderate lower tail dependence leading to a certain increased potential for joint defaults. In Figure 2.28 (lower part of figure) we graphically illustrated the effect of lower tail dependence on dependent default times: for $\eta = 2$ we expect a certain tendency of default times to take place in a (to some extent) 'coordinated earlier' way, but not as strong as we saw it in Figure 2.28 in the case $\eta = 5$.

Following Theorem 2.5.9, our Monte Carlo simulation starts with a gamma-distributed random variable

$$X \sim \Gamma(1/\eta, 1).$$

Next, we need to randomly draw a uniform random variable

$$U_i \sim U([0,1])$$

for every name $i \in I$ such that $(U_i)_{i \in I}$ is independent. The inverse generator of the Clayton copula is

$$\varphi_\eta^{-1}(v) = (1+v)^{-1/\eta}.$$

Based on Theorem 2.5.9, CWIs can be defined by

$$\mathrm{CWI}_i = \varphi_\eta^{-1}\left(-\frac{\ln U_i}{X}\right) \quad (i \in I)$$

and corresponding default times are given by

$$\tau_i = \mathbb{F}_i^{-1}(\mathrm{CWI}_i).$$

For the sake of completeness let us briefly check back that default times defined in such a way do the job, hereby re-confirming Theorem 2.5.9 for the special case of an Archimedean copula generated by φ_η:

$$\begin{aligned}
\mathbb{P}[\tau_i < t_i : i \in I] &= \mathbb{P}[\mathrm{CWI}_i < \mathbb{F}_i(t_i) : i \in I] \\
&= \mathbb{P}\left[-\frac{\ln U_i}{X} < \varphi_\eta[\mathbb{F}_i(t_i)] : i \in I\right] \\
&= \int_0^\infty \mathbb{P}\left[U_i < \exp(-x\varphi_\eta[\mathbb{F}_i(t_i)]) : i \in I\right] \gamma_{1/\eta,1}(x)\,dx \\
&= \int_0^\infty \prod_{i \in I} \exp(-x\varphi_\eta[\mathbb{F}_i(t_i)]) \, \gamma_{1/\eta,1}(x)\,dx \\
&= \mathbb{E}\left[\exp\left(-X \sum_{i \in I} \varphi_\eta[\mathbb{F}_i(t_i)]\right)\right] \\
&= \left(1 - m + \sum_{i \in I} \mathbb{F}_i(t_i)^{-\eta}\right)^{-1/\eta}
\end{aligned}$$

hereby applying the Laplace transform of a gamma-distributed random variable according to Appendix 6.1,

$$\mathbb{E}[\exp(-tX)] = (1+t)^{-1/\eta} \quad \text{for} \quad X \sim \Gamma(1/\eta, 1).$$

This shows the desired stochastics of the simulated default times: their marginal distributions are specified by the PD term structures represented by the default time distribution functions $\mathbb{F}_i$, whereas their multi-variate dependence is specified by the Clayton copula.

In the sequel, we will repeatedly simulate Gaussian, Student-t, and Clayton default times. Hereby, we always proceed in the way just described in the previous three sections.

2.6.4 Nth-to-Default Simulation Study

After the technical preparations from the previous three sections we are now ready for the 10-names default basket. In general, the nth-to-default of a basket is represented by $\tau_{(n)}$ out of the *order statistics*

$$\tau_{(1)} \leq \tau_{(2)} \leq \cdots \leq \tau_{(10)}$$

of the default times $\tau_1, ..., \tau_{10}$ of the 10 names (n ranging between 1 and 10). In our simulation study we focus on the first- and second-to-default, where the first mentioned is illustrated in Figure 2.32. Table 2.11 shows the result of a Monte Carlo simulation of $\tau_{(1)}$, whereas Table 2.12 shows the corresponding simulation result for $\tau_{(2)}$.

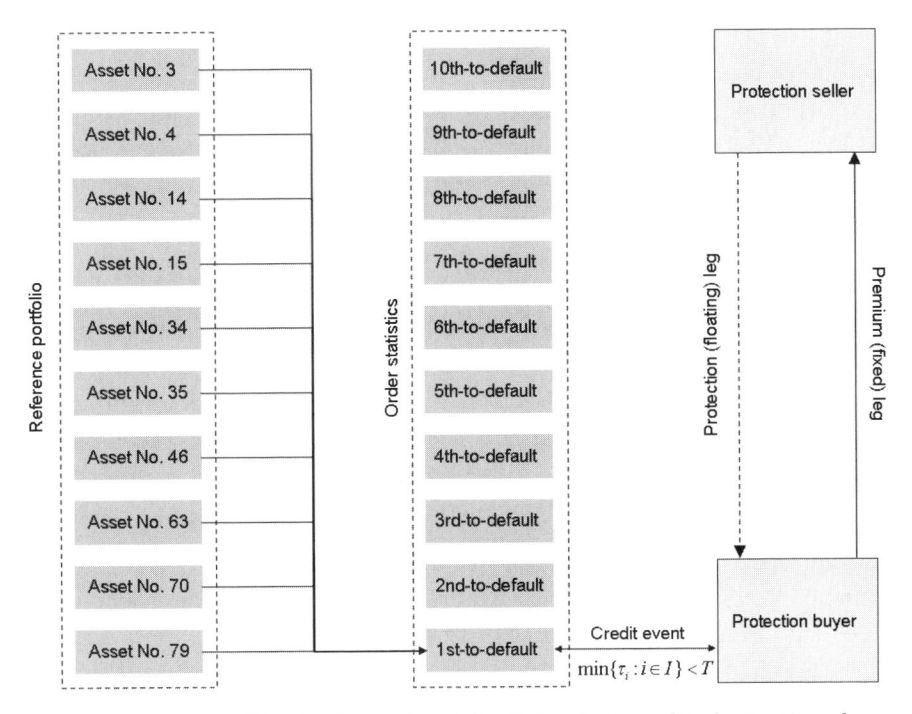

FIGURE 2.32: Illustration of a default basket credit derivative, here the first-to-default

In Figure 2.32, the *protection buyer* buys protection on the first default, or, more general, on the first *credit event*,[15] occurring in the pool. Hereby, credit events are typically defined w.r.t. standard ISDA[16] contracts. ISDA agreements significantly contribute to the standardization of credit derivatives like credit default swaps (CDS) and other instruments. Standardization of derivative contracts is a major success factor in developing liquid markets; see also Section 3.4.4. The second counterparty involved in the basket credit derivative is the *protection seller* who receives a fixed premium (the *premium leg*) from the protection buyer as compensation for the credit risk taking. Premium payments are typically made on a quarterly basis. The obligation of the protection seller is to provide credit protection against - in our case, the first loss in the reference portfolio. Hereby, settlement of protection in case of a credit event can happen in two ways: either the protection buyer has the right to sell the defaulted asset to the protection seller for a pre-specified price (*physical settlement*), e.g., at par, or the protection seller pays to the protection buyer a pre-specified amount (*cash settlement*), e.g., the realized loss of the default asset.

CDS do not necessarily have to be referenced to a single name. As in Figure 2.32, CDS can refer to a basket of assets or a larger pool or portfolio and can even have embedded structural elements. However, the more standard or 'plain vanilla' the instruments, the more liquidity can be assumed. In this section we only consider *unfunded* or *purely synthetic* instruments where no principal payments are involved (as long as no defaults occurred). Later we will also discuss *funded* instruments and also briefly compare advantages and disadvantages of funded and unfunded transactions; see Section 2.7 and Chapter 3. In an unfunded transaction as in our case in this section, the protection seller pays a premium or spread and not a full coupon (e.g., LIBOR + spread) so that no refinancing component is involved.

The terms and conditions of our first-to-default basket transaction can be summarized as follows, where 'PS' stands for protection seller and 'PB' stands for protection buyer.

- The reference portfolio or basket consists of the 10 names listed in Table 2.9. Each reference name is linked to an exposure of 10 mn USD such that the basket in total refers to 100 mn USD.

---
[15] Credit events typically include bankruptcy, failure to pay, and restructuring.
[16] International Swaps and Derivatives Association; see www.isda.org

- A credit event is triggered by a default event of a reference asset where 'default' refers to the typical payment default definition on financial obligations (90 days past due) or to bankruptcy or insolvency of a reference asset obligor; note that credit events in CDS contracts can also refer to other events like a rating downgrade to a certain pre-specified level or a relevant financial or debt restructuring (which can occur in the context of a 'bankruptcy protection' measure), etc.

- The PB pays an annualized premium in quarterly payments (in arrears) applied to the notional exposure of 10,000,000 USD, which is the effective exposure at risk insured by the PS.

- The basket has a term of 5 years in case no credit event occurs.

- The first credit event in the reference basket terminates the transaction and triggers a payment of

$$\text{assumed realized loss} = \text{exposure} \times \text{LGD} =$$
$$= 10{,}000{,}000 \times 50\% = 5{,}000{,}000 \text{ USD}$$

to be made by the PS to the PB. Note that the assumed LGD not necessarily refers to a realized recovery[17] amount but rather is pre-specified in the contract as a lump-sum fraction of the referenced exposure. Transactions of this type are called *fixed recovery transactions*.

Obviously, the protection sellers' best case is a five-year transaction without any credit event at all. Second best is a credit event at the latest possible time (namely, in quarter 20), and so on. The earlier a credit event occurs the less time the protection seller has to collect premium payments. Therefore, the *timing of defaults* is the key risk driver to be modeled via Monte Carlo simulation. In the previous three sections, we demonstrated how default times can be simulated w.r.t. the Gaussian, the Student-t, and the Clayton copula functions. We exercised the simulations and calculated the first-to-default times per copula simulation via Equation (2.63). Table 2.11 shows the result of this exercise for the first-to-default. Tables 2.12 and 2.13 show the same exercise if the transaction would focus on the second-to-default or third-to-default, respectively.

---

[17]If the PS is a bank, the PS can have a certain interest to use the upside potential

TABLE 2.11: Basket simulation study (occurrence likelihoods of first-to-default times) for different copula functions (df $= 5$; $\eta = 2$)

First-to-default time		Gauss	Student-t	Clayton
Quarter	1	0.28%	0.25%	0.14%
Quarter	2	0.47%	0.45%	0.24%
Quarter	3	0.68%	0.64%	0.36%
Quarter	4	0.86%	0.75%	0.45%
Quarter	5	0.95%	0.82%	0.57%
Quarter	6	1.02%	0.94%	0.62%
Quarter	7	1.13%	1.02%	0.70%
Quarter	8	1.25%	1.15%	0.81%
Quarter	9	1.25%	1.10%	0.83%
Quarter	10	1.27%	1.18%	0.81%
Quarter	11	1.37%	1.17%	0.82%
Quarter	12	1.33%	1.24%	0.89%
Quarter	13	1.32%	1.24%	0.80%
Quarter	14	1.35%	1.29%	0.78%
Quarter	15	1.35%	1.31%	0.80%
Quarter	16	1.33%	1.21%	0.78%
Quarter	17	1.27%	1.17%	0.85%
Quarter	18	1.31%	1.17%	0.76%
Quarter	19	1.25%	1.18%	0.70%
Quarter	20	1.39%	1.24%	0.72%
Quarters	>20	77.59%	79.49%	86.59%

TABLE 2.12: Basket simulation study (occurrence likelihoods of second-to-default times) for different copula functions (df $= 5$; $\eta = 2$)

2nd-to-default time		Gauss	Student-t	Clayton
Quarter	1	0.00%	0.02%	0.07%
Quarter	2	0.00%	0.05%	0.09%
Quarter	3	0.01%	0.06%	0.10%
Quarter	4	0.02%	0.08%	0.11%
Quarter	5	0.04%	0.11%	0.16%
Quarter	6	0.04%	0.11%	0.17%
Quarter	7	0.07%	0.12%	0.17%
Quarter	8	0.06%	0.16%	0.23%
Quarter	9	0.08%	0.18%	0.20%
Quarter	10	0.13%	0.18%	0.25%
Quarter	11	0.16%	0.24%	0.26%
Quarter	12	0.13%	0.21%	0.25%
Quarter	13	0.20%	0.22%	0.25%
Quarter	14	0.19%	0.27%	0.28%
Quarter	15	0.21%	0.28%	0.29%
Quarter	16	0.22%	0.26%	0.27%
Quarter	17	0.22%	0.28%	0.30%
Quarter	18	0.29%	0.29%	0.28%
Quarter	19	0.23%	0.30%	0.31%
Quarter	20	0.29%	0.29%	0.32%
Quarters	>20	97.41%	96.30%	95.66%

Default Baskets 133

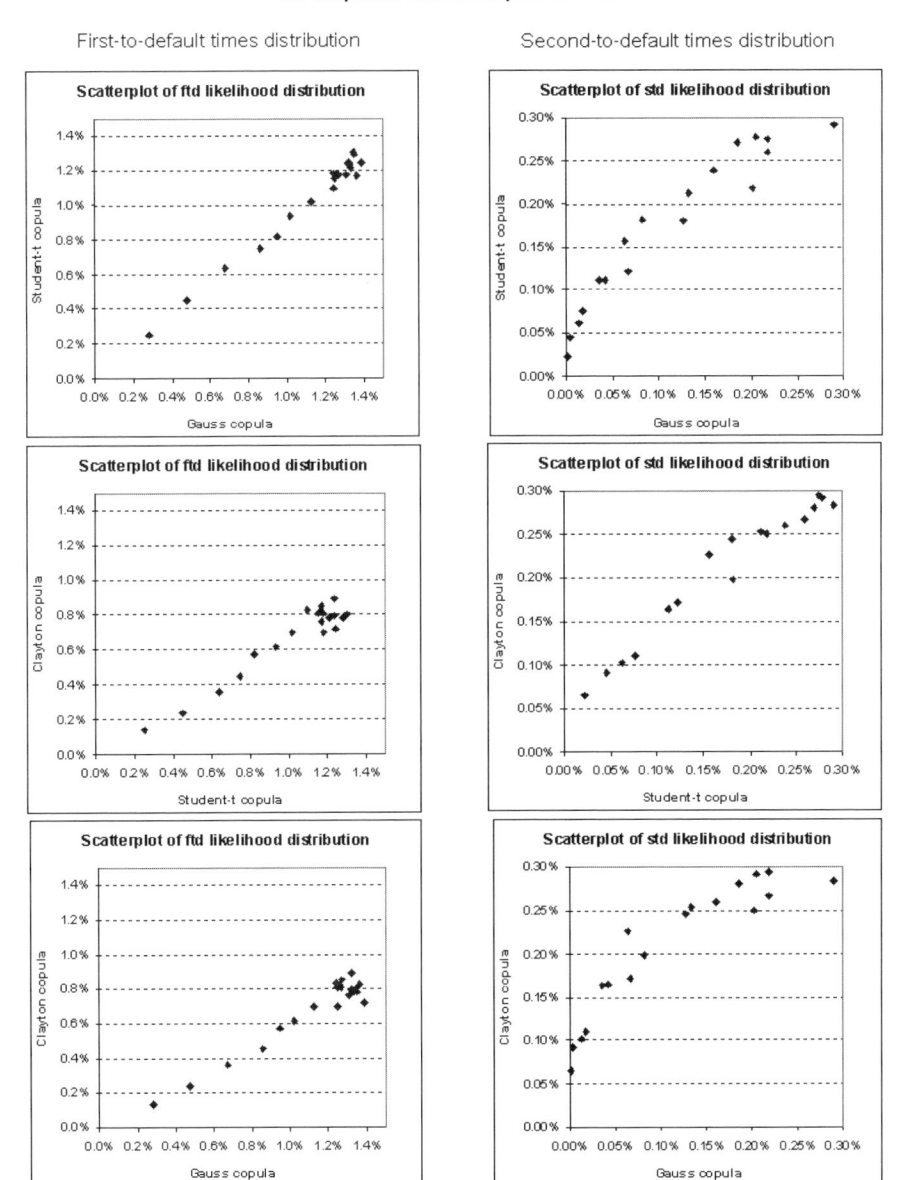

FIGURE 2.33: Scatterplot of first-to-default times (at the left-hand side) and second-to-default times (at the right-hand side) likelihoods for different copula functions (df $= 5$ for Student-t; $\eta = 2$ for Clayton); the scatterplot is an illustration of Tables 2.11 and 2.12

TABLE 2.13: Basket simulation study (occurrence likelihoods of third-to-default times) for different copula functions (df = 5; $\eta = 2$)

3rd-to-default time		Gauss	Student-t	Clayton
Quarter	1	0.00%	0.01%	0.05%
Quarter	2	0.00%	0.01%	0.07%
Quarter	3	0.00%	0.02%	0.07%
Quarter	4	0.00%	0.01%	0.06%
Quarter	5	0.00%	0.02%	0.10%
Quarter	6	0.00%	0.02%	0.13%
Quarter	7	0.00%	0.03%	0.11%
Quarter	8	0.00%	0.03%	0.16%
Quarter	9	0.00%	0.03%	0.14%
Quarter	10	0.01%	0.05%	0.17%
Quarter	11	0.01%	0.05%	0.15%
Quarter	12	0.00%	0.06%	0.18%
Quarter	13	0.02%	0.05%	0.18%
Quarter	14	0.01%	0.06%	0.16%
Quarter	15	0.02%	0.06%	0.21%
Quarter	16	0.03%	0.06%	0.16%
Quarter	17	0.02%	0.10%	0.18%
Quarter	18	0.03%	0.07%	0.18%
Quarter	19	0.03%	0.07%	0.22%
Quarter	20	0.06%	0.08%	0.21%
Quarters	>20	99.77%	99.13%	97.14%

Figure 2.33 illustrates in its upper part Table 2.11 by showing a scatterplot of likelihoods from the first-to-default time distribution of the three copula functions. Figure 2.33, in its lower part, shows the same exercise but for the second-to-default time distributions. What we see confirms our expectation when taking into account what we learned regarding tail dependencies and correlations in previous sections. Let us summarize what we find for the first-to-default time distribution.

- The probability that the protection seller is a pure payment receiver and the insurance becomes not effective equals in the Gaussian case 77.6%, in the Student-t case 79.5%, and in the Clayton case 86.6%. The higher the lower tail dependence (i.e., the potential for joint defaults), the better for the protection seller, an observation we made several times already.

- Consistent with the observation just mentioned is the fact that Student-t dependent default times lead to a probability-weighted

---

of the bank's recovery process.

later first-to-default time and Clayton dependent default times are likely to occur even later, although in the Student-t case the difference to the Gaussian case (given our particular parameterization) almost appears negligible; compare also with Figure 2.29 where we already have seen that the first-to-default is less sensitive to copula differences than the second-to-default (Figure 2.30).

A nice illustration of the difference between 'coordinated default times' and 'independent default times' was given some years ago by J. P. MORGAN via their *'correlation cat'*: imagine a (blind) cat that has to cross a room from the left-hand side to the right-hand side with mouse traps spread over the floor. A mouse trap, although conceptually intended to kill mice, can seriously hurt the cat if she accidentially steps into it. The worst case for the cat, given there are sufficiently many traps inducing a high trap density, is the case where the trap distribution shows no clusters at all, a situation that can be compared with independent or 'uncoordinated' default times. In contrast, the more clustered the traps are, the better the chances for the cat to find a safe path from one side of the room to the other side of the room, a situation that can be compared with dependent or to some extent 'coordinated' default times.

For the second-to-default and third-to-default time distributions, we find confirmed what already has been said on the relation between the timing of defaults and applied copula functions. In general, it can be expected that *the higher the correlation and the higher the lower tail dependence, the greater the potential for joint defaults at earlier times.*

We now focus on the first-to-default protection selling agreement made in our basket contract and turn our attention to the *evaluation* of the first-to-default basket transaction as an evaluation example. Analogous considerations can be applied to the second- and third-to-default in our example.

Figure 2.34 illustrates cash flows for the protection seller over time. We assume the convention that the protections seller (who receives quarterly premiums in arrears) receives a last premium at the end of the 20th payment period (no default scenario) or at the end of the payment period where the first default took place, which at the same time is the last payment period because our contract says that the occurrence of the first default triggers termination of the transaction.

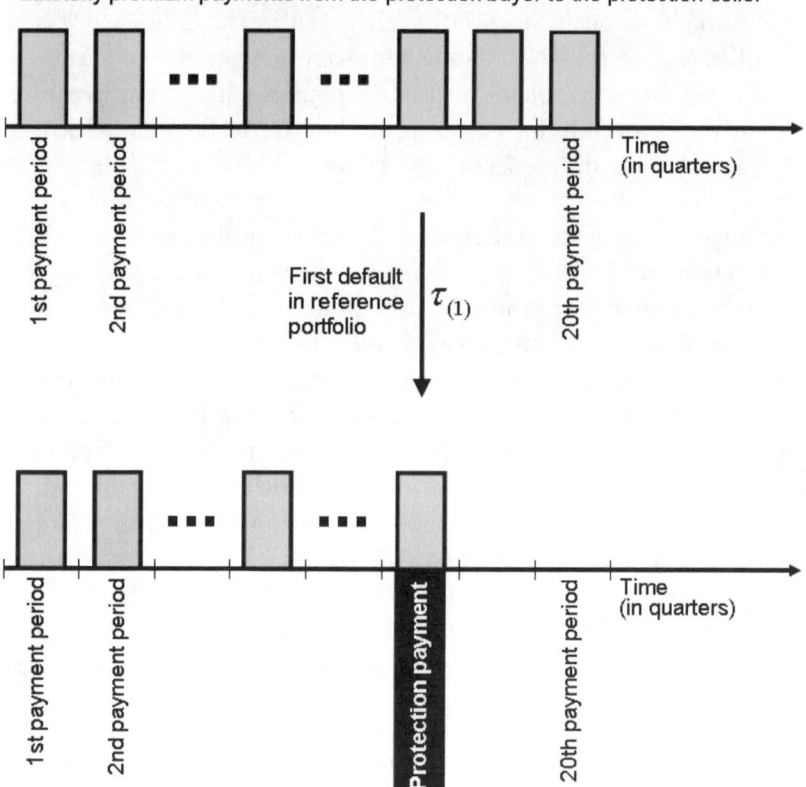

FIGURE 2.34: Illustration of cash flows for the protection seller in the first-to-default basket transaction

2.6.5 Evaluation of Cash Flows in Default Baskets

Note that in this section we count time in quarters and not in years for the sake of a more convenient notation.

The net cash flow C_{PS} for the protection seller can be written as follows. Denoting the spread to be paid from the protection buyer to the protection seller by S, we obtain

$$C_{\text{PS}}(S) = \min\{\tau_{(1)}, 20\} \times S \times [10 \text{ mn}] \times \frac{1}{4} - \mathbf{1}_{\{\tau_{(1)} \leq 20\}} \times [5 \text{ mn}] \quad (2.99)$$

where $\tau_{(1)}$ does now count time units in quarters. Figure 2.34 illustrates the cash in-flows (premiums) and potential cash out-flows (contingent protection payment) for the protection seller.

Default Baskets

Table 2.11 gives us an overview of the full distribution of possible scenarios, *discretized to a quarterly time grid*, which is a sufficiently fine resolution[18] for basket-type transactions and CDOs. For instance, w.r.t. a Gaussian copula model we have

$$\mathbb{P}[\tau_{(1)} = 7] = 1.13\%$$

meaning that with a likelihood of 113 bps the first default occurs in the 7th (quarterly) payment period. The likelihood (under a Gaussian copula assumption) that the protection seller receives the contractually promised premium for (at least) 7 payment periods is given by

$$\mathbb{P}[\tau_{(1)} \geq 7] = 1 - p_{1st}^{(6)} = 1 - \sum_{t=1}^{6} \mathbb{P}[\tau_{(1)} = t] = \quad (2.100)$$

$$= 1 - 4.26\% = 95.74\%$$

where $p_{1st}^{(t)}$ has been introduced[19] in Equation (2.30). Note that due to our premium payment convention (payments 'in arrears' per payment period) mentioned in the context of Figure 2.34 the protection seller receives a last premium at the end of the payment period in which the first default occurs. Moreover, with certainty he receives a premium payment in the first payment period. Conventions like this are subject to be negotiated in the contract of the transaction.

For calculating the expected cash flow for the protection seller we have to use the first-to-default time distribution to come to a weighting of cash flows shown in Figure 2.34. A premium payment will be made if no default occurred in previous payment periods. Therefore, the probability weights are given by Equation (2.100). We obtain

$$\mathbb{E}[C_{\text{PS}}(S)] = \sum_{t=1}^{20} \mathbb{P}[\tau_{(1)} \geq t] \times S \times 10 \times \frac{1}{4} - \mathbb{P}[\tau_{(1)} \leq 20] \times 5 \quad (2.101)$$

for the expected cash flow to the protection seller where units of money (mn USD) are dropped now to make formulas a bit shorter.

Calculation $\mathbb{E}[C_{\text{PS}}(S)]$ as a function of the spread S, we obtain Figure 2.35 showing the expected cash flow for the protection seller, for the time being ignoring the time value of money (no discounting applied).

---

[18] Because we consider a quarterly time grid, we postulate $\mathbb{P}[\tau_{(1)} \geq 1] = 1$.
[19] Again note that our usual convention is to count time in years; here we exceptionally count time in quarters.

Undiscounted cash flow for protection seller

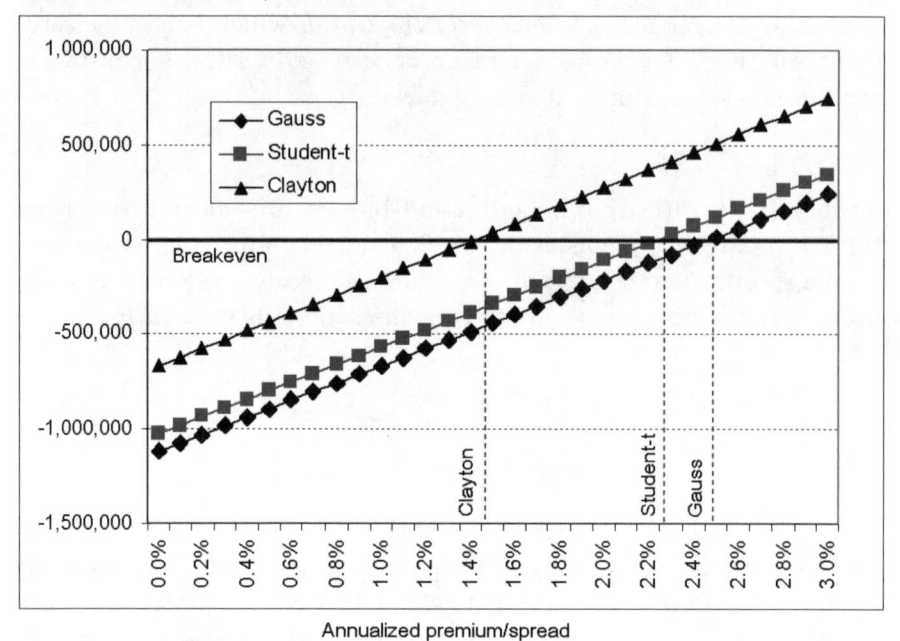

FIGURE 2.35: Expected cash flow to the protection seller in dependence of the assumed annualized premium/spread

The *breakeven spread* $S_{\text{breakeven}}$ (without discounting) can be determined according to the natural equation

$$\mathbb{E}[C_{\text{PS}}(S_{\text{breakeven}})] \stackrel{!}{=} 0 \qquad (2.102)$$

referring to the case where the protection seller on average does not make a profit nor a loss. We obtain

- $S_{\text{breakeven}} = 247$ bps for the Gaussian copula
- $S_{\text{breakeven}} = 224$ bps for the Student-t copula
- $S_{\text{breakeven}} = 142$ bps for the Clayton copula

Now, as a next step we have to take the time value of money into account. For this purpose, the probability weighting of cash flows as shown in Figure 2.34 has to be supplemented by *discounting* the single cash flows. For the calculation of the breakeven spread, now based on

Default Baskets

discounted cash flows, we adopt a comparable approach as for *risk-neutral* pricing: we look for the spread making the expected discounted cash flow for the protection seller equal to zero or 'neutral' (as in Equation (2.102)) where cash flows are discounted w.r.t. the *risk-free interest rate*. Let us assume that the risk-free interest rate equals 3%. Then, the expected discounted cash flow (where discounted cash flows are denoted by $\tilde{C}$ instead of C) for the protection seller is given by

$$\mathbb{E}[\tilde{C}_{\text{PS}}(S)] = \sum_{t=1}^{20} \frac{\mathbb{P}[\tau_{(1)} \geq t] \times S \times 10 \times \frac{1}{4} - \mathbb{P}[\tau_{(1)} = t] \times 5}{(1 + 0.75\%)^t} \qquad (2.103)$$

where 0.75% is the quarterly interest rate corresponding to an annual rate of 3%. Equation (2.103) boils down to Equation (2.101) if we set the risk-free interest rate equal to zero. In the same way, as in Equation (2.102), we can now calculate the breakeven spread, which we call the *fair spread* S_{fair} in the discounted case. It is the spread such that

$$\mathbb{E}[\tilde{C}_{\text{PS}}(S_{\text{fair}})] \stackrel{!}{=} 0. \qquad (2.104)$$

Calculating the fair spread for the three copula functions, we obtain

- $S_{\text{fair}} = 244$ bps for the Gaussian copula
- $S_{\text{fair}} = 221$ bps for the Student-t copula
- $S_{\text{fair}} = 140$ bps for the Clayton copula

The difference between fair and breakeven spreads, given a risk-free rate of 3%, is negligibly small. The reason for the low impact of discounting on the implied fair spread in our particular situation is easy to understand: the breakeven spread according to Equations (2.102) and (2.104) mainly is driven by the *balance* between positive and negative expected cash flows, namely

$$\mathbb{P}[\tau_{(1)} \geq t] \times S \times 10 \times \frac{1}{4} \quad \text{and} \quad \mathbb{P}[\tau_{(1)} = t] \times 5$$

along the quarterly time grid. The denominator in (2.104) plays only a minor role for this balance. However, if the first-to-default time distribution would be more inhomogeneous with high occurrence likelihoods close to the maturity of the transaction, then discounting would have a higher impact. In order to illustrate this, we consider an extreme case scenario where the first-to-default occurs with certainty in the last payment period of the transaction,

$$\mathbb{P}[\tau_{(1)} = 20] = 100\%,$$

which immediately implies

$$\mathbb{P}[\tau_{(1)} \geq j] = 100\% \quad \forall\, j = 1, 2, ..., 20.$$

Then, Equation (2.103) can be reduced to

$$\mathbb{E}[\tilde{C}_{\text{PS}}(S)] = -\frac{5}{(1+r/4)^{20}} + S \times 10 \times \frac{1}{4} \times \delta(r)$$

where r denotes the (risk-free) interest rate and

$$\delta(r) = \sum_{t=1}^{20} \frac{1}{(1+r/4)^t}$$

denotes the r-dependent sum of discount factors. Based on this, we obtain a very simple form of Equation (2.104) such that the fair spread, as a function of the risk-free interest rate, can be calculated as

$$S_{\text{fair}}(r) = \frac{5}{(1+r/4)^{20}} \times \frac{2}{5} \times \frac{1}{\delta(r)}.$$

For $r = 3\%$, we obtain $\delta(r) = 18.51$ (in contrast to $\delta(0) = 20$) and $(1+r/4)^{-20} = 86.12\%$ such that

$$S_{\text{fair}}(3\%) = 9.31\% \quad \text{and} \quad S_{\text{fair}}(0\%) = 10\%$$

for the considered extreme first-to-default time distribution with a 100% occurrence probability in the last payment period. With respect to this particular scenario or case, the spread difference equals roundabout 7%, which no longer is negligible as it has been the case for our *simulated* first-to-default time distribution. The back-of-the-envelope calculations just exercised lead us directly to our next topic.

2.6.6 Scenario Analysis

Besides a full Monte Carlo simulation of default times, scenario analysis as done at the end of the previous section where we considered a very special first-to-default time distribution is a mighty tool to *explore the potential solution space*, hereby catching-up with our discussion in Section 2.5.9. As protection seller in a first-to-default basket transaction as in Section 2.6.4, the Monte Carlo simulation of the first-to-default time distribution gives us a full and clear picture of what we have to expect including deviations from expectations. However, in the sequel we want to demonstrate that manually set scenarios can provide valuable insight in a transaction. Moreover, the results from such exercises can be used for *plausibility checking* on simulation-based results.

2.6.6.1 Best and Worst Case Scenarios

Good starting points are typically *worst case* and *best case* scenarios. In our example, the best case scenario is simply the case where no default occurs during the lifetime of the transaction. The protection seller then collects (nominal)

$$20 \times S \times 10,000,000 \times \frac{1}{4} = S \times 50,000,000 \text{ USD}.$$

For a spread of $S = 250$ bps this yields a total of $1,250,000$ USD of premium payments collected by the protection seller.

The worst case scenario is the situation where default occurs already in the first payment period. The protection seller then collects just one quarterly premium of

$$S \times 10,000,000 \times \frac{1}{4} = S \times 2,500,000 = 62,500 \text{ USD}$$

but has to make a payment to the protection buyer as a consequence of the protection agreement in size of $5,000,000$ USD.

Based on the two extreme scenarios we can conclude that the range of possible cash flow scenarios for the protection seller equals

$$[\text{cash range}] = [-4,937,500 \ , \ 1,250,000] \quad \text{USD} \qquad (2.105)$$

if the contractually agreed spread/premium equals $S = 250$ bps.

Best and worst case scenarios as well as the scenario example at the end of the previous section are scenarios where we adopt the point of view of *deterministic default timing* by considering situations where the first default occurs in a pre-specified payment period. However, scenario analysis can be extended to include *pre-specified default time distributions*. We call the latter-mentioned *distributional scenarios*. Figure 2.36 shows several distributional scenarios we will consider next.

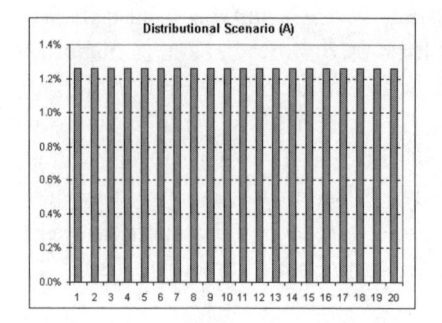

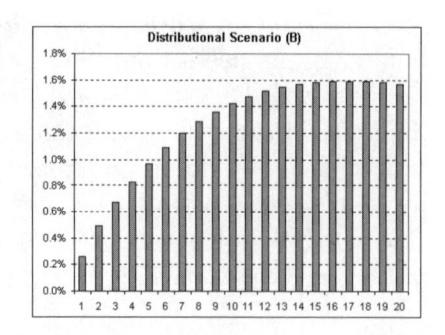

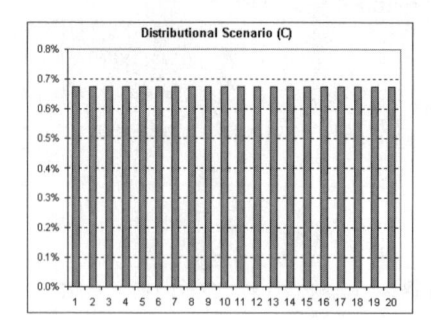

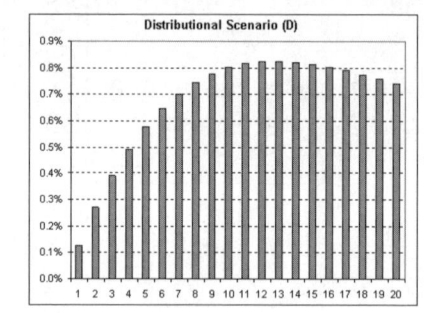

142   *Structured Credit Portfolio Analysis, Baskets & CDOs*

FIGURE 2.36: Pre-specified (not simulated but assumed) distributional scenarios for the first-to-default time; x-axis shows time in quarters; y-axis shows the probability that the first-to-default time falls into the considered payment period

2.6.6.2 Scenarios (A) and (B)

Note that we now go back to our usual convention to write time t in years, no longer in quarters as done in the previous section where the quarterly notation made formulas a bit more handy.

Distributional Scenario (A) in Figure 2.36 argues that the sum of PDs of the 10 assets in the reference basket

$$\sum_{i \in I} p_i^{(t)} = 2 \times p_{AAA}^{(t)} + 2 \times p_{AA}^{(t)} + 2 \times p_A^{(t)} + 3 \times p_{BBB}^{(t)} + p_{BB}^{(t)}$$

must be an *upper bound for the first-to-default likelihood* for time horizon t. We give it a name and denote this PD term structure by

$$(p_{upper}^{(t)})_{t \geq 0} = \left(\min \left[1, \sum_{i \in I} p_i^{(t)} \right] \right)_{t \geq 0}.$$

Default Baskets 143

TABLE 2.14: Upper cumulative PD term structure for scenario analysis calculations (Scenarios (A) and (B) in Figure 2.36)

quarter	1	2	3	4	5
PD_upper	0.26%	0.75%	1.42%	2.26%	3.23%

quarter	6	7	8	9	10
PD_upper	4.32%	5.52%	6.80%	8.17%	9.59%

quarter	11	12	13	14	15
PD_upper	11.07%	12.58%	14.13%	15.70%	17.28%

quarter	16	17	18	19	20
PD_upper	18.87%	20.46%	22.05%	23.63%	25.20%

A discretized version of $(p_{upper}^{(t)})_{t \geq 0}$ for quarterly time horizons is shown in Table 2.14 and the figure attached below.

Scenario (A) now takes the cumulative PD at the 5-year horizon

$$p_{upper}^{(5)} = 25.2\%$$

and distributes the mass of this likelihood uniformly over the relevant period of time, which are 20 quarters in our case. This implies a first-to-default time distribution of

$$\mathbb{P}[\tau_{(1)} = t] = 1.26\% \quad \text{for all} \quad t = 0.25, 0.5, ..., 5$$

and $\quad \mathbb{P}[\tau_{(1)} > 5] = 74.8\%$

(see Figure 2.36 where the first 20 payment periods are shown). The implied fair spread in this distributional scenario equals

$$S_{\text{fair}} = 285 \text{ bps}$$

for an assumed risk-free interest rate of 3%.

Scenario (B) distributes $p_{upper}^{(5)} = 25.2$ over the 20 payment periods in line with the PD term structure shown in Table 2.14. In Figure 2.36 an illustration of the scenario is given. The fair spread then equals

$$S_{\text{fair}} = 275 \text{ bps}$$

w.r.t. a risk-free interest rate of 3%. Figure 2.37 shows a comparison of the Gaussian copula simulation case (see Table 2.11) and distributional scenarios (A) and (B). It does not come as a big surprise that (A) yields a higher fair spread than (B), which yields a higher fair spread than the Gaussian copula simulation.

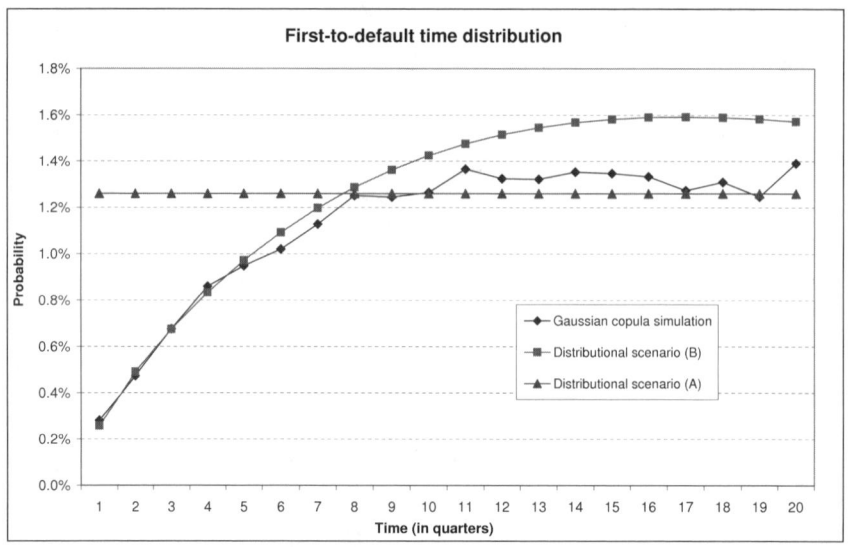

FIGURE 2.37: Comparison of first-to-default time distributions in the context of distributional scenario analysis

Note that by distributional scenarios as elaborated above we arrived at, e.g., 'reasonable' spread figures *without any simulation efforts at all*, just by a back-of-the-envelope cash flow calculation based on a certain assumed *distributional scenario*.

Default Baskets 145

2.6.6.3 Scenarios (C) and (D)

Scenarios (C) and (D) replace in the previous section the PD term structure

$$(p_{upper}^{(t)})_{t\geq 0} = \left(\min\left[1, \sum_{i\in I} p_i^{(t)}\right]\right)_{t\geq 0}.$$

by the PD term structure

$$(p_{max}^{(t)})_{t\geq 0} = \left(\max_{i\in I} p_i^{(t)}\right)_{t\geq 0}$$

which refers to a perfect dependence case. Because BB is the worst case rating in the 10-names reference pool, the max-PD term structure coincides with the BB-term structure.

Scenario (C) now assumes a uniform distribution of the 5-year BB-PD, whereas Scenario (D) distributes mass of the first-to-default time distribution over the payment periods in line with the BB-PD term structure; see Figure 2.36. For distributional Scenario (C) we obtain

$$S_{\text{fair}} = 144 \text{ bps}$$

whereas for distributional Scenario (D) we get

$$S_{\text{fair}} = 141 \text{ bps}$$

which is slightly below the uniform distribution case.

2.6.6.4 Scenario Summary and Conclusion

Altogether we have the following implied spread results:

- Monte Carlo simulation
    - Gaussian copula: $S_{\text{fair}} = 244$ bps
    - Student-t copula: $S_{\text{fair}} = 221$ bps
    - Clayton copula: $S_{\text{fair}} = 140$ bps

- Distributional scenario analysis
    - Scenario (A): $S_{\text{fair}} = 285$ bps
    - Scenario (B): $S_{\text{fair}} = 275$ bps
    - Scenario (C): $S_{\text{fair}} = 144$ bps

– Scenario (D): S_{fair} = 141 bps

where discounting always has been done w.r.t. a risk-free interest rate of 3% as in previous calculations.

Protection selling on a first-to-default basket is an *investment in credit risks*. Basis for an investment decision always should be the results from a *full Monte Carlo simulation yielding the full distribution of possible scenarios*. However, back-of-the-envelope calculations as done for Scenarios (A) to (D) can serve as

- Plausibility checks of obtained simulation results

- First-order 'quick-and-dirty' pre-evaluations

- Extreme case considerations, etc.

In our example, Scenarios (C) and (D) can be used to make plausibility checks regarding the Clayton copula simulation case because they reflect a *high dependence situation*. As we have seen, the back-of-the-envelope calculations of the fair spread nicely confirm the implied fair spread arising from Monte Carlo simulation. Scenarios (A) and (B) can be used to cross-check Gaussian and Student-t copula simulations. As we have seen, we can confirm the simulation-based fair spreads in reasonable ranges.

There are many variations of the scheme regarding basket evaluation. For instance, in arbitrage transactions with more sophisticated structural elements, basket cash flows can be designed to depend on spreads of the underlying reference assets. In such cases, it is recommendable to work with *risk-neutral* 'spread-implied' default probabilities instead of working with *historical*[20] 'ratings-based' default probabilities as we did in our example. In general, it is common market practice to back-out default probabilities from market prices of traded instruments (see Section 3.4.4) in order to avoid price arbitrage opportunities for other market players who will not adopt the bank-internal view of credit evaluations where the internal rating plays a dominant role.

In any case, the modeling techniques elaborated in the context of our first-to-default basket example are universally applicable and independent of the parameterization of default probabilities (risk-neutral

---

[20] Based on observed real-life default frequencies.

2.7 Example of a Basket Credit-Linked Note (CLN)

We now turn our attention to another important example of structured credit products, namely so-called *credit-linked notes* (CLNs) or *structured notes* as they are sometimes called. In contrast to synthetic unfunded credit derivatives like the first-to-default basket in the last section, a credit-linked note is a *funded* instrument comparable to a *bond with an embedded credit default swap*. As in all of these instruments, the performance of the CLN is linked to the performance of a single credit-risky asset (classical single-name CLN) or a basket (sometimes called an *index*) of credit-risky instruments (CLN referenced to a basket). Given the topic of this book, we only consider portfolio-referenced CLNs.

There are different ways to create CLNs, but two roles in a CLN transaction are always the same. The *issuer* of the CLN adopts the role of a *protection buyer*, whereas the *investor* in a CLN takes on the role of the *protection seller*. The investor in (buyer of) the note makes a *principal payment* at the beginning of the transaction to the issuer of the note. At maturity, the issuer pays back the full principal in case that no credit event of reference assets occurred. However, if a credit event occurred, the repayment is reduced by the realized loss or some contractually agreed proxy for the realized loss. In this way, the buyer of a CLN invests in the credit risk of reference assets. In return for the taken credit risk, the investor receives a *coupon* from the CLN issuer on a regular basis, typically in quarterly payment periods. Note that the CLN issuer not necessary represents a bank or corporate, it can also be a *special purpose vehicle* (SPV). Analogous to what has been said in Section 2.6.4, settlement of credit events of underlying reference entities can follow *physical settlement* or *cash settlement*. Due to its flexibility, CLNs are often used as building blocks in more complex structured

148   *Structured Credit Portfolio Analysis, Baskets & CDOs*

credit products like collateralized debt or synthetic/hybrid obligations.

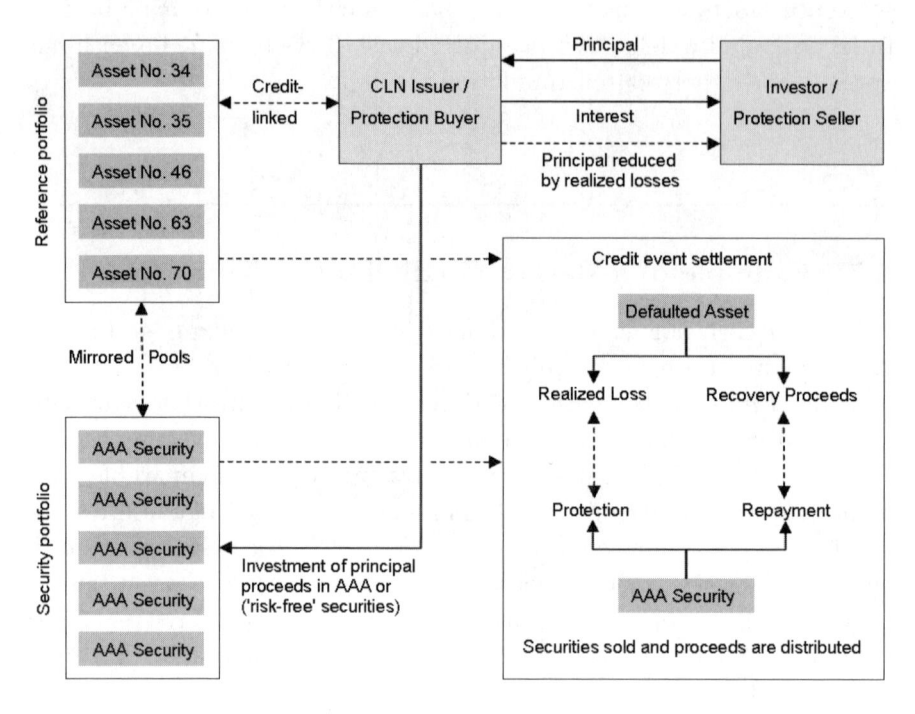

FIGURE 2.38:   Illustration of a CLN transaction referenced to a basket of 5 reference names

As indicated in the introductory remarks for this section, CLNs can be designed for the sake of different purposes in different ways. The transaction we want to model is illustrated in Figure 2.38 and can be considered as a *funded basket credit swap*, already conceptually close to CDOs, which will be discussed in Chapter 3. A summary of the term sheet of our CLN can be given as follows.

- The reference portfolio to which the performance of the CLN is linked is given by 5 reference names taken from Table 2.9. We again denote the index set for the 5 obligors by I. We set

$$I = \{4, 14, 15, 35, 46\}.$$

Every reference name carries an exposure of 10 mn USD. The assumed and contractually agreed/fixed LGD equals 50%.

Default Baskets

- The issuer of the CLN sells a note written over 50 mn USD to the investor who pays the principal amount of the note to the issuer at the beginning of the transaction.

- The issuer invests the principal proceeds in AAA-rated securities, which, for the sake of simplicity, we will consider as risk-free assets in the sequel. Following standard market practice we call these securities *collateral securities* although they are not really 'collateral' in the strict meaning of the word.

- The CLN pays an interest payment (coupon) to the investor. We assume that interest payments are floating in the form

$$\text{LIBOR} + \text{spread}$$

  where LIBOR[21] refers to the 3-month EUR-referenced risk-free rate banks use for short-term lending in the capital market. Figure 2.39 shows LIBOR rates for the last 4 years. For our model, we fix LIBOR at the 3% level.

- It is assumed that the collateral securities pay LIBOR flat so that *LIBOR fluctuations do not bother the issuer of the CLN;* see Figure 2.40.

- In case of a credit event of a reference entity, the transaction proceeds as follows:

  - The reference asset, which had a credit event, is removed from the reference pool.
  - The outstanding notional of the note is reduced accordingly, future coupon payments are applied to the reduced outstanding notional of the note.
  - Securities from the collateral security portfolio are sold in order to free-up principal proceeds in a way exactly matching the notional of reference assets, which had a credit event.
  - Proceeds from selling securities are distributed as follows. In line with our LGD = 50% assumption, half of the proceeds are paid back to the investor and the other half flows into the pocket of the issuer as protection payment for assets that had a credit event.

---

[21] London Inter Bank Offered Rate.

150   *Structured Credit Portfolio Analysis, Baskets & CDOs*

- The term of the transaction is 3 years.

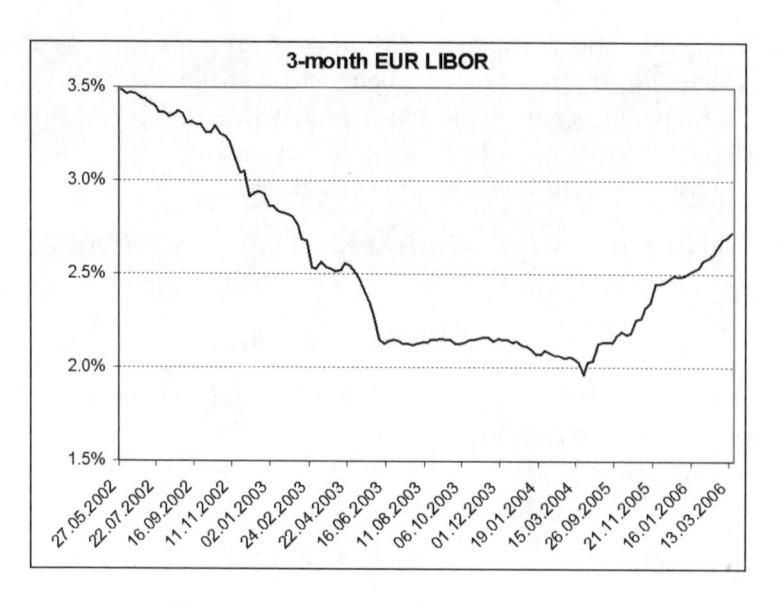

FIGURE 2.39:   Historical 3-month (EUR-referenced) LIBOR rates

Figure 2.40 illustrates the cash flows in the CLN transaction. Interest payments to be made to the investor have two components: LIBOR, which stands for 'risk-free refinancing', is collected from the collateral securities, spread has to be paid by the issuer of the CLN. Because the transaction works in a way keeping the reference pool and the collateral securities pool always at the same level of notional, LIBOR fluctuations indeed are not having any influence on the transaction from the issuer's perspective.

What is the difference between the considered CLN and a purely synthetic credit derivative, which could be structured in a way exactly matching the economics of our CLN? Now, the main difference is that the CLN involves a *principal stream* and an *interest stream* in its cash flows, whereas a comparable unfunded transaction would only involve an interest stream as long as no defaults occurred. There are advantages and disadvantages of both variants. A major advantage of our CLN is the avoidance of *counterparty risk:* protection on the reference assets is guaranteed by the collateral securities, which are sold in suffi-

Default Baskets

cient quantity in case of a credit event on reference assets. Therefore, the issuer has no risk that the protection seller for some reason does not pay the contractually agreed protection payment. However, this advantage is bought at the expense of market risk of the collateral securities because it could be the case that for reasons not known today they cannot be sold to the market at par. In some transactions, market risk of collateral securities is mitigated by a structured *repo agreeement*. As always in structured credit products, variations of the scheme are thinkable and can be found in the market: in practical applications it is always possible to structure a CLN in a way best matching investors and issuers demands.

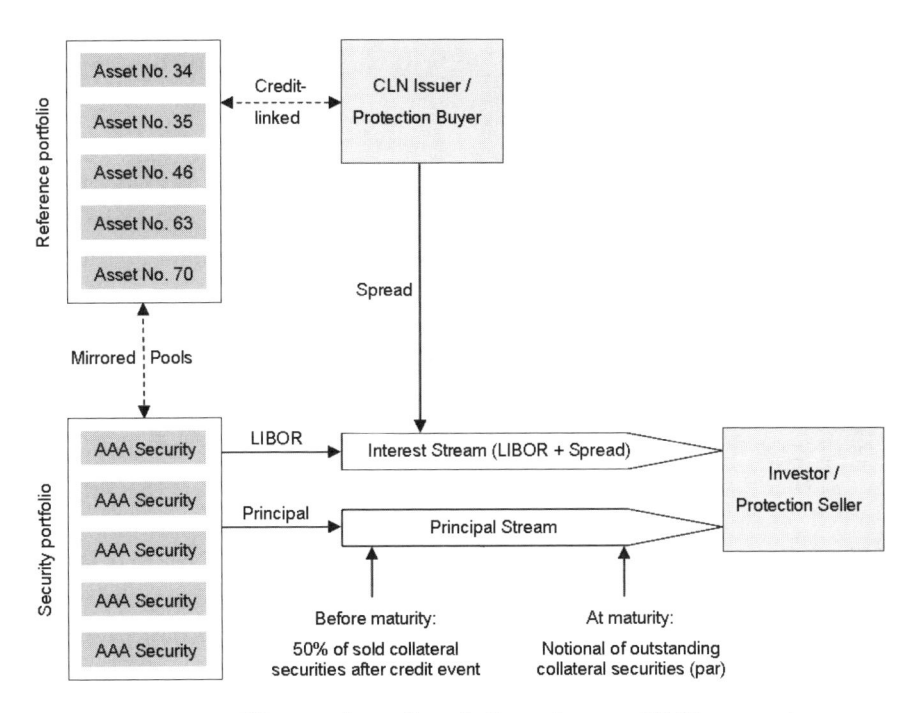

FIGURE 2.40: Illustration of cash flows in our CLN example

We now want to model the CLN from issuer's and protection seller's perspective. The first step in such models always is to gain an understanding of the involved credit portfolio. Here, we are facing 5 already known names: in Section 2.6 we modeled a portfolio for a first-to-default basket, which contained the 5 names constituting our CLN reference basket. The default time distributions of the 5 reference assets are, therefore, already known to us. Again we consider the Gaussian copula, the Student-t copula with 5 degrees of freedom, and the same correlation matrix as the Gaussian copula, and the Clayton copula with $\eta = 2$. Table 2.15 shows the distributions of the nth default times where n ranges from 1 to 5. But before we turn our attention to Table 2.15 as outcome of a default time Monte Carlo simulation we want to find a reasonable range of what we have to expect, comparable to our scenario analysis for the first-to-default example in Section 2.6.6.

Given our NHCTMC term structures, the (cumulative) 3-year PDs of our reference assets are given by

$$p_4^{(3)} = 0.04\%, \quad p_{14}^{(3)} = 0.10\%, \quad p_{15}^{(3)} = 0.10\%,$$

$$p_{35}^{(3)} = 0.29\%, \quad p_{46}^{(3)} = 1.52\%.$$

First, let us provide a very rough estimate of an upper bound for the worst case loss the investor could face. In the same way as in Section 2.6.6 we argue that the probability of at least one default in the reference portfolio, which we call the *hitting probability* of the reference pool, is bounded from above by the sum of the reference assets default probabilities, where we have to consider the PD time horizon matching the term of the transaction, which is 3 years in our example:

$$\mathbb{P}\left[\sum_{i \in I} \mathbf{1}_{\{\tau_i \leq 3\}} > 0\right] \leq \sum_{i \in I} p_i^{(3)} = 2.05\%. \qquad (2.106)$$

This means that in roughly maximal 2% of the cases we have to expect a 'hit' of the reference pool resulting in a loss of principal for the investor. In this sense, *the hitting probability represents the default probability of the reference portfolio*. Now, a worst case hit of the pool is a complete *wipe-out* of the pool. A kind of worst case expectation is to consider the hitting probability and attach to it the wipe-out event of the reference pool. Given the contractual LGD of 50%, we can now combine the pool PD with the contractual LGD applied to the portfolio wipe-out scenario yielding a *worst case expected loss*

$$\mathrm{EL}_{pool}^{(wc)} = \mathrm{PD}_{pool}^{(wc)} \times \$\mathrm{LGD}_{pool}^{(wc)} = \qquad (2.107)$$

$$= 2.05\% \times (50\% \times 50,000,000) = 512,500 \text{ USD}$$

TABLE 2.15: Default time distributions (nth default time for the 3 considered copula functions for the 5 reference assets underlying the CLN (rounded to basispoint level)

Gauss	1st	2nd	3rd	4th	5th
1	0.05%	0.00%	0.00%	0.00%	0.00%
2	0.09%	0.00%	0.00%	0.00%	0.00%
3	0.09%	0.00%	0.00%	0.00%	0.00%
4	0.14%	0.00%	0.00%	0.00%	0.00%
5	0.15%	0.00%	0.00%	0.00%	0.00%
6	0.15%	0.00%	0.00%	0.00%	0.00%
7	0.17%	0.01%	0.00%	0.00%	0.00%
8	0.22%	0.00%	0.00%	0.00%	0.00%
9	0.19%	0.00%	0.00%	0.00%	0.00%
10	0.22%	0.00%	0.00%	0.00%	0.00%
11	0.32%	0.01%	0.00%	0.00%	0.00%
12	0.24%	0.01%	0.00%	0.00%	0.00%
>=13	97.99%	99.97%	100.00%	100.00%	100.00%

Student-t	1st	2nd	3rd	4th	5th
1	0.07%	0.00%	0.00%	0.00%	0.00%
2	0.09%	0.00%	0.00%	0.00%	0.00%
3	0.09%	0.01%	0.00%	0.00%	0.00%
4	0.12%	0.01%	0.00%	0.00%	0.00%
5	0.15%	0.01%	0.00%	0.00%	0.00%
6	0.14%	0.02%	0.00%	0.00%	0.00%
7	0.16%	0.02%	0.01%	0.00%	0.00%
8	0.20%	0.02%	0.00%	0.00%	0.00%
9	0.20%	0.02%	0.00%	0.00%	0.00%
10	0.22%	0.02%	0.01%	0.00%	0.00%
11	0.23%	0.02%	0.00%	0.00%	0.00%
12	0.25%	0.03%	0.01%	0.00%	0.00%
>=13	98.12%	99.84%	99.98%	100.00%	100.00%

Clayton	1st	2nd	3rd	4th	5th
1	0.06%	0.01%	0.00%	0.00%	0.00%
2	0.07%	0.01%	0.00%	0.00%	0.00%
3	0.07%	0.01%	0.01%	0.00%	0.00%
4	0.06%	0.02%	0.01%	0.00%	0.00%
5	0.12%	0.02%	0.01%	0.01%	0.00%
6	0.13%	0.02%	0.01%	0.00%	0.00%
7	0.13%	0.03%	0.01%	0.01%	0.00%
8	0.14%	0.04%	0.01%	0.01%	0.01%
9	0.14%	0.02%	0.02%	0.01%	0.00%
10	0.17%	0.03%	0.02%	0.01%	0.01%
11	0.19%	0.05%	0.02%	0.01%	0.01%
12	0.17%	0.02%	0.03%	0.01%	0.01%
>=13	98.57%	99.74%	99.87%	99.92%	99.97%

where the superscribed 'wc' stands for 'worst case' and '$LGD' stands for LGD in units of money. In this way, we obtain an *upper bound for the worst case loss expectation* the investor has to face.

Now let us consider Table 2.15, which gives us the full distribution of scenarios arising from Monte Carlo simulation. As expected, we obtain an increasing potential for joint defaults when replacing the Gaussian copula by the Student-t copula and the Student-t copula by the Clayton copula. The *hitting probabilities* are given by

$$\text{PD}_{pool} = 1 - \mathbb{P}[\tau_{(1)} > 3 \text{ years}]$$

where $\tau_{(k)}$ again denotes the kth default time in the notation of the order statistics of the default times of the 5 reference assets. We obtain[22]

- $\text{PD}_{pool} = 201$ bps for the Gaussian case
- $\text{PD}_{pool} = 188$ bps for the Student-t case
- $\text{PD}_{pool} = 143$ bps for the Clayton case

Table 2.15 provides much more information to us. For instance, the *wipe-out probabilities* for the Gaussian and Student-t case are zero but

- $\mathbb{P}\left[\prod_{i \in I} \mathbf{1}_{\{\tau_i \leq 3\}} = 1\right] = 3$ bps for the Clayton[23] case

In other words, the upper tail dependence of the Clayton copula for $\eta = 2$ is strong enough to yield a non-negative wipe-out likelihood (at measurable basispoint level).

Before we say a few more words regarding the Monte Carlo simulation of the CLN, we want to briefly discuss a natural question arising in the context of pool PDs calculated above. Based on Table 2.15, we found

- $\text{PD}_{pool} = 201$ bps for the Gaussian case

whereas for our (back-of-the-envelope calculated) upper estimate of the pool PD we found

- $\text{PD}_{pool}^{(wc)} = 205$ bps

---

[22] For verification via a pocket calculator note that figures in Table 2.15 are rounded to basispoint level.

[23] For verification via a pocket calculator note that figures in Table 2.15 are rounded to basispoint level.

Default Baskets

As expected from the attribute 'worst case', we have

$$\text{PD}_{pool} < \text{PD}_{pool}^{(wc)}$$

but is the small distance of only 4 bps between the two pool PDs really *plausible*? Once more, catching-up with our plausibility and scenario analysis discussion in Section 2.6.6, we want to provide an argument why there is no need to be concerned about only 4 bps difference (Gaussian case!) between the simulated pool PD and a worst case pool PD.

The 205 bps worst case pool PD are based on the sum of 3-year PDs of reference assets. If the default events of the 5 reference assests would be disjoint, Inequality (2.106) would become an Equation,

$$\mathbb{P}\Big[\sum_{i \in I} \mathbf{1}_{\{\tau_i \leq 3\}} > 0\Big] = \sum_{i \in I} p_i^{(3)} = 2.05\%,$$

and we would obtain $\text{PD}_{pool} = \text{PD}_{pool}^{(wc)}$. In our case, they are not disjoint. Then, the corresponding equation is the following:

$$\mathbb{P}\Big[\sum_{i \in I} \mathbf{1}_{\{\tau_i \leq 3\}} > 0\Big] = \sum_{i \in I} p_i^{(3)} - \mathbb{P}\Big[\sum_{i \in I} \mathbf{1}_{\{\tau_i \leq 3\}} > 1\Big]. \quad (2.108)$$

Figure 2.41 illustrates Equation (2.108) for a situation with 3 assets.

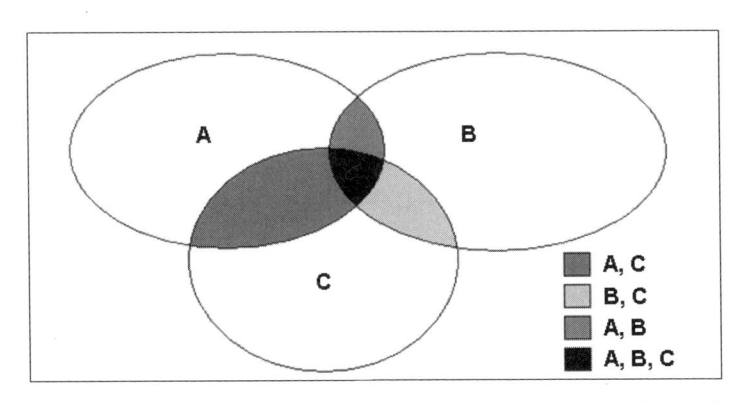

FIGURE 2.41: Venn diagram illustration of Equation (2.108)

- Event A refers to a default of name A, Event B refers to a default of name B, and event C refers to a default of name C, all w.r.t. a pre-specified evaluation horizon.

156   *Structured Credit Portfolio Analysis, Baskets & CDOs*

- The likelihood of the event $A \cup B \cup C$ refers to the union of the area of the three ellipses. This 'union event' corresponds to the left-hand side of Equation (2.108).

- The right-hand side of Equation (2.108) is the union of the three ellipse areas corrected by overlapping areas or, as we would say in our credit risk terminology, *joint default events* in order to avoid double and triple counting of (sub)areas.

After these preparations, we are now ready to do a *plausibility cross-check* of the 4 bps difference between PD_{pool} and $\text{PD}_{pool}^{(wc)}$ via a very rough (upper) estimate of the joint default probabilities subtracted at the right-hand side of Equation (2.108). We proceed as follows.

- The probability for joint defaults in Equation (2.108) can be decomposed into *elementary events*. Denoting (as always) defaults with a '1' and non-defaults with a '0', the space of possible portfolio default scenarios w.r.t. any fixed time horizon (3 years for our CLN) can be described by the sequence/code space

$$\mathcal{S} = \{0,1\}^5 = \{(i_i, i_2, i_3, i_4, i_5) \mid i_i, i_2, i_3, i_4, i_5 \in \{0,1\}\}.$$

For instance, we have

$$\mathbb{P}[(i_i, i_2, i_3, i_4, i_5) = (0,1,0,0,0)] = p_{14}^{(3)} = 0.10\% \quad \text{and}$$

$$\mathbb{P}[(i_i, i_2, i_3, i_4, i_5) = (0,0,0,0,1)] = p_{46}^{(3)} = 1.52\%.$$

The wipe-out probability of the pool can be written as

$$\mathbb{P}[(i_i, i_2, i_3, i_4, i_5) = (1,1,1,1,1)]$$

and, as can be seen in Table 2.15, strongly depends on the chosen copula function. We can write the code space $\mathcal{S}$ in the following form as a partition of subspaces,

$$\mathcal{S} = \mathcal{S}_0 + \mathcal{S}_1 + \mathcal{S}_2 + \mathcal{S}_3 + \mathcal{S}_4 + \mathcal{S}_5$$

where $\mathcal{S}_k$ denotes the space with codes referring to exactly k defaults. For instance,

$$\mathcal{S}_2 = \{(i_i, i_2, i_3, i_4, i_5) \in \mathcal{S} \mid i_i + i_2 + i_3 + i_4 + i_5 = 2\}$$

which is the space of events with exactly 2 joint defaults.

- The number of codes in the space $\mathcal{S}_{\geq 2}$ of all joint default events
$$\mathcal{S}_{\geq 2} = \mathcal{S}\backslash(\mathcal{S}_0 \cup \mathcal{S}_1)$$
can be calculated as $\#\mathcal{S}_{jd} = 2^5 - 1 - 5 = 26$ due to
    - $\#\mathcal{S}_0 = 1$
    - $\#\mathcal{S}_1 = 5$
    - $\#\mathcal{S}_2 = 10$
    - $\#\mathcal{S}_3 = 10$
    - $\#\mathcal{S}_4 = 5$
    - $\#\mathcal{S}_5 = 1$

    The symmetry in the number of events arises from the perspective of defaults versus survivals, e.g., 3 joint defaults correspond to 2 joint survivals, and so on.

- Based on what has been said regarding joint default probabilities in previous sections, we can estimate from above every of the probabilities $\mathbb{P}[S]$ of joint default events $S \in \mathcal{S}_{\geq 2}$ by rough upper bounds in the following way. We are considering the Gaussian copula case so that we need to know pairwise CWI correlations to do this exercise. Table 2.16 shows the pairwise CWI correlations for the reference portfolio of our CLN; see Section 2.6.1 for a demonstration on how to calculate CWI correlations in our example. In addition, Table 2.17 shows the calculation of JDPs for the 10 events (upper triangle matrix) referring to exactly 2 joint defaults. Summing up the upper triangle matrix of Table 2.17, we get
$$\mathbb{P}[\mathcal{S}_2] = \sum_{S \in \mathcal{S}_2} \mathbb{P}[S] = 2.46 \text{ bps}$$
as the probability for exactly two joint defaults.

- In the same way, we can find conservative estimates for the probabilities of 3, 4, and 5 joint defaults in the reference pool. It turns out (in line with intuition and expectations) that $\mathbb{P}[\mathcal{S}_3]$, $\mathbb{P}[\mathcal{S}_4]$, and $\mathbb{P}[\mathcal{S}_5]$ are safely negligible so that altogether one can say that the probability of joint default events,
$$\mathbb{P}\left[\sum_{i \in I} \mathbf{1}_{\{\tau_i \leq 3\}} > 1\right],$$

TABLE 2.16: CWI correlation matrix for the CLN reference portfolio

CWI Corr	4	14	15	35	46
4	100%	23%	21%	19%	17%
14	23%	100%	13%	21%	11%
15	21%	13%	100%	24%	12%
35	19%	21%	24%	100%	10%
46	17%	11%	12%	10%	100%

TABLE 2.17: Matrix of JDPs referring to joint default events with exactly 2 names defaulting in the CLN reference portfolio

JDPs	4	14	15	35	46
4	0.0363%	0.0004%	0.0003%	0.0006%	0.0022%
14	0.0004%	0.1032%	0.0004%	0.0019%	0.0037%
15	0.0003%	0.0004%	0.1032%	0.0023%	0.0040%
35	0.0006%	0.0019%	0.0023%	0.2901%	0.0087%
46	0.0022%	0.0037%	0.0040%	0.0087%	1.5171%

appearing with a negative sign at the right-hand side in Equation (2.108) indeed can be expected to be in the range of a few basispoints only.

Therefore, *the 4 bps difference between the simulated pool PD (based on the Gaussian copula function) and the worst case pool PD is completely in line with our findings from the discussion above on joint default events and their likelihoods.* Based on the *tail dependence of the Student-t and Clayton copulas* it is not surprising that the difference between simulated pool PD and worst case pool PD is higher for Student-t and Clayton copulas than w.r.t. the Gaussian copula simulation.

The discussion we just had on challenging the derived 4 bps difference of two different model outcomes provides another example for 'model thinking'. Looking at the same object from different angles can help to better understand the effect of model choices and different parameterizations. In this particular example, the discussion helped us to better understand joint default events and their occurrence likelihoods.

We conclude this section by some further findings from the Monte Carlo simulation of the transaction. Figure 2.40 shows the so-called *principle* and *interest streams* of the transaction. For an understanding of the credit risk of structured credit products is is important to *explicitly model all cash flows relevant for the transaction*. We will come

Default Baskets 159

back to this remark in Section 3. Therefore, we should now consider the principle and interest stream arising in our CLN transaction.

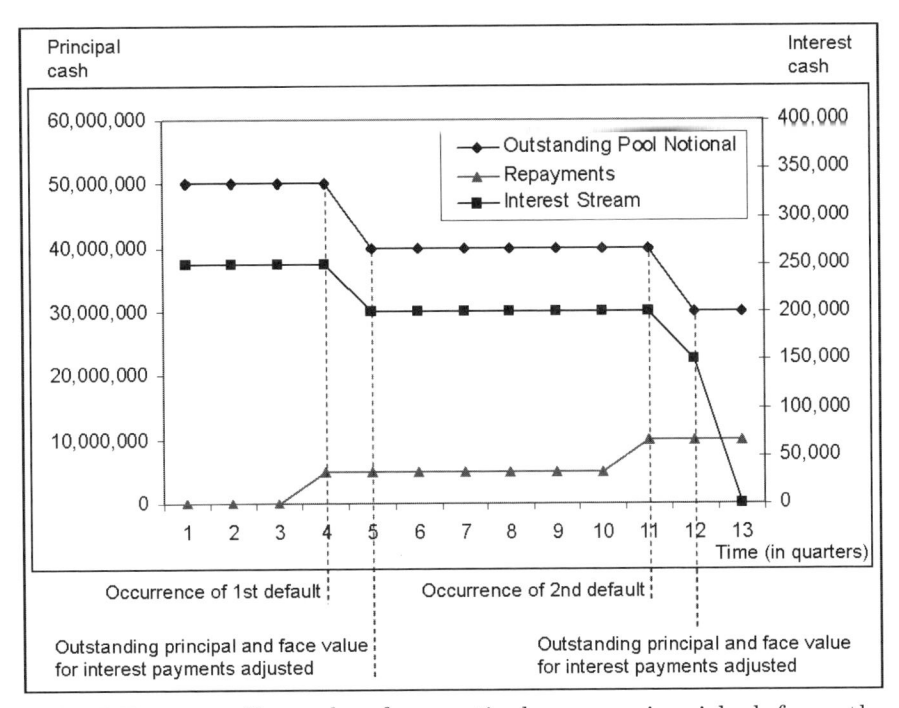

FIGURE 2.42: Example of a particular scenario picked from the Monte Carlo simulation of the CLN

Figure 2.42 shows one particular scenario picked from the Monte Carlo simulation of the transaction. The left y-axis shows principal cash, whereas the right y-axis shows interest cash at a different scale.

- The first default occurs in the 4th quarter and triggers a repayment[24] of 5,000,000 USD to the investor. For the following payment period (quarter 5), the outstanding notional of the CLN is reduced by the defaulted gross exposure of 10,000,000 USD. Because our contract prescribes an LGD of 50%, the realized loss stream (not shown in Figure 2.42 equals the repayment stream in a time congruent way.

---
[24]LGD×[outstanding notional on defaulted name].

160   *Structured Credit Portfolio Analysis, Baskets & CDOs*

- For the interest stream, we assume a spread on top of LIBOR of 200 bps. LIBOR as base component of the interest stream has not been considered because it is paid by the underlying collateral securities and transferred to the investor to cover the LIBOR part of the CLN coupon. Therefore, we can restrict the interest stream to spread payments. For the originally outstanding 50,000,000 USD we obtain a quarterly spread payment of 250,000 USD, which can be read off from the right y-axis in Figure 2.42.

- For quarter 5, the first default triggered a reduction of the CLN's outstanding notional. This impacts the interest stream because the 200 bps spread are now applied to 40,000,000 USD instead to the original 50,000,000 USD. Quarterly spread payments are, therefore, reduced from 250,000 USD down to 200,000 USD.

- The next and last default in this scenario occurs in quarter 11, which leads to another repayment of 5,000,000 USD to the investor, and another 5,000,000 USD realized loss to be covered by the investor, which has the role of the protection seller. For the following payment period, the outstanding notional of the CLN is reduced down to 30,000,000 USD such that quarterly spread payments drop by another 50,000 USD.

In this way, considering single scenarios can serve as a 'first order check' that cash flows are modeled in line with the *term sheet* or, later in CDOs, *offering circular* of the transaction.

Figure 2.43 shows expected cash flows from averaging over all scenarios in the Monte Carlo simulation for the Gaussian copula case. Let us illustrate by means of an example that expected cash flows are in line with Table 2.15. According to Figure 2.43, the average cumulative loss after 3 years equals roundabout 100,000 USD. Looking at Table 2.15 we find that the probability for more than one default is negligible in the Gaussian case ($\approx$ 3 bps). We also find that with a probability of roundabout 2% the reference pool experiences one default over the lifetime of the transaction. The net exposure at risk equals 5,000,000 USD such that Table 2.15 suggests a cumulative expected loss of

$$2\% \times 5,000,000 = 100,000 \text{ USD}.$$

This is ca. 1/5 of the worst case expected pool loss estimated in Equation (2.107), where 'worst case' was associated with a 'complete wipeout' of the reference portfolio. Table 2.15 shows that in the Gaussian

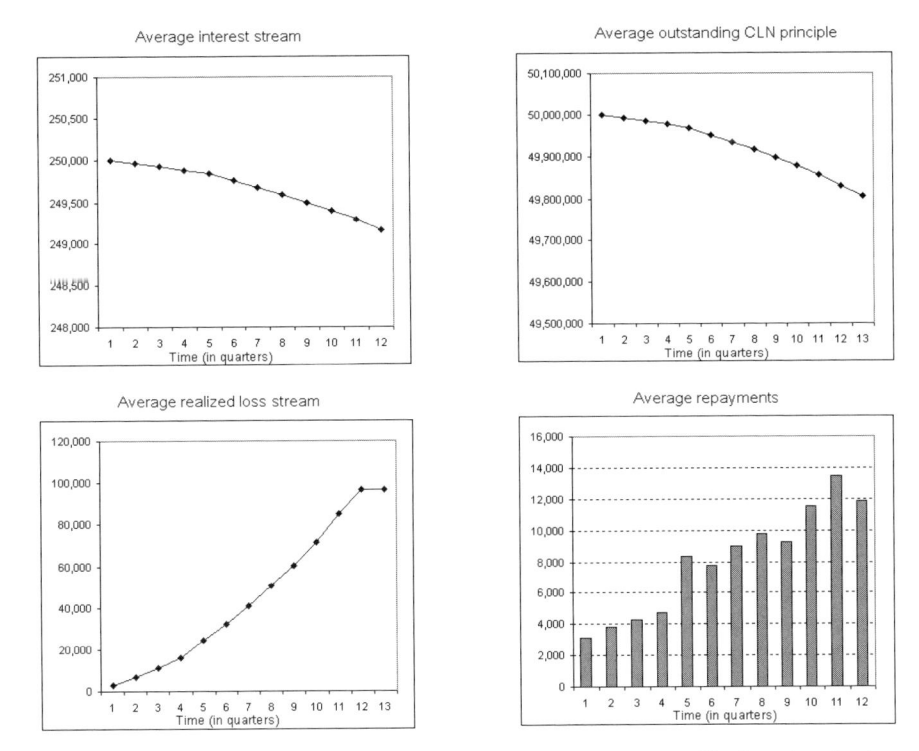

FIGURE 2.43: Average cash flows in the Monte Carlo simulation of the CLN transaction (interest: quarterly payments; realized losses: cumulative path; repayments: quarterly payments; all w.r.t. Gaussian copula simulation)

case the potential for more than one default is basically zero so that we are in a much better position than assumed in the worst case pool loss scenario.

For the Student-t and the Clayton copula the situation is slightly but not dramatically different. Following the notation developed in our discussion above, we obtain in the Student-t case

- $\mathbb{P}[\mathcal{S}_0] = 98.12\%$

- $\mathbb{P}[\mathcal{S}_1] = 172$ bps

- $\mathbb{P}[\mathcal{S}_2] = 14$ bps

- $\mathbb{P}[\mathcal{S}_3] = 2$ bps

162   *Structured Credit Portfolio Analysis, Baskets & CDOs*

- $\mathbb{P}[S_4] = 0$
- $\mathbb{P}[S_5] = 0$

and in the Clayton copula case

- $\mathbb{P}[S_0] = 98.57\%$
- $\mathbb{P}[S_1] = 117$ bps
- $\mathbb{P}[S_2] = 14$ bps
- $\mathbb{P}[S_3] = 5$ bps
- $\mathbb{P}[S_4] = 4$ bps
- $\mathbb{P}[S_5] = 3$ bps.

These probabilities[25] are in line with probabilities we find in Table 2.15. Let us make an example for illustration purposes. In Table 2.15, Student-t copula part, row '>= 13', column '2nd' we find

$$\mathbb{P}[\tau_{(2)} \geq 13] = 99.84\%.$$

This likelihood must coincide with (Student-t copula case!)

$$1 - (\mathbb{P}[S_2] + \mathbb{P}[S_3] + \mathbb{P}[S_4] + \mathbb{P}[S_5]) = 1 - 0.0014 - 0.0002$$

which is the case.

If we would impose a *capital structure* on the CLN's reference portfolio as done in our first-to-default basket or as will be done in Chapter 3, then the *timing of defaults* would become an important model result. Moreover, even in our CLN example the timing of defaults plays a certain role for the *interest stream* of the transaction: *the earlier defaults occur, the less interest can be collected over the lifetime of the deal*. As seen in previous discussions on default times, the timing of defaults is a *blend of effects from underlying PD term structures and the chosen copula function*. Our calculations before illustrate the copula effect on default timing, providing a good example why it is important to *derive*

---
[25] For verification via a pocket calculator note that figures in Table 2.15 are rounded to basispoint level; note also that results from Monte Carlo simulations are always subject to random/simulation fluctuations.

the distribution of defaults over time from a model and avoid making assumptions regarding default timing as sometimes done by market participants.

This concludes our example of a CLN transaction. One can, in the same way as done in our discussion on the first-to-default basket, extend the model to evaluation and/or pricing issues. However, the main work is represented by the modeling of default times because the performance of cash flows depends one-on-one on the *timing of defaults* and the *materiality of losses*. If these two issues are under control, evaluation of cash flows or pricing a deal is not more than a routine exercise. In addition, one can say that the *deal price* in such transactions typically depends on more than just credit risk aspects. For instance, liquidity and the complexity of the transaction as well as other factors play a role in the determination of the deal price.

Chapter 3

Collateralized Debt and Synthetic Obligations

This chapter deals with another main topic of this book, namely *collateralized debt obligations* (CDOs). As we will find out soon, default baskets as introduced in the previous chapter and CDOs have much in common. For instance, the protection seller in a first-to-default basket is in a position comparable to an *equity* or *first loss piece* investor in a CDO transaction. We will come back to this and other analogues soon.

Analogues between CDOs and baskets make it possible to carry-over most of the already discussed modeling techniques from the basket world into the CDO world. However, CDOs provide more flexibility and can be rather complicated regarding cash flows at the liability side of the transaction. Therefore, cash flow modeling has a major part in any discussion on CDOs. Besides Monte Carlo simulation, additional tools like analytic and semi-analytic techniques (see Section 3.3.6.2) and other modeling approaches will be discussed in the sequel.

The CDO market is steadily growing and this growth can be expected to continue over the next years. Not only large players but also smaller banks discovered CDOs as meaningful and profitable tools for *tailor-made* long and short positions in credit risk. In addition, the standardization of *index tranches* provides very liquid structured credit products, that are actively traded in the market; see Section 3.4.4.

Given the variety of products and applications, we had to make compromises regarding product coverage and items for discussion. In the same way as in Chapter 2, our main intention is the development of modeling tools and techniques. Hereby, we do not need to develop everything newly from scratch: techniques from basket modeling (Chapter 2) can be carried-over to CDOs and work well in this context too. A second intention we have in mind with this chapter is to provide to the reader a 'flavor' of what CDOs are all about. Therefore, we start our discussion with an exposition on certain applications and CDO types one finds in the market today and then continue with a discussion on various modeling techniques.

3.1 A General Perspective on CDO Modeling

This section provides a 'mini course' on CDOs as structured credit instruments. Motivations and applications will be discussed as well as some 'real life' examples from the CDO market. Because the main topic of this book is 'modeling' and not 'market overview', we will keep the exposition short. A nice market overview including lots of examples and providing a strong background on structured credit products including CDOs and related instruments can be found in the book by CHOUDRY [33]. Our discussion in this section is an updated version of the survey [26], supplemented by some recent developments in the market.

What makes CDOs as an asset class so successful? This question will bring us right at the heart of CDOs. We will consider four main motivations market participants have for engaging in CDO transactions:

- Risk transfer, or, more generally speaking, tailor-made short and long positions in credit risks and the trading of such risks, e.g., via secondary markets, see Section 3.1.2

- Spread and rating arbitrage opportunities; see Section 3.1.3

- Funding benefits, see Section 3.1.4

- Regulatory capital relief; see Section 3.1.5

Transactions can be of advantage from the perspective of more than just one of these motivations; for instance, the first and second mentioned have much in common. However, these four motivations constitute the major drivers of the CDO market and, therefore, should be discussed in the sequel. Modeling approaches for the four motivations are basically the same although certain value drivers enjoy stronger emphasis in one or the other case. For instance, transactions with a focus on risk transfer need to be evaluated based on the bank-internal quantification of transferred credit risks, being in line with risk measurement standards of the bank, whereas in arbitrage-motivated transactions market spreads and risk-neutral valuation naturally play a dominant role; see Section 3.4.4.

The four mentioned motivations or drivers of the CDO market will be covered in four sections in the sequel. Before that, we need a few entry remarks on CDOs in general.

3.1.1 A Primer on CDOs

Active management of credit portfolios is a core competence of advanced banks. Instead of reducing the credit business to a pure buy and hold strategy, banks want to actively improve the risk/return profile of the credit portfolio by going long (buying credit risk, selling protection) and short (selling credit risk, buying protection) in credit positions optimizing the return on economic capital as well as the overall P&L situation of the bank.

Going short (buying protection) in credit risk is sometimes also called *hedging*. This can be done for single names as well as for whole portfolios. In this book, we focus on transactions at portfolio level. In Section 2, we already extensively discussed *risk transfer* via basket transactions. For instance, buying protection on the first default in a basket can be realized via a first-to-default, buying protection in the second default can be done via a second-to-default, etc. In Sections 2.6 and 2.7 we considered short and long positions in structured credit risk, where a long positions refers to the *investor* in a transaction and the short position refers to the *originator* of a transaction.

Figure 3.1 shows a generic example of a CDO transaction, which serves us as an example to explain some basics of CDOs. In general, a CDO has two legs. The left-hand side in Figure 3.1 is called the *asset side* of the CDO and refers to the underlying *reference* pool or the *securitized* portfolio. The right-hand side of Figure 3.1 is called the *liability side* of the CDO. It consists of securities issued in the market by an *issuer*, which often is a so-called *special purpose vehicle* (SPV) whose only reason for existence is the issuance of notes of the particular transaction. The liability side is *linked* to the asset side of the CDO such that the *performance of securities at the liability side is a function of the performance of the underlying credit portfolio.*

CDOs can be categorized in various different ways. For instance, the type of underlying credits is one way to divide CDOs in different classes. Most common underlying asset classes include

- Loans: *collateralized loan obligations* (CLOs)

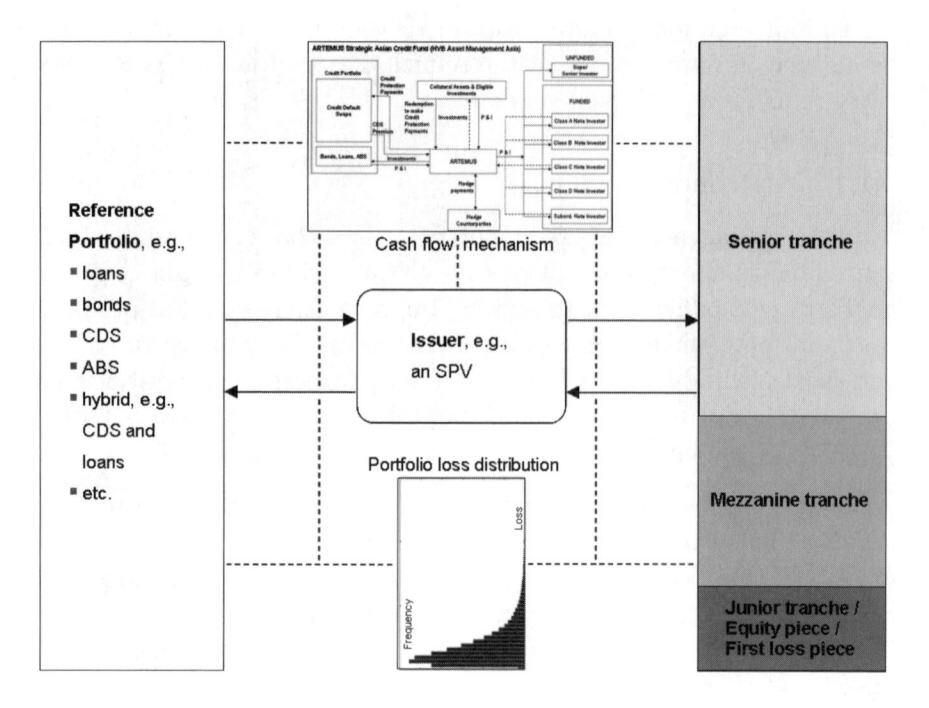

FIGURE 3.1:   Illustration of a CDO transaction

- Bonds: *collateralized bond obligations* (CBOs), including ABS bonds as underlying

- Mortgages[1]: *mortgage-backed securities* (MBS) including *residential mortgage-backed securities* (RMBS) and *commercial mortgage-backed securities* (CMBS)

- Credit default swaps (CDS): *collateralized swap/synthetic obligations* (CSOs)

- CDO tranches: *CDO of CDOs* or *CDO-squared* structures (CDO^2)

Another way of CDO categorization is a clustering w.r.t. *cash flows* at the liability side of the CDO. For instance, CSOs with CDS as underlying reference are naturally structured as a pure *credit derivative*

---
[1]Note that MBS are sometimes not counted under the headline of CDOs, but because they can be modeled based on the same principles as CDOs we consider them, from modeling perspective, as a subclass of CDOs.

without involving any *principal cash flows*. The first-to-default basket in Section 2.6 is an example for a comparable *unfunded* transaction focussing only on the first-to-default, which corresponds to a special type of *equity tranche* as shown in Figure 3.1. Such transactions are completely *unfunded* and cash flows consist of an *interest* or *spread stream* only. Other transactions involve also a *principal stream* where funding and repayments play an integral part of the *cash flow waterfall*. For example, the CDO explained in Section 3.1.3 is a *partially funded* transaction consisting of funded lower tranches and an unfunded *super senior swap* on top for the most senior part of the portfolio. Funded or partially funded transactions often arise from two motivations. If *funding* or *refinancing* is an issue for the originator, then a principal stream is required in the structure of the CDO. A second main motivation, often in the context of *partially* funded transactions, for funded CDO tranches is the function of principal as *collateral funding* in a comparable way as explained in Section 2.7 where principal proceeds from issuance of the CLN were used to buy risk-free or low risk securities functioning like *collateral* to guarantee for required cash for contingent payments on the credit protection agreement embedded in the note.

Another major difference CDOs can exhibit are *cash flow* versus *synthetic* CDOs, although the two notions are differently used in the market and can be expected to be not fully consistent applied among different banks and market participants. One way to think about cash flow structures versus purely synthetic structures is the following. Synthetic structures can be considered as pure credit derivative type transactions where one party in the deal adopts the role of the protections seller for a certain part in the *capital structure* of a portfolio in exchange for a *premium*. In contrast, cash flow structures typically involve a certain mechanism of cash flow distribution, often involving an interest *and* a principal stream, at the liability side of the CDO. In recent years, transactions were structured in a way combining typical synthetic with typical cash flow characteristics. Such transactions are sometimes called *hybrid*; see Section 3.1.3 for an example of such a structure.

Another aspect shown in Figure 3.1 is the *tranching* at the liability side. The illustration is kept in a fairly generic way, only showing the three main positions in the *capital structure* of a CDO, namely a *junior tranche* (also called *equity* piece or *first loss* piece/tranche), a *mezzanine tranche*, and a *senior tranche*. In general, it can be expected that

- 'Good' cash flows like interest and repayments flow *top-down* through the liability side of a CDO, essentially meaning that senior tranches receive interest and repayments before mezzanine tranches and mezzanine tranches get interest and repayments before junior or equity note holders receive any payments

- 'Bad' cash flows like losses are allocated *bottom-up* at the liability side of the CDO such that equity note holders bear the first loss, mezzanine investors carry the second-tranche loss, and senior investors or swappers have to pay for losses not before all other, more junior, investors lost all of their invested capital or paid for all protected amounts subordinated to the senior tranche, respectively

Let us illustrate the mechanism of loss allocation to tranches by means of a very simple illustrative CDO example. Assume we have a tranching of some underlying (high yield, due to the assumed tranching levels) credit portfolio as follows:

- The equity tranche ranges from 0% to 10%

- The mezzanine tranche occupies 30% in the capital structure above equity

- The senior tranche refers to the remaining upper 60% of the reference portfolio's notional

The capital below a considered tranche is always called the capital *subordinated* to that tranche. For instance, the mezzanine tranche in our example has 10% of subordinated capital, whereas the senior tranche is protected by 40% subordinated capital.

Figure 3.2 illustrates the allocation of losses to the tranched securities at the liability side of the CDO. At the lower left-hand side of Figure 3.2 some cumulative loss sample paths from a Monte Carlo simulation of the underlying reference pool are plotted. In the example, some paths remain within the equity tranche, a majority of paths runs into the mezzanine tranche, and a few paths even eat into the senior tranche. For example, assume the following cumulative loss scenarios.

- If the underlying portfolio over the lifetime of the transaction had a cumulative realized loss (net of recovery) of 5% of the notional of the pool, then equity holders would suffer a loss of

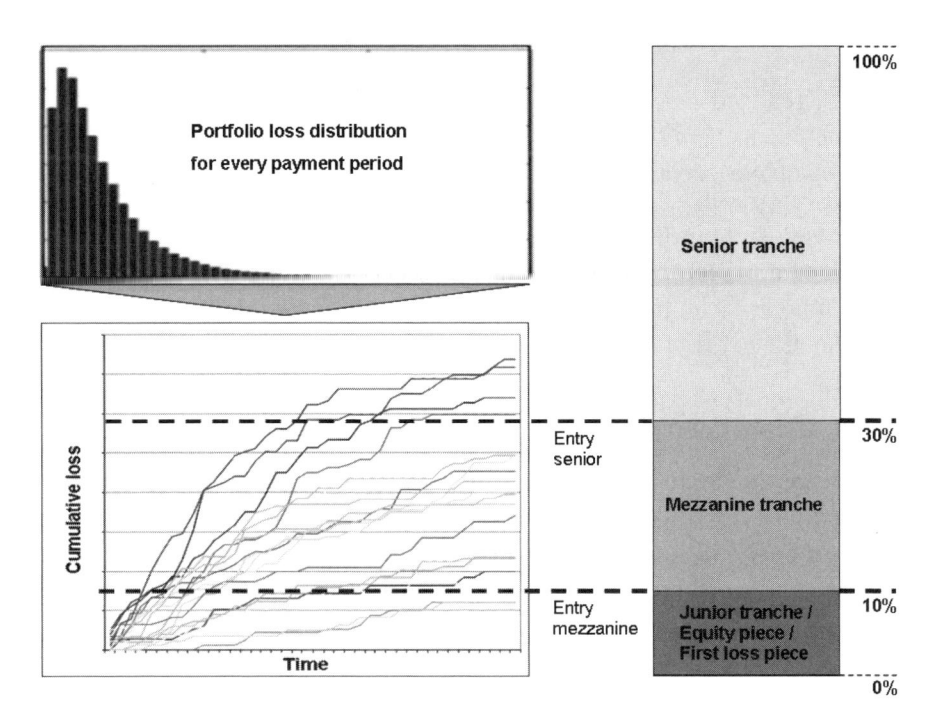

FIGURE 3.2: Illustration of the tranching of losses at the liability side of a CDO

50%, mezzanine and senior note holders would be loss-free. In a pure swap-type credit derivative transaction this would mean that the equity investor in his role as protection seller would have to pay the realized loss of 50% of the equity piece volume to the protection buyer. In case of a funded equity tranche, the equity holder gets only 50% of his invested capital repaid at maturity of the deal. Note that equity positions are not necessarily in one hand. Often the originator sells part of the equity to get rid of some part of the first loss but at the same time keeps part of the equity in his own book, maybe to demonstrate confidence and trust regarding the structured deal, maybe to collect *excess spread* as outlined in the deal described in Section 3.1.3.

- If the total cumulative realized loss would be equal to 25%, then equity note holders would have a realized loss of 100% of their invested capital or referenced notional, mezzanine investors would suffer a loss of 50% of their tranche and senior note holders would be loss-free again.

- A realized loss scenario exceeding 40% would mean a complete *wipe-out scenario* for the junior and mezzanine tranches and would involve a certain loss for the senior tranche.

The lower and upper boundaries of a tranche are also called the *attachment* and *detachment* points[2] of the tranche. For example, the mezzanine tranche in Figure 3.2 has an *attachment point* of 10% and a *detachment point* of 40%.

3.1.2 Risk Transfer

Transfer of credit risks is our first-mentioned motivation. In the same way as for baskets (see Section 2), risk transfer means that there is also a party willing to take the risk in exchange for a certain credit risk premium. Therefore, we can speak broader about *short* and *long positions* in a pre-specified part of the capital structure of a portfolio. To explore this, we supplement Figure 3.2 by Figure 3.3, which shows the loss distribution of a credit portfolio with an illustrative tranching attached to it. In some way, possibly in addition also involving certain cash flow elements like spread re-direction triggers and so on, any CDO refers to a *tranched loss distribution* in a wider sense.

Figure 3.3 shows an illustration of a tranching of a credit portfolio's loss distribution. More explicitly, denote the notional exposure of some underlying reference portfolio in percent. Any tranching of the portfolio consisting of N tranches can be represented as a *partition* of the unit interval

$$[\alpha_0, \alpha_1) \cup [\alpha_1, \alpha_2) \cup \cdots \cup [\alpha_{N-1}, \alpha_N]$$

for a strictly increasing sequence of α's, where $\alpha_0 = 0$ and $\alpha_N = 1$. Let us now focus on one particular tranche

$$T = T_{\alpha,\beta} = [\alpha_n, \alpha_{n+1})$$

for some $n \in \{0, 1, 2, ..., N-1\}$. Let us assume[3] for a moment that the loss of the tranche T is a function of the loss of the underlying reference

---
[2] The detachment point sometimes is also called *exhausting point*.
[3] We will later see that in simple synthetic CDOs indeed this assumption makes perfect sense.

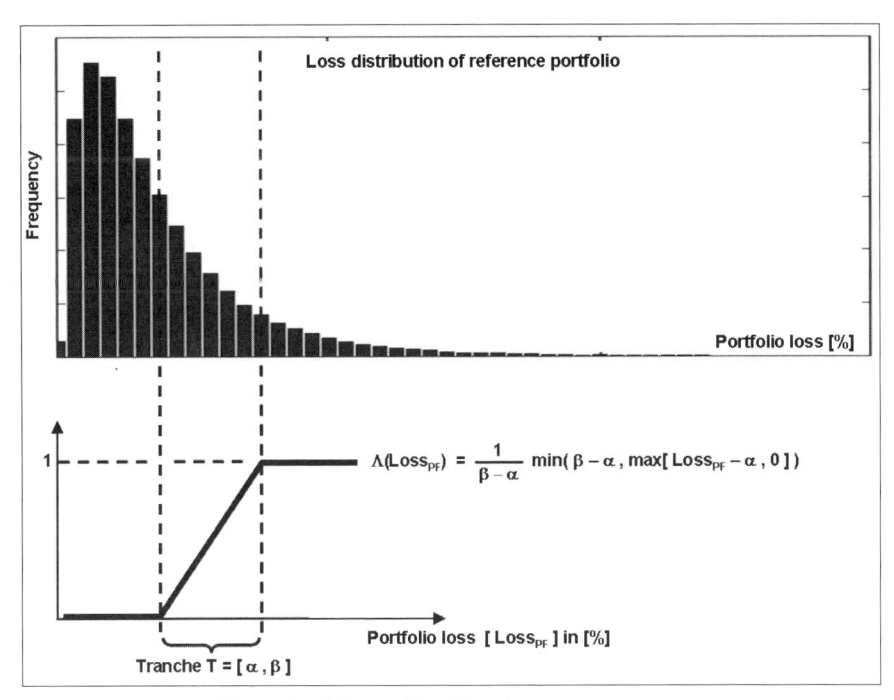

Collateralized Debt and Synthetic Obligations

FIGURE 3.3: Tranched loss distribution

portfolio only, considered w.r.t. an evaluation time horizon matching the term of the CDO, e.g., t years. Denoting the cumulative portfolio loss at time t by $L^{(t)}$ we get

$$\Lambda(L^{(t)}) = \Lambda_{\alpha,\beta}(L^{(t)}) = \frac{1}{\beta - \alpha} \min\left[\beta - \alpha, \max[0, L^{(t)} - \alpha]\right] \quad (3.1)$$

for the loss allocated to tranche T, normalized to the tranche's size. This is illustrated in Figure 3.3.

By choosing α and β in Equation (3.1), market players can engage in their preferred risk position in the capital structure of the underlying reference portfolio. Here are some examples:

- The bank owning the portfolio might want to get rid of medium to tail risks and seek protection for all portfolio losses exceeding a threshold of 4%; they could then buy protection in the market on a tranche ranging from $\alpha = 4\%$ to $\beta = 100\%$. Obviously, *the larger α, the lower the premium to be paid on such an insurance* because of a corresponding increase in *subordinated capital* protecting the tranche investor up to the attachment point of the

174   *Structured Credit Portfolio Analysis, Baskets & CDOs*

tranche. The bank would then be *long* in the capital structure of the portfolio ranging from 0% up to 4% and *short* in all risks exceeding the 4% loss threshold.

- If the same bank feels uncomfortable with the expected loss of the portfolio, it could try to find an equity investor in the market taking the first, e.g., 2% of portfolio losses. The bank would then be *short* in an equity tranche on the portfolio ranging from 0% to 2% and *long* in a combination of tranches (mezzanine and senior risk) capturing all portfolio losses exceeding the 2% threshold. The insurance premium to be paid on an equity tranche typically is very high because equity tranches are not protected by subordinated capital.

These simple examples show that parts of the credit risk of the underlying portfolio can be cut-out and transferred to some risk taker in a tailor-made way. Investors in the corresponding deals take on the opposite position in the transaction by functioning as a protection seller who needs a certain premium as compensation for the taken risk.

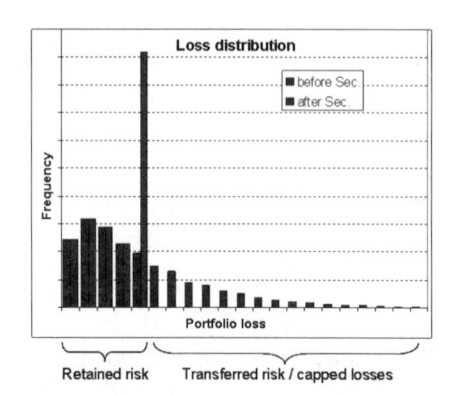

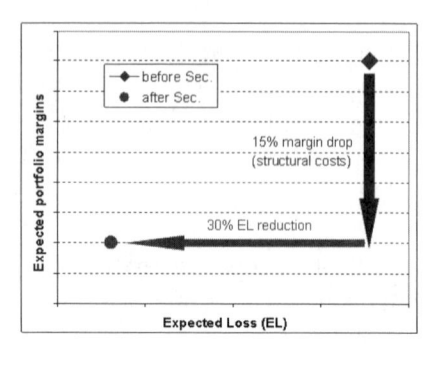

FIGURE 3.4:   Effective risk transfer in a synthetic CLO transaction

Banks indeed use CDOs for risk transfer purposes. Figure 3.4 shows an example of a securitized portfolio where some part of the portfolio's margins are used to pay spreads on tranches at the liability side of a synthetic CLO. Here, the bank as originator is the protection buyer and keeps the first loss in form of the equity tranche. At the left-hand side of Figure 3.4 the portfolio loss distribution is shown. Before securitization, the originating bank is exposed to the full risk of the portfolio. After securitization, all portfolio losses exceeding the size of the kept equity piece are effectively transferred to tranche investors: the loss distribution is capped at the detachment point of the kept part of the capital structure of the portfolio. At the right-hand side of Figure 3.4, the profitability of the transaction is sketched. Here we have a lucky situation: the expected loss of the portfolio *after* synthetic securitization is reduced by 30%, whereas the overall portfolio margins are only reduced by 15%. This results in a positive leverage of funds compared to transferred risks.

Let us comment on some questions sometimes asked by people new in the field of CDOs. Readers familiar with CDOs should skip the following pages until the beginning of the next section.

A typical question sometimes asked is:

If expected losses of the securitized portfolio are located within the equity tranche and the equity tranche is kept by the originator, can there be some effective risk transfer at all?

The answer to this question is hidden in the well-known *insurance paradigm* on which the concept of credit risk costs is based. Every insurer includes the amount of expected losses arising from the considered business as one out of several additive components in their pricing. In the same way, expected losses in credit risk have to be an integral part of the pricing concept of the bank. Now, the expected loss of a portfolio can be written as the *mean value* of all potential losses in the credit portfolio. A good way to think about this is in terms of a Monte Carlo simulation of portfolio losses. We can write

$$\mathrm{EL}^{(t)} \approx \frac{1}{n} \sum_{i=1}^{n} \hat{L}_i^{(t)}$$

where $\mathrm{EL}^{(t)}$ denotes the expected loss of the portfolio and $L_i^{(t)}$ refers to the simulated portfolio loss from scenario i out of n scenarios in the simulation. Let us assume the portfolio is structured in two tranches, namely an equity tranche $[0, \alpha)$ kept by the originating bank and an upper tranche $[\alpha, 1]$ hedged in the capital market. The protection agreement on the upper tranche now changes the expected loss to

$$\mathrm{EL}^{(t)}_{capped} \approx \frac{1}{n} \sum_{i=1}^{n} \min[\hat{L}_i^{(t)}, \alpha]$$

because losses exceeding α are only to be born by the originating bank up to $\alpha\%$ of the notional of the portfolio, whereas losses $\max[\hat{L}_i^{(t)} - \alpha, 0]$ are taken by the tranche investor. Therefore, we have

$$\text{EL}_{capped}^{(t)} \overset{!}{<} \text{EL}^{(t)}$$

if and only if in the Monte Carlo simulation the *likelihood for at least one cumulative loss path* $\hat{L}^{(t)}$ *exceeding the detachment point of kept tranches* is positive (given we have sufficient time and computational power for the simulation); see also Figure 3.2. So even on expected losses, a CDO-based hedge of upper tranches can have a positive effect considering the business case through the glasses of the credit insurance paradigm: because the potential for high losses is capped after securitization, the insurance premium should be a little lower too. The same rationale applies in an even stronger way to tail losses.

Note that such EL transfer or tail risk transfer looks positive if seen from the angle of the whole securitized portfolio. Normalized to, or seen from, the perspective of a kept first loss piece, the *percentage risk obviously is much higher after transferring the senior risk than before*. It really depends on the view a bank has and on the need to transfer risks if an overall business case for a risk transfer transaction looks positive or negative.

Another question often discussed concerns the *balance between costs and benefits* of portfolio-based risk transfers:

Given the average weighted spread to be paid on tranches at the liability side of a CDO plus required upfront payments for paying lawyers, rating agencies, structuring teams and the underwriting unit: are the benefits (e.g., risk transfer) really justifying to spend the money?

There is no standard answer to this question, transactions have to be evaluated case by case. Based on what already has been said on risk transfer, one can say that most securitization transactions actually lead to some risk transfer. The challenge here is to come to a reasonable cost/benefit relation as shown in Figure 3.4 where the reduction in risk costs dominates the charged spreads at the liability side of the transaction. In an evaluation of securitization costs versus securitization benefits one in addition has to take into account that upfront costs like agency fees and lawyer costs are immediately P&L effective in contrast to risk cost savings, which have a calculatory character. It really depends on the scope of the transaction, the bank's way to quantify *tail*

Collateralized Debt and Synthetic Obligations 177

risk costs, the willingness of the senior management of the bank to pay opportunity costs for risk transfer if necessary, and the structure of the transaction, whether a securitization involves not only an effective but also an economically reasonable risk transfer; see also our discussion in Section 3.5.

A last question we want to briefly comment on is:

If the bank securitizes a subportfolio out of the overall credit portfolio of the bank, is the effect on Economic Capital (EC) always positive because the assets are removed from the loan book?

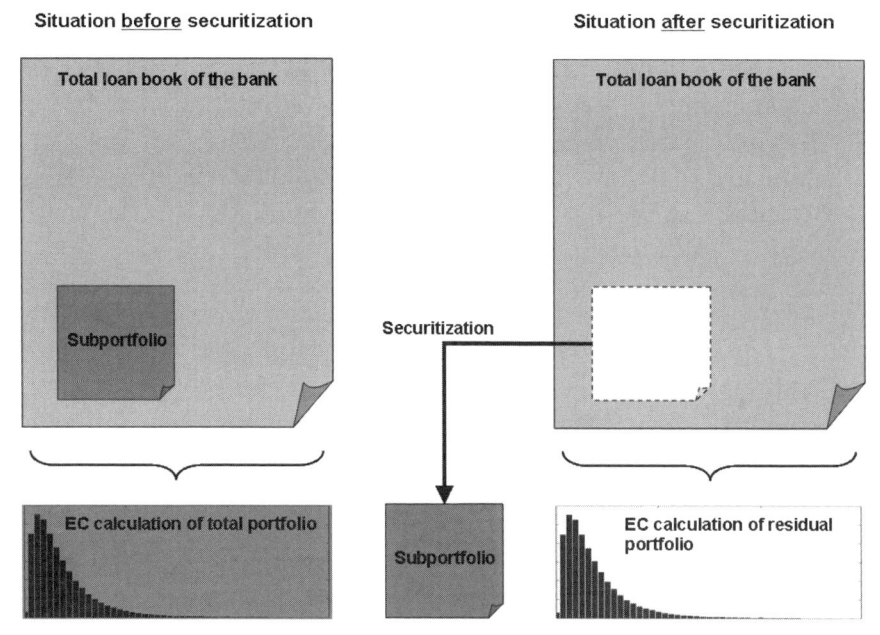

FIGURE 3.5: Effect of a securitization transaction on the loan book of the bank

Figure 3.5 illustrates the situation before and after a securitization transaction, seen from the perspective of the total credit portfolio of the bank. Before securitization (the left-hand side in Figure 3.5), the EC of the portfolio is a conglomerate of all credit-risky instruments including the assets from the considered subportfolio. Securitizing the subportfolio means removing certain credit risks from the loan book of the bank, which in principal can have a positive impact if considered in isolation from the overall credit portfolio of the bank. However, in situations where the assets in the securitized subportfolio (the right-hand side of Figure 3.5) have a strong diversification effect on the overall loan book, it can indeed happen that the EC calculation of the total loan book yields an inconvenient surprise in that the EC situation after securitization is worse than before securitizing the subportfolio. Obviously, the size of the securitized subportfolio compared to the size of the loan book of the bank plays a certain role for the outcome of such a calculation as well as the *risk contribution measure* applied in the calculation. To make an example, we consider the following situation. A universal bank has half of its business in well diversified corporate and retail banking and the other half of its business in investment banking. If the bank decided to securitize large parts of its corporate and retail portfolio it would be very likely that the overall EC required after the transaction is higher than before securitizing the corporate and retail portfolio due to the *removed diversifying credit risks*. In the same way as mentioned in the discussion on cost/benefit relations of risk transfer measures, the impact of a securitization transaction on the bank's Economic Capital has to be evaluated case by case.

One keyword mentioned in the introduction not covered so far is the *trading* of credit risks. For this topic, we ask the reader for patience: in Section 3.4.4 we will find nice examples for credit risk trading in today's CDO market. We now turn our attention to the second motivation for CDOs mentioned in the introductory section of this chapter.

3.1.3 Spread and Rating Arbitrage

This section is dedicated to *spread arbitrage*, but *rating arbitrage* is closely related and often involved. We will come back to this topic later in the text. In order to give our exposition a 'real life flavor', we introduce spread arbitrage by means of an example taken from the Asian structured credit market. We focus on the transaction described in [26]. This deal has been issued in 2003 already and incorporates lots of structural and cash flow features we want to use as an example for 'state of the art' financial engineering in structured credit products. Note that a comparable transaction with the same name and same manager/arranger (namely, HVB Asset Management Asia) came out in 2005; see [4] and [5]. Also note that the following discussion serves only as an illustration for sophisticated structured credit products, there are many comparable transactions by other banks out there in the market. It is by no means our intention to judge in any way about the profitability of the transaction, no matter if seen from the perspective of investors or seen from the perspective of the arranger.

We begin this section by quoting some passages from an article that appeared in the web at www.FinanceAsia.com (written by Rob Davies, March 20, 2003):

HVB Asset Management Asia (HVBAM) has brought to market the first ever hybrid collateralized debt obligation (CDO) managed by an Asian collateral manager. The deal, on which HVB Asia (formerly known as HypoVereinsbank Asia) acted as lead manager and underwriter, is backed by 120 *million of asset-backed securitization bonds and* 880 *million of credit default swaps ... Under the structure of the transaction, Artemus Strategic Asian Credit Fund Limited, a special purpose vehicle registered in the Cayman Islands, issued* 200 *million of bonds to purchase the* 120 *million of cash bonds and deposit* 80 *million into the guaranteed investment contract (GIC), provided by AIG Financial Products. In addition, the issuer enters into credit default swap agreements with three counterparties (BNP Paribas, Deutsche Bank and JPMorgan) with a notional value of* 880 *million. On each interest payment date, the issuer, after payments of certain senior fees and expenses and the super senior swap premium, will use the remaining interest collections from the GIC accounts, the cash ABS bonds, the hedge agreements, and the CDS premiums from the CDS to pay investors in the CDO transaction ... The transaction was split into five tranches, including an unrated* 20 *million junior piece to be retained by HVBAM. The* 127 *million of A-class notes have triple-A ratings from Fitch, Moody's and S&P, the* 20 *million B-notes were rated AA/Aa2/AA, the* 20 *million C bonds were rated A/A2/A, while the* 13 *million of D notes have ratings of BBB/Baa2 and BBB.*

Figure 3.6 shows a structural diagram of the transaction, whereas Figure 3.7 provides a cash flow diagram for the deal. As mentioned in the quotation from FinanceAsia before, the three major rating agencies analyzed the transaction. The reader can find the results of their analysis in the three presale reports [1], [2] and [3] of the transaction.

In the sequel, we discuss the transaction step by step. Hereby we focus on certain relevant aspects only and simplify structural elements as much as possible. For example, hedge agreements regarding inter-

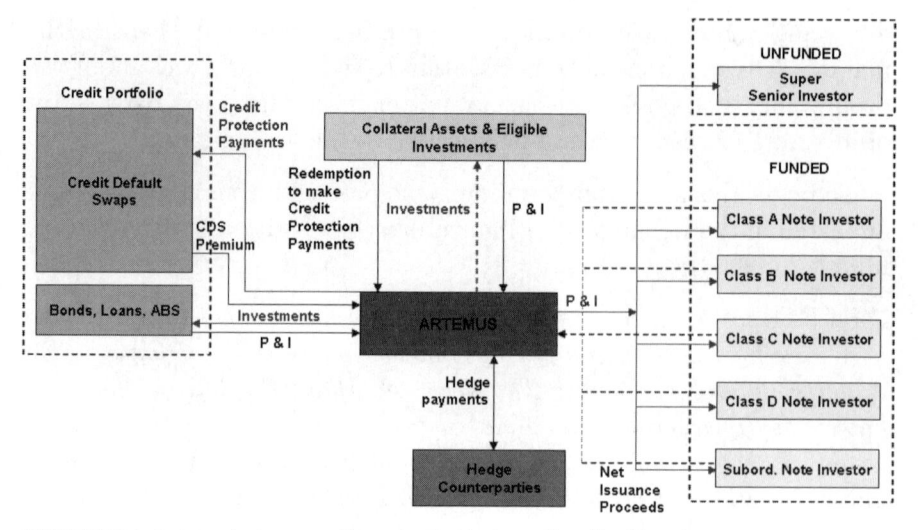

180 *Structured Credit Portfolio Analysis, Baskets & CDOs*

FIGURE 3.6: Artemus Strategic Asian Credit Fund

est rates and currencies will be excluded from our discussion. In this section all amounts of money refer to USD.

3.1.3.1 Liability Side of the Structure

The issuer (Artemus Strategic Asian Credit Fund Limited, an SPV at Cayman, from now on shortly called 'Artemus') issued 200 mm of bonds, split in five tranches reflecting different risk/return profiles. Artemus (as protection buyer) also entered into a CDS agreement (super senior swap) on a notional amount of 800 mm with a *super senior swap* counterparty. Such counterparties (protection sellers on super senior swaps) are typically OECD-banks with excellent credit quality. Because the liability side has a *funded* (200 mm of notes) and an *unfunded* (800 mm super senior swap) part, the transaction is called *partially funded*.

3.1.3.2 Asset Side of the Structure

The proceeds of the 200 mm issuance have been invested in a guaranteed investment contract (GIC account; 80 mm in eligible collateral assets) and asset-backed securities (ABS bonds; 120 mm). Additionally, the issuer sold protection on a pool of names with an aggregated notional amount of 880 mm. Because the asset side consists of a mixture of debt securities (ABS bonds) and synthetic assets (CDS), the

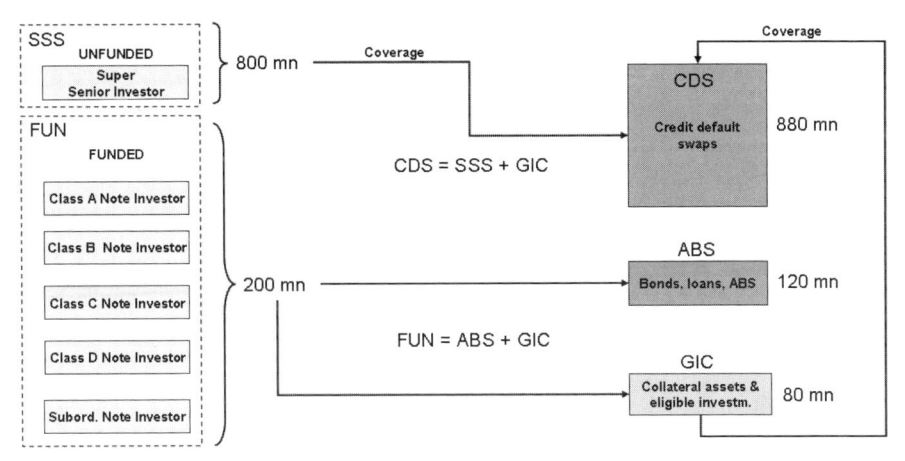

FIGURE 3.7: Cash flow summary for Artemus Strategic Asian Credit Fund ('FUN' stands for 'funded volume', 'GIC' for the notional in the 'GIC account', 'SSS' for the super senior swap volume, 'ABS' for the ABS bond volume, and 'CDS' for the total CDS notional)

transaction is called *hybrid*. Note that the GIC is considered as 'approximately risk-free' (AAA-rated liquid securities, equivalent to a cash account).

3.1.3.3 Settlement of Credit Events

If credit events happen on the 880 mm CDS agreement (Artemus is protection seller), a settlement *waterfall* takes place as follows.

- Proceeds from the GIC account are used by Artemus to make payments on the CDS agreement.

- If proceeds from the GIC are not sufficient to cover losses, principal proceeds from the debt securities are used to pay for losses.

- If losses exceed the notional amount of the GIC and principal proceeds, then ABS securities are liquidated and proceeds from such liquidation are used for payments on the 880 mm CDS agreement.

- Only if all of the before mentioned funds are not sufficient for covering losses, the super senior swap will be drawn (Artemus bought protection from the super senior swap counterparty).

Note that (at start) the volume of the GIC plus the super senior swap notional amount exactly match the 880 mm CDS agreement, and that the 120 mm ABS Securities plus the 880 mm CDS volume match the 1 bn total tranche volume on the liability side; see Figure 3.7. However, these coverage equations refer only to principal and swap notionals outstanding. But there is much more *credit enhancement* in the structure, because additional to the settlement waterfall, interest proceeds, mainly coming from the premium payments on the 880 mm CDS agreement and from the ABS bonds, mitigate losses as explained in the following section.

3.1.3.4 Distribution of Proceeds

Principal proceeds (repayment/amortization of debt securities) and interest proceeds (income on ABS bonds, the GIC, hedge agreements and premium from the 880 mm CDS agreement) are generally distributed *sequentially top-down* to the note holders *in the order of their seniority*. On top of the *interest waterfall*, fees, hedge costs and other senior expenses, and the super senior swap premium have to be paid. Both, principal and interest payments are subject to change in case certain *coverage test* are broken. There are typically two types of coverage tests in such structures:

- **Overcollateralization** tests (O/C) take care that the available (principal) funds in the structure are sufficient for a certain coverage (encoded by O/C-ratios greater than 100%) regarding repayments due on the liability side of the transaction.

- **Interest coverage** tests (I/C) make sure that any expenses and interest payments due on the liability side of the structure and due to other counterparties involved, e.g., hedge counterparties, are (over)covered (encoded by I/C-ratios greater than 100%) by the remaining (interest) funds of the transaction.

If a test is broken, cash typically is redirected in a way trying to bring the broken test in line again. In this way, the interest stream is used to mitigate losses by means of a changed waterfall. It is beyond the scope of this survey to dive deeper into such cash flow mechanisms, but interested readers find in [25], Chapter 8, a detailed working example for a cash flow CDO model.

3.1.3.5 Excess Spread

As already mentioned before, interest proceeds are distributed top-down to the note holders of classes A, B, C, and D. All *excess* cash left over after senior payments and payments of coupons on classes A to D is paid to the subordinated note investors. Here, HVB Asset Management Asia (HVBAM) retained part of the subordinated note (which corresponds to the equity piece introduced in Section 3.1.1). Such a constellation is typical in *arbitrage-motivated* structures: most often, the originator/arranger keeps some part of the most junior piece in order to *participate in the excess spread of the interest waterfall*. In addition, retaining part of the 'first loss' of a CDO to some extent 'proves' to the market that the originator/arranger itself trusts in the structure and the underlying *embedded credit enhancements*. As indicated before in our discussion on coverage tests, *if tests are broken excess cash typically is redirected* in order to *protect senior note holder's interests*. Here, the *timing of defaults is essential* in the same way as for our basket examples in Chapter 2: if defaults occur at the end of the lifetime of the deal (*backloaded*), subordinated notes investors had plenty of time to collect excess spread and typically will achieve an attractive overall return on their investment even if they lose a substantial part of their invested capital. In contrast, if defaults occur at an early stage of the transaction (*frontloaded*), excess cash will be diverted as a credit enhancement for senior tranches and no longer distributed to the equity investors. This is a bad scenario for equity investors, because they bear the first loss (will lose money) but now additionally miss their (spread) upside potential because excess cash is trapped.

3.1.3.6 Where Now Does the Arbitrage Come From?

The key observation is that on the 880 mm CDS agreement and on the 120 mm ABS securities on the asset side premiums are collected on a *single-name* base, whereas premium/interest payments to the super senior swap counterparty and the note holders refer to a *diversified pool* of ABS bonds and CDS names. Moreover, the tranching of the liability side into risk classes contributes to the spread arbitrage in that tranches can be sold for a comparably low spread if sufficient *credit enhancement* (e.g., subordinated capital, excess cash trapping, etc.) is built up for the protection of senior notes.

The total spread collected on single credit-risky instruments at the asset side of the transaction exceeds the total 'diversified' spread to be paid to investors on the tranched liability side of the structure. Such a mismatch can create a significant arbitrage potential leading to an attractive excess spread for equity or subordinated note holders.

There are many such transactions motivated by spread arbitrage opportunities in the CDO market. In some cases, structures involve a so-called *rating arbitrage* mentioned in the headline and at the beginning of this section, which arises whenever *spreads increase quickly and rapidly and the corresponding ratings do not react fast enough to reflect the increased risk* of the instruments. Rating arbitrage as a phenomenon is an important reason why typically a *serious analysis of arbitrage CDOs should not rely on tranche ratings but on a fully fledged Monte Carlo simulation of risks and chances of the transaction.*

Looking at arbitrage structures from an economic point of view, a well-structured transaction like Artemus Strategic Asian Credit Fund has, due to the arbitrage spread involved, *a potential to offer an interesting risk/return profile to note investors as well as to the originator/arranger holding (part of) an unrated junior piece.* It is certainly possible that the incorporated spread arbitrage is sufficiently high to compensate both groups of people adequately for the risk taken. In case of the Artemus Strategic Asian Credit Fund, the echo in the market and from rating agencies has been very favorable so far.

3.1.3.7 Super Senior Swap Positions

Regarding super senior swaps one can say that in most transactions the likelihood that the super senior tranche gets hit by a loss will be close to zero. *Scenarios hitting such a tranche typically are located far out in the tail of the loss distribution of the underlying reference pool.* Looking at super senior swaps from a business (non-mathematical) perspective, one can say that in order to cause a hit on a super senior tranche the economy has to turn down so heavily that it is very likely that problems will have reached a level where a super senior swap hit is just the tip of the iceberg of a catastrophic global financial crisis.

3.1.4 Funding Benefits

There is not much to say about this third motivation banks have for CDO issuance. In so-called *true sale* transactions, the transfer of assets is not realized by means of derivative constructions but rather involves a true sale 'off balance sheet' of the underlying reference assets to an SPV, which then issues notes in order to refinance/fund the assets purchased from the originating bank. The advantage for the originator is the receipt of cash (funding). A typical example for such transactions in the market today are the already-mentioned *commercial mortgage-backed securities* (CMBS) where banks refinance *income-producing real estates* like office buildings, shopping malls, etc., by means of CDO-type structures in the capital market. Such deals have several advantages for the originating bank. First, funding via the capital market backed by collateral securities can, under certain circumstances, be cheaper than the usual refinancing of the bank. Second, the credit risk arising from the credit approval of the commercial real estates is distributed back to the market, hereby involving an effective risk transfer for the lending institute. Third, regulatory and economic capital required for financing the commercial real estate transactions is freed up as soon as the assets are securitized in the market. Investment banks strongly rely on securitization-based lending in order to *increase capital velocity and efficiency*.

In general, funding can be an issue for banks whose rating has declined to a level where funding from other sources is expensive. In fact, the very first securitizations in the market were motivated by funding problems of the originator. The advantage of refinancing by means of securitizations is that resulting *funding costs are mainly related to the credit quality of the securitized assets and not so much to the rating of the originating (lending) institute.* Nevertheless, typically there remains a certain link to the originator's rating if the originating bank also acts as the *servicer* of the securitized portfolio, which sometimes naturally is the case. Then, investors and rating agencies will take into account the so-called *servicer risk* inherent in the transaction.

As a last remark, note that transactions can be *funded* without involving a *funding benefit* because the principal collected from the issuance of notes is not used for refinancing but for collateral-like security reasons. We discussed an example for such a situation in Section 2.7. There, we introduced a basket credit-linked note economically comparable to an unfunded basket derivative but, due to the principal proceeds from the CLN, with the advantage of completely eliminated counterparty risk. In CDOs, one often finds the following two categories:

- Partially funded transactions, where all tranches below the super senior tranche are funded and the super senior tranche is an unfunded pure portfolio swap contract
- Fully funded transactions, where all tranches are funded with principal collected into a collateral-like security depot, which func-

tions like a cash guarantee on the (contingent) default leg of the transaction, independent and remote from the protection seller; such transactions are eligible for a full regulatory capital relief if the bank sells all tranches to the market

Today, we see an increasing market share of fully funded transactions, where even the most senior tranche involves the issuance of notes, but, according to the fast evolving CDO market, such trends can change very quickly.

3.1.5 Regulatory Capital Relief

Regulatory capital relief has been a major motivation for securitizations under the old capital accord. An example for a typical pattern in so-called synthetic *regulatory arbitrage* CLOs is the following:

The originating bank bought protection on a pool of loans, e.g., corporate loans or residential mortgage-backed loans, by issuing credit-linked notes (CLNs) in form of some lower tranches (funded part) and entering into a super senior swap as protection buyer on a large unfunded upper part (90% or more) of the overall reference portfolio, which in some cases could contain several billion of Euro. The super senior swap counterparty (protection seller) had to be an OECD bank to make the transaction meaningful from a capital relief point of view.

In the Basel I capital accord, loan pools required regulatory capital in size of 8%×RWA where RWA denotes the *risk-weighted assets* of the reference pool. Without collateral, this means that 8% of the total notional had to be available as regulatory capital. Taking into account (eligible) collateral, could substantially reduce the RWA. For instance, residential mortgages required only a 50% instead of a 100% risk weight.

After securitization, the RWA of the pool could be reduced. The super senior protected volume required only a 20% risk weight and volumes protected by CLNs were eligible for a zero risk weight due to the funded collateral-type securities, which typically were bought from proceeds arising from the issuance of notes; see Section 2.7. Often, originators could not place the equity piece in the market or considered such a placement to be too expensive. Then, they kept and held the equity piece in their own books, which typically required a full capital deduction. Nevertheless, *regulatory capital could be substantially reduced as a consequence of a well-structured securitization.* Banks with a tight capital situation used regulatory arbitrage as a tool for freeing up capital urgently needed somewhere else in the bank. As 'opportunity costs' for capital relief, the originating bank had to pay interest/spreads to note investors, a super senior swap premium, upfront costs (rating agencies, lawyers, structuring and underwriting costs) ongoing administration costs, and possibly some other expenses. The business case seen from the capital situation of the bank is the essential decision making instrument to judge about such transactions.

Now, *capital relief as a driver for securitizations* probably will continue to play a certain role in the new regulatory framework (see [15]) but the focus on asset classes is likely to change in the course of Basel II. It is beyond the scope of this book to provide a survey on the securitization framework in the new capital accord, but a few remarks can give introductory guidence what we have to expect under the new capital rules regarding securitizations and capital relief.

First of all, we should mention that in contrast to Basel I (the old framework) the new capital rules are much more risk sensitive. As already mentioned, the standard risk weight under Basel I was 100%, which led to a regulatory capital of

$$[\text{risk weight}] \times [\text{solvability coefficient}] = 100\% \times 8\% = 8\%$$

for various assets originated and later on securitized by banks. The new accord works completely different. In the *internal ratings-based approach* (IRB), banks calculate the risk weight of an asset in the following way[4] (see [15], §271-272):

$$\text{RWA} = 12.5 \times \text{EAD} \times \text{LGD} \times K(\text{PD}) \times M(\text{PD}, \text{MAT})$$

$$K(\text{PD}) = N\left[\frac{N^{-1}[\text{PD}] + \sqrt{\varrho(\text{PD})}\, q_{99.9\%}(Y)}{\sqrt{1 - \varrho(\text{PD})}}\right] - \text{PD}$$

$$\varrho(\text{PD}) = 0.12 \times \frac{1 - e^{-50 \times \text{PD}}}{1 - e^{-50}} + 0.24 \times \left(1 - \frac{1 - e^{-50 \times \text{PD}}}{1 - e^{-50}}\right).$$

---

[4] Note that the function K is the quantile function (here, with respect to a confidence level of 99.9%) of the VASICEK distribution described in Appendix 6.7. Another related reference is Section 3.3.6.1.

Note that different parameterizations have to be applied for different asset classes. For instance, for SMEs[5] some firm size adjustment is applied ([15], §273-274); for retail exposures, e.g., residential mortgages or revolving retail exposures, other correlation parameterizations are used ([15], §327-330), and so on. The RWA formula shown before refers to a *standard corporate loan* where

- $M(\text{PD}, \text{MAT})$ stands for an adjustment factor depending on the effective maturity of the asset and its PD.

- $q_{99.9\%}(Y)$ denotes the 99.9%-quantile of a standard normal random variable Y.

- The formula for the correlation parameter ϱ is an interpolation between 12% and 24% quantifying the systematic risk ('R-squared'; see Section 1.2.1) of an asset as a function of its PD; the lower the PD of the firm, the higher its R-squared or ϱ.

The rationale underlying the correlation parameter $\varrho = \varrho(\text{PD})$ as a function of the rating or PD is that one expects on average to have better ratings for large (multi-national) firms and 'large' often is associated with higher systematic risk. From a mathematical as well as an empirical point of view this simplification (reducing a two-parameter model to a one-parameter model) is questionable but might serve its purpose in a regulatory framework.

Looking closer at the RWA formula for standard corporate loans, we find (see also Equation (2.12) and Section 6.6) that the the parameter ϱ can be interpreted as a *correlation parameter*, but, in fact, the Basel II model itself lacks any portfolio effects, because *regulatory capital allocated to a client is independent of the surrounding reference portfolio*; see the before-mentioned RWA formula. This constitutes a major issue of criticism on the new capital accord: Basel II does neither reward well diversified nor penalize low diversified portfolios.

Altogether we can, based on the few remarks we made, easily summarize the dilemma of the securitization framework in the new capital accord: *a portfolio effect-free model never can appropriately treat correlation products like CDOs and other basket-type transactions.*

---

[5] Small- and medium-sized enterprises.

The Basel Committee on Banking Supervision (the authors of the new accord) faced the difficult task to find work-arounds regarding the just-mentioned dilemma in order to treat correlation products and securitizations in the new regulatory capital framework. In [15], §538-643, one finds the securitization framework, which is binding from 2007/2008 on. Main components are the so-called *ratings-based approach* (RBA) which must be applied to rated tranches (to cut a long story short) and the so-called *supervisory formula* (SFA) which provides a capital charge figure for tranches in securitizations; see [15], §624. Explaining both approaches and their application is beyond the scope of this section. Nevertheless, we want to give two examples for consequences we think one can safely conjecture.

- Residential mortgage-backed securities (RMBS): risk weights under Basel II fall substantially (from 50% to 35% in the standardized approach and even more under the IRB approach; see [15], §72 and §328). Therefore, we expect that RMBS motivated by capital relief are no longer really attractive for most banks. However, true sale transactions involving funding benefits have a potential to remain of interest even under Basel II.

- Corporate loan securitizations: because the new risk weights under Basel II are rather risk sensitive, securitizations for high quality (good ratings) corporate loans for the purpose of capital relief are not really efficient. However, subinvestment grade names and lower rated risks are heavily risk-weighted under Basel II. Because such (e.g., high-yield) assets typically pay attractive margins/profits compensating for the taken risk, banks might still want to originate such assets and then securitize the assets in order to get rid of heavy capital charges. In addition, economic risk transfer and spread arbitrage as outlined before will remain a driver for securitizations.

A major challenge in securitizations will be the management and sale of first loss pieces. In extreme cases, risk weights on low-rated or unrated lower tranches (like equity pieces) can go up to 650% such that banks will do a lot to get rid of these pieces. It is very likely that even more than as of today non-regulated players like hedge funds will be happy buyers of highly risk-weighted high spread-carrying lower tranches arising from securitizations. This concludes our brief collection of remarks on regulatory arbitrage and capital relief via securitizations.

3.2 CDO Modeling Principles

In this section, we briefly comment on general principles of CDO modeling before we actually demonstrate modeling techniques by means of an example. The modeling scheme we are going to present is close to the modeling principle applied in the context of baskets and CLNs (see, e.g., Section 2.7). A major difference is that in certain parts of the CDO world one typically has to spend an additional (significant) amount of time with cash flow modeling at the liability side of the transaction, as we will see in a moment.

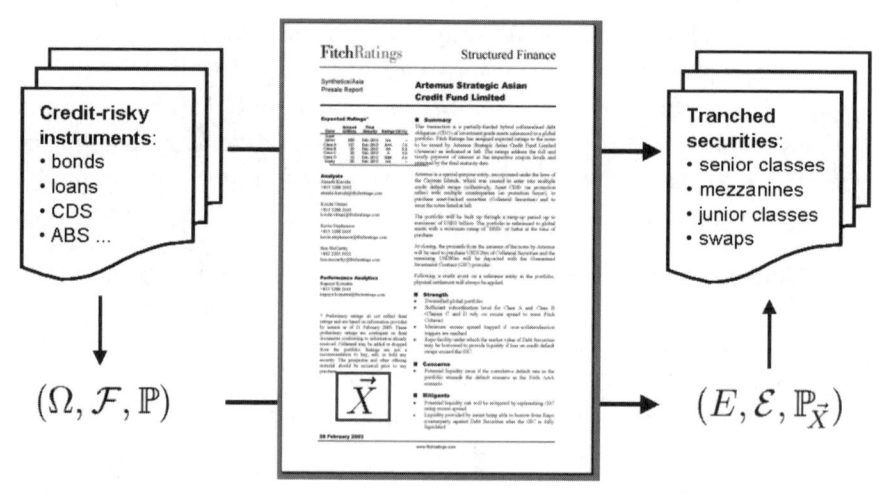

FIGURE 3.8: CDO modeling scheme - cause and response

Figure 3.8 shows a generic scheme of a CDO model. At the left-hand side we find the already-mentioned *asset side* of the transaction. Randomness typically only occurs via uncertainties at the asset side. Therefore, we can in an abstract way attach a probability space $(\Omega, \mathcal{F}, \mathbb{P})$ to the asset side representing the probabilistic mechanism steering all random effects at the asset side including payment defaults, prepayments, collateral management, and everything that is not known today and subject to random fluctuations. Recall that a probability space consists of

- A space Ω of elementary outcomes

- A set ('σ-algebra') of sets (events) in Ω to which a probability can be assigned

- A probability measure $\mathbb{P}$, which assigns probabilities to events

In one or the other way every CDO model relies implicitly or explicitly on the modeling of such a probability space.

At the right-hand side we have the *liability side* of the structure consisting of tranched securities. The liability side is linked to the asset side by means of a random variable or vector $\vec{X}$ *translating asset scenarios into liability scenarios*, hereby inducing a probability space $(E, \mathcal{E}, \mathbb{P}_{\vec{X}})$ at the liability side, carrying the image measure of the random variate $\vec{X}$. For instance, payment defaults can be seen as 'events' in $(\Omega, \mathcal{F}, \mathbb{P})$, which will be transformed via $\vec{X}$ into cash flow scenarios at the liability side of the structure leading to a bottom-up allocation of losses to tranches, and so on. Figure 3.8 represents a fundamental way of thinking about CDO models: *given an asset scenario $\omega \in \Omega$ is determined, the corresponding cash flow scenario $\vec{X}(\omega)$ immediately also is determined and follows deterministically the pre-specified rules of the transaction described in the term sheet or offering circular of the deal.*

Based on Figure 3.8, it is obvious that *Monte Carlo simulation is the preferred tool for CDO evaluation* due to its intrinsic scenario orientation. Any CDO can be modeled in the following way:

1. Step: Monte Carlo simulation of the underlying reference portfolio at the asset side of the transaction; the mathematical counterpart of this first step in the model is the construction of the probability space $(\Omega, \mathcal{F}, \mathbb{P})$.

2. Step: Modeling of the random variate $\vec{X}$ translating asset scenarios into cash flow and liability side scenarios. The mathematical result of this is the $\vec{X}$-induced probability space $(E, \mathcal{E}, \mathbb{P}_{\vec{X}})$. For instance, this part of the model includes the modeling of the cash flow waterfall in cash flow CDOs: given an asset scenario in form of, e.g., a realization $\omega = (\tau_1, ..., \tau_m) \in \Omega$ of default times for m underlying reference assets, $\vec{X}(\omega)$ collects all relevant, default time adjusted, cash flows into a liability side scenario. In Figure 3.9, default time adjusted cash flows are illustrated.

3. Step: Evaluation of simulation results, e.g., hitting probabilities and expected losses for tranches, distributions of the internal rate of return for tranches, expected interest and principal streams, risk adjusted duration of tranches, and so on.

Besides Monte Carlo simulation, which works always and for any CDO, we will discuss *analytic*, *semi-analytic*, and *comonotonic approximations*, which enable a quicker evaluation of CDOs in certain cases, e.g., in cases with many underlying names such that limit distributions for the portfolio loss can be applied (analytic approximation) or in cases where the underlying names are *exposure-exchangeable* (comonotonic approximation); see Sections 3.3.6.2 and 3.3.6.3.

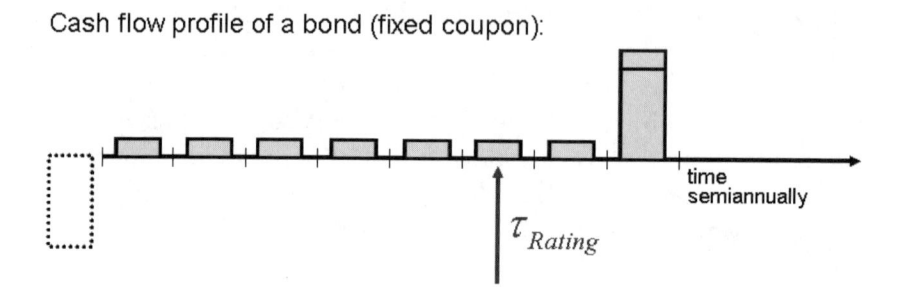

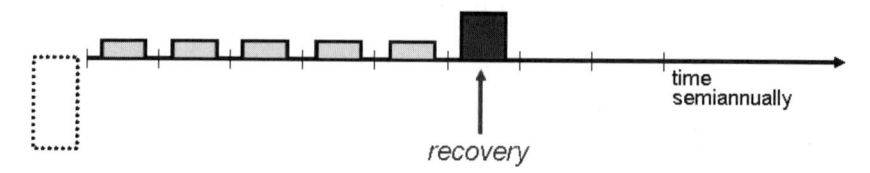

FIGURE 3.9: Transformation of cash flows by (simulated) default times in a Monte Carlo simulation of CDO-type structures

Modeling the asset side (1. Step in our discussion before) is, from a mathematical point of view, the most challenging part of any CDO model. However, modeling the liability side, namely the structure and cash flows of a CDO can be far from being trivial too. First of all the modeler needs to carefully study all available documentation, starting from presales and ending at a (typically voluminous) offering memorandum. It also involves a careful balance between modeling the 'full range' of all cash flow elements and simplifying structural elements for the benefit of a better handling. Taking shortcuts regarding cash flow elements can be dangerous, of course, and has to be done with great care on a case-by-case basis. To illustrate this, we give two examples.

Example 1: A Harmless Simplification

Assume that a CDO-SPV issues tranches where one tranche is split and issued in two currencies, e.g., 40% of the tranche are Euro denominated (Euro LIBOR as reference interest rate) and 60% of the tranche are Sterling denominated (Sterling LIBOR referenced). Lets say the underlying assets are all Sterling denominated and (Sterling LIBOR) floating rate notes. Obviously, there is some currency mismatch inherent in the CDO, which typically is hedged by means of a basis swap and a currency swap. The good news regarding the CDO model is that as long as the hedges are in place, there is no need to model the randomness underlying the currency risk. Instead, the hedge costs can just be implemented as another (senior) deduction from available funds in the interest waterfall of the CDO.

Example 2: A Dangerous Simplification

A less harmless shortcut is the following situation. An investor considers buying a mezzanine tranche of a cash flow CDO. In order to come to a quick decision, the investor only 'tranches-up' the loss distribution of the underlying pool (see Figure 3.3) in order to get an estimate of the mezzanine tranche's default probability and expected loss. Now, if the CDO is not just a 'plain vanilla' structure but incorporates some redirection of cash flows based on credit events in the asset pool (which will be the case in almost all cases of cash flow deals), such a 'tranched loss' approach (ignoring cash flow triggers and waterfall changes) is very likely to be misleading. For more sophisticated structures at least a semi-analytic approach (see 3.3.6.2), or even better, a 'full' Monte Carlo simulation approach must be recommended.

This concludes our introductory survey on CDOs and their applications. We now turn our attention to the modeling of CDOs.

3.3 CDO Modeling Approaches

We start our exposition on CDO models by an application of the *dependent default times approach* as introduced in Section 2.5 to a CDO example outlined below. Default times approaches to CDO evaluation are current market standard but have, as all modeling approaches, advantages and disadvantages. To name one particular disadvantage, default times cannot adequately handle situations where rating triggers are involved. For instance, rating triggers can be used in the structural definition of a deal to enforce a removement of assets from the portfolio underlying the deal if their rating falls below a critical threshold. Default times by definition are not following a *multi-period* or *dynamic* logic so that rating triggers can only 'artificially' be included in the deal evaluation via a separate side calculation in the waterfall model.

Better suitable for structural elements like rating triggers are so-called *multi-step* models where the evolution of the transaction 'payment period by payment period' is modeled. In Section 3.3.5.1, we explain how such models can be implemented.

Section 3.3.5.2 indicates a model approach based on diffusion-based *first passage time* models. From a probabilists point of view, first passage time models appear as very natural candidates for time-dynamic portfolio models. There, CWIs, which were considered w.r.t. different fixed time horizons, are replaced by *ability to pay processes* (APPs), which are based on a time-dynamic stochastic process concept, as indicated in the context of Figure 1.3.

The last modeling tools we will discuss in this section are *analytic*, *semi-analytic*, and *comonotonic* simulation techniques, which can speed-up CDO evaluations enormously, given a structure and the corresponding pool are 'eligible' for an application of such techniques. We will come back to criteria supporting an application of semi-analytic techniques in Sections 3.3.6.1, 3.3.6.2, and 3.3.6.3.

3.3.1 Introduction of a Sample CSO

Application of default times in CDO modeling will be demonstrated by means of a sample CDO where the underlying reference portfolio is the one described in Appendix 6.9. Note that subportfolios out of

Collateralized Debt and Synthetic Obligations

the 100-name portfolio from 6.9 were already used as reference pools in Sections 2.6 and 2.7. The transaction we want to model is an illustrative (!) *partially funded cash flow CSO*; see Figure 3.10.

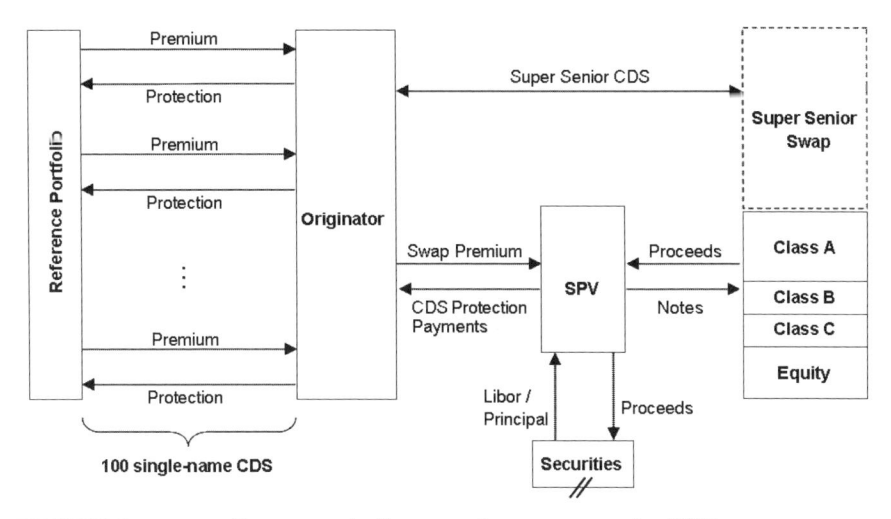

FIGURE 3.10: Structural diagram for our sample CSO

The transaction is structured as follows.

- The originator of the transaction sells protection on the 100 names in the reference portfolio (Appendix 6.9) by means of 100 single-name CDS with a total swap volume of 1 bn USD, 10 mn USD swap volume on each of the 100 names. This constitutes the asset side of the transaction.

- At the same time, the originator, via an SPV, buys protection on the 1 bn USD swap volume he now is exposed to. This constitutes the liability side of the structure.

- The SPV is the issuer in the transaction and issues 4 notes, namely, class A, class B, class C and a non-rated class E (the equity piece) securities in the capital market. Principal proceeds received from issuing the notes are invested in an account of risk-free (cash-equivalent) collateral-like securities paying USD-referenced 3-month LIBOR flat; see also Section 2.7 where a comparable mechanism is described. We fix USD-referenced 3-

month LIBOR at 1.5%. Class A, B, C notes, and the equity piece constitute the funded part of the transaction.

- Interest paid to notes investors can be written as

$$\text{LIBOR} + \text{spread}$$

where the spread of a tranche depends on

– Its seniority and the amount of subordinated capital below the considered tranche

– The risk characteristics of the underlying CDS portfolio

– The current state of the market regarding liquidity and complexity premiums for transactions of this type

see the interest waterfall decribed later in this section.

- Additionally, the issuers enters into a super senior swap agreement with an OECD[6] bank in order to buy protection against tail events in the CDS pool. The super senior swap constitutes the unfunded part of the transaction.

- In case of a default in the CDS pool, the realized loss is paid on the protection selling agreement by liquidating securities from the collateral-like account and using the proceeds to make the corresponding contingent payments. If losses exceed the funded volume (i.e., the collateral account dried-up), the super senior swap counterparty has the obligation to pay for the residual losses not covered by available securities. Following a default, the outstanding swap volume is reduced 'top-down' by recovered amounts on defaulted asset.

- At maturity (after 5 years), proceeds from liquidating the remaining assets in the collateral-like account are distributed 'top-down' to funded notes investors, i.e., class A has first priority, class B second priority, class C third priority and equity investors receive all of the remaining cash after a full pay-down of the three more senior classes. From an economic point of view this means that losses eat 'bottom-up' into tranches at the liability side.

---

[6] In order to get capital relief.

Collateralized Debt and Synthetic Obligations 197

Note the difference in the interest stream between the CLN transaction in Section 2.7 and the transaction modeled here: the CLN pays interest on the volume of the reference portfolio such that losses that reduce the volume of the reference portfolio *directly*, result in a reduced interest payment to the protection seller (CLN holder). In the transaction here, the mechanism is different because the deal is structured like a *cash flow* transaction: the interest stream collects cash on top of the waterfall described below via the total collected interest income, which consists of LIBOR flat paid by collateral-like securities plus spreads paid by non-defaulted CDS. In the course of defaults, collateral-like securities as well as CDS are removed from the transaction so that a reduced cash flow feeds the interest stream for the next payment period. Because beneficial cash flows like interest and principal are distributed top-down in the transaction (see the waterfall below), tranches suffer from defaults in an *indirect* way by means of a reduced interest stream in the cash flow waterfall distributing *available cash* to note holders.

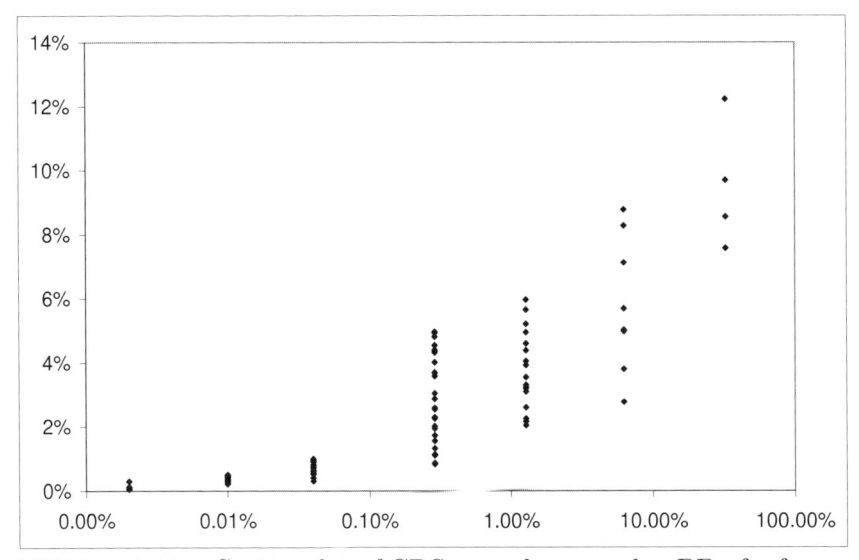

FIGURE 3.11: Scatterplot of CDS spreads versus log-PD of reference names in the portfolio described in Appendix 6.9

Figure 3.11 shows the assumed CDS spreads collected on single-name basis on the 100 names in the underlying portfolio. As expected, there is a clear trend of increased spreads in case of increased PDs. However, for lower-rated names, spreads are more deviating from the trend than in case of better-rated names in absolute terms, whereas in relative terms (good ratings have PDs at basispoint level) the deviation can be more significant for good credit quality names. Reasons for spread deviations can be liquidity and/or different levels of *rating arbitrage* referring to situations where the spread on a name already reflects the market view a of higher risk than suggested by the rating of the name.

Table 3.1 on page 204 shows the tranching of the deal, corresponding ratings of the CSO tranches and the spreads paid (on top of LIBOR for funded notes) to investors and the super senior swap counterparty. Equity pays no contractually promised spread but only excess spread, paying out the remaining cash in the interest stream after paying all contractually agreed payments to more senior tranches in the deal.

Next, we should describe the cash flow waterfall of the transaction. Let us consider the principal stream first. It is restricted to amounts in the collateral-like account funded by the issuance of notes by the SPV in order to guarantee the contingent payments of the SPV to the protection buyers at the left side of Figure 3.10 under the 100-names CDS agreements, covering losses below the super senior tranche. Therefore, we can turn our attention immediately to the interest stream of the transaction. Regarding interest payments (CDS premium collections; contributing to the *interest stream*) at the asset side, an amount of

$$0.25 \times \text{spread}_i \times \text{notional}(\text{CDS}_i) \qquad (i = 1, ..., 100)$$

is collected in every (quarterly) payment period. The collateral-like account contributes an amount of

$$0.25 \times \text{LIBOR} \times \text{volume}(\text{securities})$$

to the available proceeds of the interest stream. The total collected cash will be distributed at the liability side of the structure according to the following waterfall:

- Payment of structural/transaction fees (including administrative costs) in size of 7 bps on the total outstanding pool volume
  quarterly payment: 7/4 bps

- Payment of the super senior swap premium from remaining interest funds
  quarterly payment: 10/4 bps

Collateralized Debt and Synthetic Obligations 199

- Payment of interest/coupon to class A note holders from remaining interest funds
  quarterly payment: (LIBOR + 30 bps)/4

- Payment of interest/coupon to class B note holders from remaining interest funds
  quarterly payment: (LIBOR + 100 bps)/4

- Payment of interest/coupon to class C note holders from remaining interest funds
  quarterly payment: (LIBOR + 350 bps)/4

- Payment of all of the remaining interest cash to equity tranche holders.
  quarterly payment: excess cash to equity note holders

This concludes are description of the CSO and its cash flow waterfall.

3.3.2 A First-Order Look at CSO Performance

Before we evaluate the transaction by Monte Carlo simulation, let us do some 'back of the envelope' calculations roughly indicating the expected performance of the transaction as a first-order approximation to its 'true' performance, which, as indicated in our introductory survey in the context of Figure 3.8, typically requires Monte Carlo simulation for an offering memorandum-adequate translation of asset into cash flow scenarios at the liability side of the deal.

Looking at summary statistics of the deal, we find the following.

- All names in the reference portfolio reference the same amount of exposure (10,000,000 USD) so that the average PD in the pool can be calculated as the arithmetic mean. We have

$$PD^{(1)} = 211 \text{ bps}$$

for the average pool PD at the one-year horizon. Combined with an (assumed uniform) LGD of 50% this results in a pool EL of

$$EL^{(1)} = 106 \text{ bps}$$

at the one-year horizon. At the 5-year horizon, corresponding to the maturity of the transaction, we obtain

$$PD^{(1)} = 837 \text{ bps}$$

$$EL^{(1)} = 419 \text{ bps}$$

which can be obtained by a term structure derived from weighting the 100 single-name term structures in the pool with the portfolio-determined weights for the different rating classes; see Figure 6.4. Note that the 5-year EL in our sample portfolio actually is lower than 5 times the one-year EL, which confirms that a 211 bps PD already is in a risk area where forward PDs are decreasing over time; see Figure 2.7. Expressed in amounts of money, this yields a 5-year EL of

$$4.19\% \times 1,000,000,000 = 41,867,440 \text{ USD}. \qquad (3.2)$$

This figure for the cumulative expected loss can later be used in the Monte Carlo simulation as a first-order check if the simulation yields plausible results.

- Let us now consider the interest stream. From averaging the spreads in Figure 3.11, we get

$$\text{spread}_{in} = 263 \text{ bps p.a.}$$

for the average spread to be expected annually, which constitutes the basis for a solid cash income justified by a certain number of quite risky assets in the portfolio. In terms of money, this implies a 5-year spread income of

$$2.63\% \times 1,000,000,000 \times 5 = 131,563,570 \text{ USD} \qquad (3.3)$$

given that no defaults reduce the number of performing CDS (no-defaults case). As we will see, this assumption is overly optimistic.

- Next, we should figure out what we have to pay in terms of transactional costs. As we have seen, on top of the interest waterfall (highly prioritized) the transaction requires

$$\text{fees} = 7 \text{ bps p.a.}$$

as running costs for administration and maintenance. In terms of money, this deducts

$$0.07\% \times 1,000,000,000 \times 5 = 3,500,000 \text{ USD} \qquad (3.4)$$

Collateralized Debt and Synthetic Obligations

from the income stream over the whole lifetime of the transaction. In addition, the issuer (SPV) has to pay spreads on notes at the liability side of the structure. Contractually promised are spreads for classes A, B, C, and a premium for the super senior swap. The weighted average liability spread equals

$$\text{spread}_{out} = 3\% \times 3.5\% + 3\% \times 1\% + 6\% \times 0.3\% +$$
$$+ 83\% \times 0.1\% = 24 \text{ bps p.a.}$$

which in amounts of money sums up to[7]

$$0.24\% \times 1,000,000,000 \times 5 = 11,800,000 \text{ USD} \qquad (3.5)$$

over the 5 years of the term of the transaction.

In the same way as described in our CLN example in Section 2.7, LIBOR does not really affect the economics of the transaction (making the same simplifying assumptions as in Section 2.7) because LIBOR is paid in exact required amounts from the collateral-like securities arising from investing the proceeds from note issuance by the SPV.

As a first-order approximation of what we have to expect from the transaction, we can now make the following simple calculation:

+	131,563,570 USD	by income from CDS spreads (3.3)
−	3,500,000 USD	structural fee payments (3.4)
−	11,800,000 USD	liability/tranche spread (3.5)
−	41,867,440 USD	expected losses on CDS names (3.2)
=	74,396,130 USD	expected excess cash on equity

In other words, over the whole lifetime of the transaction, a total of *74.4 mn USD* of excess spread cash can be expected for equity note (*first loss piece*) holders, given the transaction follows the before assumed *average path*. This would mean an annual return of

$$\text{return}_{flp} = \frac{74.4 \text{ mn USD}}{5 \times 50 \text{ mn USD}} \approx 30\%. \qquad (3.6)$$

---
[7] Note rounding errors: liability spread = 0.236, not 0.24, etc.

These kind of 'back-of-the-envelope' calculations are free from waterfall mechanics and the evaluation of default scenarios. In other words, whether this expectation is justified or not strongly depends on the structure of the waterfall. As an example, we compare in Figures 3.13 and 3.14 two different waterfalls, one without excess cash conversion in principal payments and one with excess cash conversion in principal in case losses in the reference portfolio exceed a certain predefined threshold. The situation for equity investors changes dramatically based on whether such a mechanism of conversion of proceeds from the interest stream in principal proceeds is included in the waterfall or not; see also [25], Chapter 8. We come back to this later.

3.3.3 Monte Carlo Simulation of the CSO

After these preparations, we now come to the Monte Carlo simulation of the transaction. For this, we simulate *dependent default times* and evaluate cash flows exactly as in Figure 3.9, following our remarks in Section 3.2. In the same way as in Sections 2.6 and 2.7, we apply Gaussian default times, Student-t default times with 5 degrees of freedom and the same linear correlation structure as in the Gaussian case, and Clayton default times with $\eta = 2$. Simulation results we are interested in are the following:

- An important quantity regarding the risk of a tranche is the *hitting probability of tranches*, which we shortly call the PD of the tranche. Mathematically speaking, this is the likelihood

$$\mathrm{PD}_T = \mathbb{P}[L^{(t)} \geq \alpha] \tag{3.7}$$

where $T = [\alpha, \beta)$ denotes a tranche specified by its attachment point α and its detachment point β and $L^{(t)}$ refers to the total cumulative loss arising from defaults in the portfolio over the lifetime t (here, 5 years) of the deal. The hitting probability of tranches addresses a *loss on the principal stream* for a tranche, whereas another default a tranche could suffer from are *payment defaults on the interest stream*, which arise if a promised coupon or spread payment can no longer be made because the interest stream is dried-up when it comes to an interest payment for the considered tranche in the cash flow waterfall. In a Monte Carlo simulation, *principal as well as interest payment defaults can be analyzed*. We will evaluate both streams in our analysis.

- After knowing how likely it is that our tranche gets hit by losses or missed/reduced interest payments, we need to know how much losses in terms of principal and interest we have to expect on average on each tranche. In its most simple form, the *expected loss on principal* can be written in terms of the tranche loss profile function as illustrated in Figure 3.3 via

$$\mathrm{EL}_T = \mathbb{E}[\Lambda_{\alpha,\beta}(L^{(t)})] \qquad (3.8)$$

where $\Lambda_{\alpha,\beta}$ is the tranche loss profile function from Equation (3.1) equal to

$$\Lambda_{\alpha,\beta}(L^{(t)}) = \frac{1}{\beta - \alpha} \min\left[\beta - \alpha, \max[0, L^{(t)} - \alpha]\right]$$

where T is defined as before via attachment and detachment points α and β, respectively. However, in the same way as indicated in the discussion on expected excess cash, the expected loss of a tranche deviates from the pattern in Equation (3.8) as soon as non-linear cash flow elements are part of the structure of the transaction. Then, only Monte Carlo simulation, where *losses effective for a tranche can simply be accounted*, can fully reveal the loss potential of a tranche. In addition, Monte Carlo simulation will provide information regarding expected losses on the interest stream of the tranche.

- The hitting probability and the expected loss of a tranche can be combined to derive the *loss severity* of a tranche by writing

$$\mathrm{LGD}_T = \frac{\mathrm{EL}_T}{\mathrm{PD}_T} \qquad (3.9)$$

for the *loss given default of a tranche*. It is worthwhile to think about this LGD notion for a moment. Let us consider a very thin tranche where attachment and detachment points are very close, $\beta - \alpha = \epsilon > 0$ but $\epsilon \approx 0$. In such a situation, as soon as a loss hits the tranche, losses eat-up the whole volume of the tranche such that the tranche's loss severity equals 100% time-simultaneously with the hitting event. The other extreme is a tranche where attachment and detachment points are far from each other. For instance, the super senior swap in our example references a volume of $\beta - \alpha = 0.83$. In such cases, it needs a

204   *Structured Credit Portfolio Analysis, Baskets & CDOs*

TABLE 3.1:   Results of a Monte Carlo simulation of the sample CSO

Full MCS	Equity	Class C	Class B	Class A	Super Senior	Total
Vol. [USD]	50,000,000	30,000,000	30,000,000	60,000,000	830,000,000	1,000,000,000
Vol. [%]	5%	3%	3%	6%	83%	100%
Rating	NR	BB	A	AA	AAA	BBB to BB
Maturity	March 2011	March 2011	March 2011	March 2011	March 2011	March 2011
Spreads	Excess Spread	3.50%	1.00%	0.30%	0.10%	2.63%
Gaussian copula							
PD (cumul.)	99.39%	27.49%	5.88%	1.21%	0.05%	
EL (cumul.) [%]	71.65%	16.08%	3.33%	0.41%	0.00%	
EL (cumul.) [USD]	35,826,083	4,824,083	998,333	243,333	9,167	41,901,000
LGD	72.09%	58.49%	56.56%	33.38%	2.21%	
Student-t copula							
PD (cumul.)	99.18%	25.83%	8.18%	3.02%	0.57%	
EL (cumul.) [%]	68.52%	16.82%	5.57%	1.56%	0.03%	
EL (cumul.) [USD]	34,260,083	5,045,583	1,670,417	935,083	241,083	42,152,250
LGD	69.08%	65.11%	68.06%	51.58%	5.08%	
Clayton copula							
PD (cumul.)	80.96%	23.87%	15.59%	10.15%	4.24%	
EL (cumul.) [%]	46.51%	19.99%	13.28%	6.59%	0.50%	
EL (cumul.) [USD]	23,256,583	5,998,250	3,983,333	3,954,500	4,170,667	41,363,333
LGD	57.45%	83.75%	85.14%	64.92%	11.86%	

lot more losses subsequent to the loss leading to a tranche hit in order to completely wipe-out the considered tranche. Summarizing, it is obvious that the LGD of tranches is heavily driven by the *thickness* of the tranche in addition to the usual risk drivers like the amount of subordinated capital, the risk profile of the reference portfolio, and the cash flow mechanism defined in the waterfall of the transaction.

Table 3.1 shows the results of the Monte Carlo simulation of the transaction w.r.t. Gaussian, Student-t, and Clayton copula functions.

Before we do further analysis on our sample transaction, we want to make a few comments on Table 3.1.

- In Equation (3.2) we estimated a pool EL of $41,867,440$ USD just by application of the average portfolio PD term structure multiplied with an LGD of 50%. In Table 3.1 we find that the simulation-based EL pretty well matches this figure, neglecting minor differences at basispoint level from random fluctuations.

- According to Section 2.5.7 the Gaussian copula has a tail dependence of zero. Table 3.1 shows that the Gaussian copula yields by far the lowest risk profile for senior tranches. In contrast to the Gaussian copula, the Student-t and Clayton copulas exhibit tail dependence and lower tail dependence, respectively. This has consequences for the hitting probabilities and loss distribution over the capital structure of the transaction. In line with observations we made in Chapter 2 we find that equity risk declines but senior tranche's risk increases substantially when switching from Gaussian to Student-t and to Clayton copulas. Regarding first loss pieces, we again note that first-to-defaults and equity pieces have much in common.

- Class C notes show a kind of 'hybrid' behavior. Their hitting probabilities slightly decrease when switching from the Gaussian to the Student-t copula and to the Clayton copula but their expected losses increase at the same time. This becomes very explicit when looking at the corresponding tranche LGDs where we see a substantial increase with increasing (lower) tail dependence. In general, mezzanine notes like class C notes require an especially careful evaluation for investors because they are *close to the first loss piece but typically do not benefit from the upside potential of excess spread* as equity investors. In other words, mezzanine can be quite risky but has no upside potential. Actually, model risk also increases with increasing seniority because the more we go into the tail the more plays dependence modeling a role.

- Classes B and A as well as the super senior swap are located in the tail of the loss distribution. Figure 3.12 compares the cumulative portfolio loss distributions for the 3 different copula functions. Thinking in terms of a simple tranching of loss distributions as in Figure 3.3, it does not come much as a surprise that Table 3.1 comes to quite different conclusions for different copula functions. The peak at the very tail end of the Clayton copula loss distribution in Figure 3.12 collects the tail mass for losses exceeding 200 mn USD. The figure clearly shows the *extreme lower tail dependence of the Clayton copula*. We think of the Clayton copula in terms of a tool for *stress testing extreme tail behavior* rather than as a tool for representative evaluation and pricing.

- Another remark regarding Table 3.1 concerns tranche LGDs. The simulation results clearly show the interplay of LGD drivers mentioned in our discussion on tranche LGD. Class C and B notes have the same thickness such that their LGDs should be comparable. Minor differences we see in the table arise from the additional 3% subordination class B has in comparison to class C notes. In addition, the shape of the loss distribution makes a difference:

for the two copula functions with lower tail dependence, LGD of class B notes is slightly higher than for class C notes, whereas in the Gaussian copula case the class C note LGD slightly exceeds class B note LGD.

- As a last remark, we should comment on the *ratings of tranches*.[8] What we find in Table 3.1 is quite different from what we expect for single-name ratings like the ones in Table 1.1. Typically, all tranche PDs or hitting probabilities are too high to justify the rating given to the tranche. The reason is that in a real-life transaction *one would never allow equity investors to collect ongoing excess cash even in cases where already several losses occurred*. Instead, the market would enforce the incorporation of an *excess spread trap* as explained later in this section. As we will see, such a structural cash flow feature will change the risk situation of mezzanine and senior tranches substantially.

First, let us discuss some more information collected from the Monte Carlo simulation of the CSO. Hereby, we focus on the Gaussian copula case, other copulas can be evaluated analogously.

We start with *loss distributions for single tranches*. As an example, take class B notes. In Table 3.1 we find for the Gaussian case an EL of 998,333 USD. The EL is only part of the summary statistics, but Monte Carlo simulation gives us the full distribution of losses allocated to the B-tranche. We find the following loss scenario probabilities for the B-tranche:

- $\mathbb{P}[L_B^{(5)} = 0] = 94.12\%$

- $\mathbb{P}[L_B^{(5)} = 5,000,000] = 1.40\%$

- $\mathbb{P}[L_B^{(5)} = 10,000,000] = 1.06\%$

- $\mathbb{P}[L_B^{(5)} = 15,000,000] = 0.81\%$

---

[8] In our example, the ratings are assigned based on the tranche's risk, hereby taking an excess spread trapping into account; see Figure 3.2 and the discussion in the context of that figure. In practice, tranchings and ratings on tranches are a compromise of at least two, if not three, rating agency approaches to the modeling of portfolio credit risk and the evaluation of structured products.

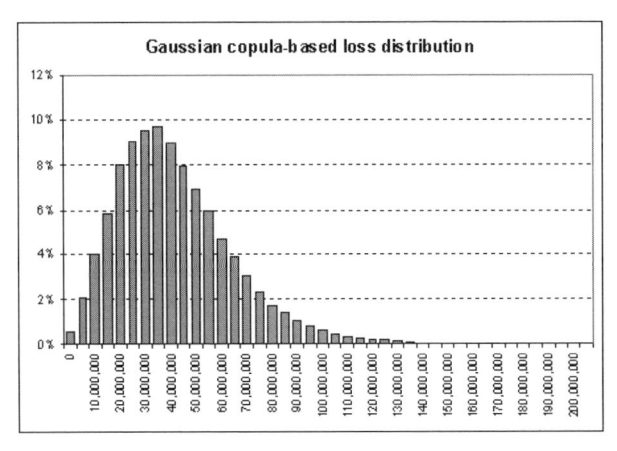

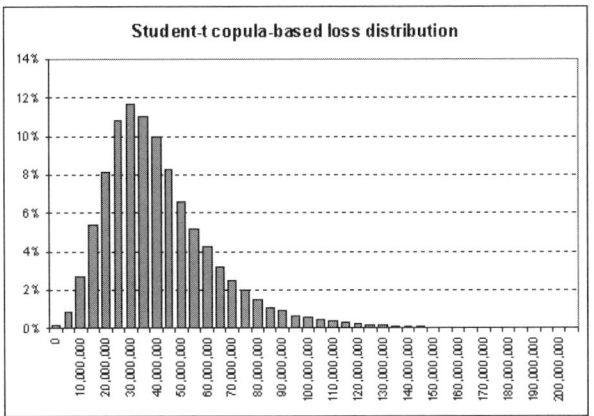

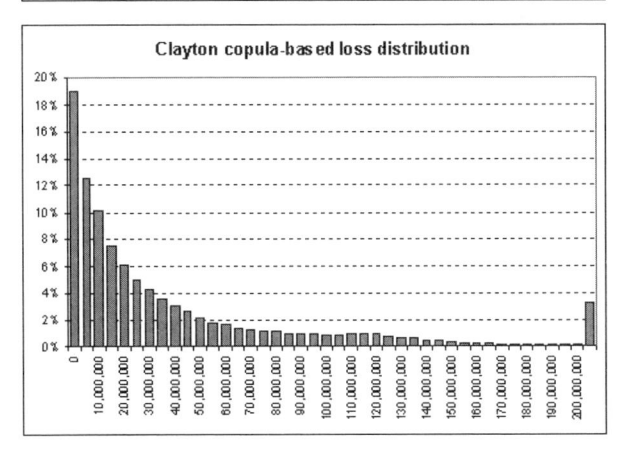

FIGURE 3.12: Comparison of 5-year cumulative loss distributions for the reference portfolio underlying our sample CSO w.r.t. the 3 applied copula functions

- $\mathbb{P}[L_B^{(5)} = 20,000,000] = 0.63\%$
- $\mathbb{P}[L_B^{(5)} = 25,000,000] = 0.42\%$
- $\mathbb{P}[L_B^{(5)} = 30,000,000] = 1.57\%$

where the superscript '5' refers to the maturity of 5 years for the CSO. The probability

$$\mathbb{P}[L_B^{(5)} = 30,000,000] = 1.57\%$$

aggregates all loss scenarios severe enough to completely wipe-out the whole tranche. Weighting each tranche loss scenario by its probability and summing up we get back the tranche EL of 998,333 USD, and

$$1 - \mathbb{P}[L_B^{(5)} = 0] = 1 - 94.12\% = 5.88\%$$

confirms the hitting probability of the B-tranche as calculated in Table 3.1. *Considering the full distribution of loss scenarios for a tranche is an important information for investors.* For instance, the tranche loss distribution reveals that with less than 2% probability class B note investors will face a loss exceeding 20 mn USD (2/3 of the invested capital), and so on.

We now come back to the equity note. By definition of the cash flow waterfall, *the performance of the equity note depends on the timing of defaults* in the following way. In case of *backloaded defaults* (late in time, close to maturity or beyond maturity), equity note holders have plenty of time to collect excess spread such that even in case of a complete wipe-out of the first loss piece at the end of the transaction a solid performance can be achieved; see also our discussion in Section 3.1.3. In case of *frontloaded defaults* (early in time, close after launching the transaction), the cash to equity note holders is reduced early due to non-performing/removed CDS premium payments. Figure 3.13 shows a scatterplot of cumulative portfolio losses versus collected excess spread for equity note holders. Pictures like Figure 3.13 are quite meaningful for investors or, as it often will be the case, arrangers of a CSO keeping part of the equity piece. The only way to generate such information is via Monte Carlo simulation. Figure 3.13 indeed illustrates in vertical direction the dependence of equity performance on the *timing of defaults*. Hereby, the maximum possible excess cash payment to equity note holders equals roundabout 120 mn USD, based on

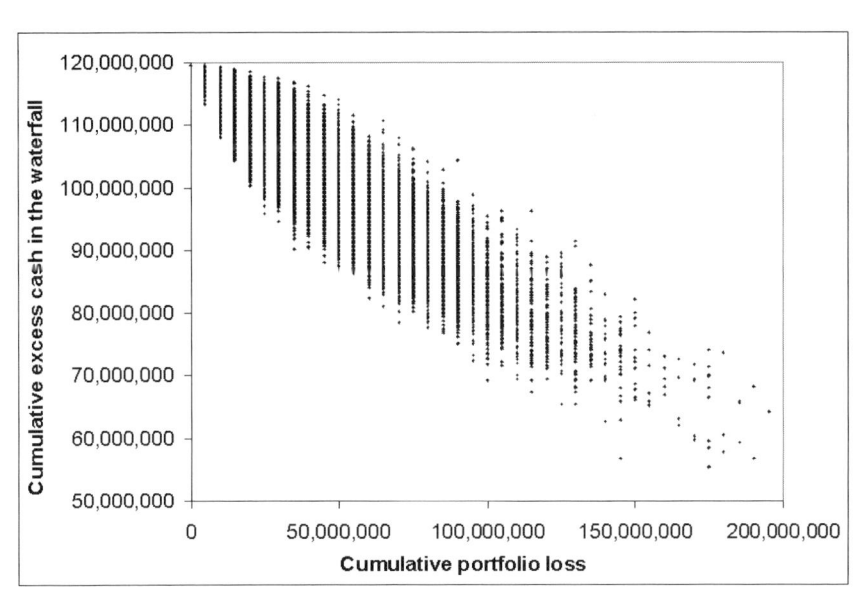

Collateralized Debt and Synthetic Obligations

FIGURE 3.13: Scatterplot of losses versus excess cash, illustrating the dependence of the performance of the equity note on the timing of defaults in the underlying reference portfolio

Equations (3.3), (3.4), and (3.5), hereby counting-in a LIBOR payment component of

$$1.5\% \times 50,000,000 \times 5 = 3,750,000 \text{ USD}.$$

Coming back to our discussion on the performance of an equity investment, Figure 3.13 graphically illustrates that equity investors are actually better off than guessed from the waterfall-free quick calculation in Equation (3.6).

However, a change in the definition of the waterfall can strongly impact the performance of the equity piece to the worse, as we will see later on page 213 in Figure 3.14.

The *stability of the interest stream* in general is a question relevant to all tranche investors. Based on the waterfall description, interest/coupon payments to tranches A, B, C, and to the super senior swap counterparty depend on available cash in the interest stream. According to Equation (3.3), roughly 130 mn USD can be collected in terms of CDS premiums in best case into the interest stream. Defaults typically reduce the available cash over time. Hereby, Figure

3.11 indicates a potential problem along these lines: more risky CDS naturally pay higher spreads such that names defaulting with higher likelihood and on average earlier in time have a more severe impact on the amount of available cash in the interest stream. However, 130 mn USD is a rich interest stream and we expect that interest payments are made in full in most of the scenarios in the simulation. Indeed, evaluating the scenarios with a focus on the interest stream we find that *all 4 tranches above equity get paid their full coupons/interest at all times in 100% of the scenarios.* As a cross-reference, we look at the equity piece income and find a positive cash flow at all times in 100% of the simulated scenarios. Regarding the equity note, the *distribution of equity tranche returns* can easily be derived from the results of the Monte Carlo simulation; see also [25], Chapter 8.

3.3.4 Implementing an Excess Cash Trap

Now, in order to illustrate the effect of non-linear waterfall features, let us now extend the definition of the waterfall by an additional element of credit enhancement for senior tranches. So far, credit enhancement for senior tranches comes from *subordination* and *cash payment priorities*. Now, we introduce in addition an *excess spread trap* triggered by cumulative losses exceeding a certain threshold:

- Still, all of the remaining cash from the interest stream after payment on all prioritized payments from the cash flow waterfall is distributed to equity tranche holders, but

  if cumulative realized losses $> \gamma \times \text{Volume}|_{t=0}$, $(0 \leq \gamma \leq 1)$,

  then in all periods of the remaining lifetime of the deal, excess cash is invested in additional (cash-equivalent) securities deposited in the collateral-like account. Accordingly, the super senior swap reference volume is reduced in size of the invested excess cash. The quantity $\text{Volume}|_{t=0}$ refers to the original notional of the pool, which equals 1 bn USD in our example.

- At maturity, the collateral-like account is cleared out and all proceeds from selling securities are distributed in line with the cash flow waterfall, paying down funded notes in their order of seniority and distributing all of the remaining excess cash to equity note holders. In this way, the loss trigger works like an excess cash *trap*, comparable to a *sinking fund*: if the re-direction of excess cash is triggered and no further losses occur, the *trapped* cash in the collateral-like account is paid to equity investors at maturity of the transaction. So the cash payment to equity investors is delayed but paid if the excess cash stored in the collateral-like account is not needed to mitigate further losses.

As mentioned, we consider only the Gaussian copula case, the other two copulas can be considered as variations of the scheme.

For our calculation, we set $\gamma = 0.05$ and $\gamma = 0.02$ such that excess spread re-direction becomes effective as soon as cumulative losses on the CDS portfolio exceed 50,000,000 USD and 20,000,000 USD, respectively. Table 3.2 shows the impact of the trigger on the performance of the tranches.

We consider the case $\gamma = 0.05$ first. Because the critical loss volume triggering excess cash diversion matches with the notional of the equity tranche, the loss potential of tranches senior to the equity piece significantly declines. The cumulative pool EL drops by roundabout 7 mn USD due to a *waterfall-enforced conversion of spread payments into principal payments*. Note that the 7 mn USD EL decline refers to a mean value aggregating all simulated scenarios into one number, there are single scenarios where the impact of the loss trigger is much heavier than can be guessed from considering mean values. The loss trigger generates additional cash support for the principal stream, which leads to reduced tranche ELs as can be seen by a direct comparison of the Gaussian copula-based part of Table 3.1 with Table 3.2. Note that even the EL of the equity piece declines by roundabout 2 mn USD. However, again this is due to a **conversion of funds from the interest stream into principal proceeds**. Loss mitigation has to be *funded from the interest stream such that excess cash payments to equity investors decline* too. From an aggregated point of view the benefit of lower equity principal losses is *off-set by lower excess cash proceeds*.

As expected, the impact is much stronger in case of $\gamma = 0.02$. Coming back to our comments on tranche ratings in our discussion of Table 3.1, we find in case of $\gamma = 0.02$ the risk of tranches *much better reflects the assigned ratings*. As already mentioned, common market practice will always enforce such a loss trigger (or some comparable feature) as an *additional credit enhancement* for tranches senior to the equity piece. The so defined structure is 'fair' also for equity investors, which in many

TABLE 3.2: Monte Carlo simulation based on the Gaussian copula of the sample CSO with effective loss triggers with $\gamma = 5\%$ and $\gamma = 2\%$ as threshold quotes ; results for other copulas can be obtained analogously

	Full MCS	Equity	Class C	Class B	Class A	Super Senior	Total
Vol. [USD]	50,000,000	30,000,000	30,000,000	60,000,000	830,000,000	1,000,000,000
Vol. [%]	5%	3%	3%	6%	83%	100%
Rating	NR	BB	A	AA	AAA	BBB to BB
Maturity	March 2011	March 2011	March 2011	March 2011	March 2011	March 2011
Spreads	Excess Spread	3.50%	1.00%	0.30%	0.10%	2.63%
			Gaussian copula ($\gamma = 1$)			
PD (cumul.)	99.39%	27.49%	5.88%	1.21%	0.05%	
EL (cumul.) [%]	71.65%	16.08%	3.33%	0.41%	0.00%	
EL (cumul.) [USD]	35,826,083	4,824,083	998,333	243,333	9,167	41,901,000
LGD	72.09%	58.49%	56.56%	33.38%	2.21%	
			Gaussian copula ($\gamma = 0.05$)			
PD (cumul.)	99.39%	9.79%	0.56%	0.10%	0.01%	
EL (cumul.) [%]	67.90%	2.44%	0.25%	0.03%	0.00%	
EL (cumul.) [USD]	33,950,709	731,532	75,449	16,794	1,635	34,776,119
LGD	68.32%	24.92%	44.64%	27.99%	3.94%	
			Gaussian copula ($\gamma = 0.02$)			
PD (cumul.)	53.76%	0.88%	0.19%	0.05%	0.01%	
EL (cumul.) [%]	15.03%	0.46%	0.11%	0.01%	0.00%	
EL (cumul.) [USD]	7,513,321	137,899	31,805	8,337	1,470	7,692,833
LGD	27.95%	52.23%	56.29%	26.89%	3.54%	

cases will coincide with the arranger of the transaction: if the reference portfolio performs well, excess spread is paid to equity investors. If the loss trigger is hit, but thereafter the pool performs ok, remaining trapped excess spread is paid to equity investors at maturity of the deal. If the trigger is hit and the pool continues to perform badly, then equity investors will not receive excess spread and all excess cash is used for the protection of higher prioritized tranches. This shows that the rationale underlying excess cash trapping makes perfect sense.

Figure 3.14 shows how the excess spread re-direction triggering based on losses (case $\gamma = 0.05$) changes the picture in Figure 3.13. Note the different scaling of the y-axis in the two figures: without trigger, the cloud of points has its lowest points roughly between the 50 and 60 mn USD level, whereas with trigger in place (Figure 3.14) the y-axis needs room down to an almost zero excess cash level. Up to a cumulative loss level of 50 mn USD the pictures in Figures 3.13 and 3.14 are similar. From 50 mn USD loss on, the trigger becomes effective such that Figure 3.14 exhibits a much wider range of possible excess spread scenarios as a function of the portfolio loss. The *timing of defaults* much stronger impacts the performance of the equity piece in case of an effective trigger, an observation that is intuitively reasonable. Obviously, the picture will look much worse for equity investors if $\gamma = 0.05$ is replaced by $\gamma = 0.02$. Then, excess cash is much earlier used for senior tranche

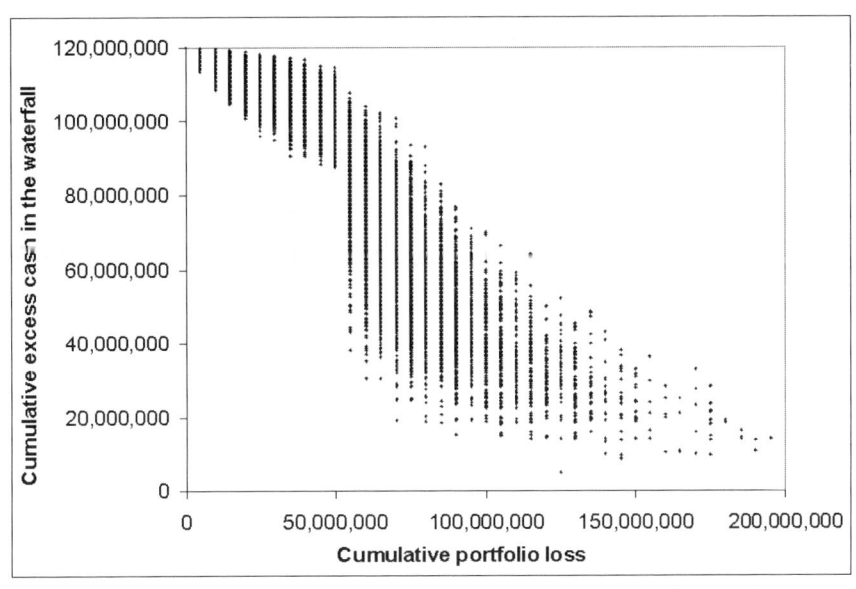

[FIGURE: Scatterplot with x-axis "Cumulative portfolio loss" from 0 to 200,000,000 and y-axis "Cumulative excess cash in the waterfall" from 0 to 120,000,000]

FIGURE 3.14: Scatterplot of losses versus excess cash, analogous to Figure 3.13, but with effective excess spread re-direction in case of losses exceeding 50,000,000 USD

protection such that in high loss scenarios the equity piece exhibits a very poor performance. Then, Equation (3.6) is a wrong conclusion.

3.3.5 Multi-Step and First Passage Time Models

For the sake of completeness, we want to use this section for a brief introduction to so-called *multi-step* models. So far we focussed on default time models, which can be considered as *static* models. To see this, consider again Figure 2.22, which shows that default times are based on a *static, non-dynamic* portfolio model, which (heuristically speaking!) *stretches static random variables over time* via the credit curves (PD term structures) of the underlying single-name credit risks. Dependencies among default times are induced by dependencies arising from the one-period portfolio model. In contrast, multi-period models aim to follow the granularity of the time axis required by the particular modeling problem and model the time evolution of the portfolio risk *period by period*. For instance, in our structured credit models, we worked on a quarterly time grid, which, in a multi-period model, requires 20 model steps for a 5-year time horizon. Multi-step models are very natural model approaches because they try to best possible capture the time evolution of credit risk w.r.t. the same logic as the future actually evolves, namely payment period after payment period.

The transaction we have in mind when describing multi-step modeling in this section is the cash flow CSO from the previous section. However, there is not much one can learn much from modeling the waterfall again such that we restrict our discussion to a more generic level of describing the model approach for the underlying reference names over the lifetime of the transaction. Modeling the liability side in dependence on the performance of the reference portfolio then follows the usual routine in the sense of Figure 3.8.

3.3.5.1 Models Based on Brownian Motion

There are different ways one can follow to come to a multi-period model. Our brief exposition on a first approach of this kind is based on papers by FINGER [48] and MOROKOFF [88]. Note that Moody's KMV as described in [88] is in the comfortable position of being able to base their multi-step modeling on the *distance to default* dynamics empirically measured from firm data over long time periods. Here, we are not in such a comfortable position, which forces us to remain in a fixed parametric framework. However, the overall modeling principles are nevertheless comparable.

The model works as follows. For the 5-year maturity of the CSO, we need to take into account 20 quarterly payment periods. In every payment period, the names defaulting in this period have to be determined via Monte Carlo simulation. Then, cash flows at the liability side of the transaction can be adjusted/transformed accordingly. Instead of CWIs as latent variables w.r.t. to *fixed time horizons*, we now choose standard Brownian motion processes $(B_t^{(i)})_{t \geq 0}$, defined on a common probability space $(\Omega, \mathcal{F}, \mathbb{P})$, for all names $i = 1, ..., m$ in the reference portfolio. The Brownian motions are assumed to be correlated with a correlation matrix $\Gamma \in [0, 1]^{m \times m}$. The interpretation of values $B_t^{(i)}$ for times t is in terms of an *asset value log-return* of firm i in the same way as described in Section 2.5.4 but in a more time-dynamic manner. If the process $(B_t^{(i)})_{t \geq 0}$ falls below a critical threshold $c_t^{(i)}$, company i is in default. For any obligor i, the *default time* τ_i (in continuous time) can now be specified as the *first passage time* of the Brownian motion

$(B_t^{(i)})_{t\geq 0}$ w.r.t. the (non-random) barrier $(c_t^{(i)})_{t\geq 0}$:

$$\tau_i = \inf\left\{t \geq 0 : B_t^{(i)} < c_t^{(i)}\right\}. \qquad (3.10)$$

In order to get back the credit curve of obligor i, the condition,

$$p_i^{(t)} \stackrel{!}{=} \mathbb{P}[\tau_i \leq t], \qquad (3.11)$$

where $(p_i^{(t)})_{t\geq 0}$ denotes the credit curve (PD term structure) of firm i (see also Remark 2.5.1), has to hold for all times t. Because the dynamics of the Brownian motion process is well-known and given, the challenge indicated by Equation (3.11) is the determination of barriers $(c_t^{(i)})_{t\geq 0}$ in line with (3.11). Another way to enforce (3.11) is to modify the process underlying the first passage time definition. In Section 3.3.5.2, we introduce such an approach.

Considering the CSO multi-step modeling problem in discrete time, we arrive at the following task. The Brownian motion process is sampled at discrete times, once in every payment period, such that instead of the full process we obtain a vector $(B_1^{(i)}, ..., B_{20}^{(i)})$ for the 20 payment periods. For every payment period $j = 1, ..., 20$, we have to find thresholds $c_j^{(i)}$ such that

$$p_i^{(j)} = 1 - \mathbb{P}\left[B_1^{(i)} > c_1^{(i)}, ..., B_j^{(i)} > c_j^{(i)}\right] \qquad (3.12)$$

where now $p_i^{(j)}$ denotes the cumulative default probability w.r.t. the first j payment periods. At the right-hand side of Equation (3.12) we find the survival probability of obligor i for the first j payment periods involved. Solving Equation (3.12) is far from being trivial. Therefore, it makes perfect sense to follow the logic of the word 'multi-step' and use the fact that the (discrete-time sampled) Brownian motion value $B_j^{(i)}$ in period j has a recursive representation based on the fact that Brownian motion can be obtained as a limit of random walks; see [31], p. 251. This enables us to recursively solve for thresholds $c_j^{(i)}$ via

$$\frac{p_i^{(j)} - p_i^{(j-1)}}{1 - p_i^{(j-1)}} = \mathbb{P}\left[B_j^{(i)} < c_j^{(i)} \mid B_{j-1}^{(i)} > c_{j-1}^{(i)}, ..., B_1^{(i)} > c_1^{(i)}\right] \qquad (3.13)$$

comparing the term structure induced *conditional* probability of default in period j given survival in previous periods with the corresponding process induced conditional probability of default given no default event occurred in previous payment periods. The starting point in the first payment period is straightforward, all subsequent periods follow by recursion and application of (3.13) w.r.t. given PD term structures; see again the paper by FINGER [48].

3.3.5.2 Models Based on Time-Changed Brownian Motion

In the previous section, we said that one approach to first passage time modeling is the definition of a suitable process such that the first passage time distribution matches given credit curves. In this section, we follow up on this remark. Starting point for our discussion is the already-introduced *Brownian first passage time*

$$\tau_i = \inf\left\{t \geq 0 \;:\; B_t^{(i)} < c_t^{(i)}\right\}.$$

from Equation (3.10) in the previous section. There, we used Brownian motion as underlying *ability to pay process* (APP) in the sense of Figure 1.3 and the corresponding discussion in Chapter 1.

The question arises, and is treated in OVERBECK and SCHMIDT [97], if we can find a better suitable process than just Brownian motion such that the corresponding first passage time $\tilde{\tau}_i$ w.r.t. a (to be defined) barrier matches a given credit curve,

$$p_i^{(t)} = \mathbb{P}[\tilde{\tau}_i \leq t] \qquad (3.14)$$

The answer given in [97] is 'yes': we can find *strictly increasing time scale transformations*

$$T_i : [0, \infty) \to [0, \infty)$$

applied to the underlying Brownian motions,

$$\tilde{B}_t^{(i)} = B_{T_i(t)}^{(i)} \qquad (t \geq 0;\; i = 1, ..., m),$$

and thresholds c_i, such that the first passage times of the so-defined APPs match given credit curves in line with Equation (3.14). In addition, the calibration can be done in a way generating prescribed *joint default probabilities* JDPs w.r.t. a fixed time horizon, e.g., one year,

$$\text{JDP}_{ij}^{(1)} = \mathbb{P}[\tilde{\tau}_i \leq 1, \tilde{\tau}_j \leq 1] \qquad (i, j = 1, ..., m;\; i \neq j).$$

Recall that, given the PDs of two names, JDPs and CWI/APP correlations w.r.t. a fixed time horizon represent the same information in a Gaussian world; see Remark 2.2.3 as well as Figure 2.5. Therefore, *JDP matching can be interpreted as a certain way of pairwise correlation matching*.

The time-transformed Brownian motion APPs provide an intuitive way to derive dependent default times by means of a reasonable *barrier diffusion* model. In HULL and WHITE [59], a comparable analysis can be found. Regarding default barrier diffusion models we also refer to ALBANESE ET AL. [7] and AVELLANEDA and ZHU [12].

For the sake of completeness, we now want to sketch the derivation of the transformations T_i. Let us start with the *hitting time* for Brownian motion $(B_t^{(i)})_{t \geq 0}$ w.r.t. a given number $c_i \in \mathbb{R}$,

$$\tau_i^{(c_i)} = \inf\left\{t \geq 0 \,:\, B_t^{(i)} = c_i\right\}.$$

A reformulation of the *reflection principle* for Brownian motion (see [109], Chapter 5) yields

$$\mathbb{P}\left[\max_{0 \leq s \leq t} B_s^{(i)} \leq c_i\right] = \mathbb{P}\left[\tau_i^{(c_i)} \geq t\right] = 2N\left[\frac{c_i}{\sqrt{t}}\right] - 1. \quad (3.15)$$

From (3.15) we immediately obtain

$$\mathbb{P}\left[\min_{0 \leq s \leq t} B_s^{(i)} \leq c_i\right] = 2N\left[\frac{c_i}{\sqrt{t}}\right]. \quad (3.16)$$

Given a *strictly increasing* time transformation T_i, we obtain for

$$\tilde{\tau}_i = \inf\left\{s \geq 0 \,:\, B_{T_i(s)}^{(i)} \leq c_i\right\}$$

the following equation,

$$\begin{aligned}
\mathbb{P}[\tilde{\tau}_i \leq t] &= \mathbb{P}\left[\min_{0 \leq s \leq t} B_{T_i(s)}^{(i)} \leq c_i\right] \quad (3.17) \\
&= \mathbb{P}\left[\min_{0 \leq s \leq T_i(t)} B_s^{(i)} \leq c_i\right] \\
&= 2N\left[\frac{c_i}{\sqrt{T_i(t)}}\right].
\end{aligned}$$

In other words, the distribution function $\tilde{\mathbb{F}}_i$ of $\tilde{\tau}_i$ can be written as

$$\tilde{\mathbb{F}}_i(t) = \mathbb{P}[\tilde{\tau}_i \leq t] = 2N\left[\frac{c_i}{\sqrt{T_i(t)}}\right]. \tag{3.18}$$

Solving Equation (3.18) for the transformation T_i yields[9]

$$T_i(t) = \left(\frac{c_i}{N^{-1}[\tilde{\mathbb{F}}_i(t)/2]}\right)^2. \tag{3.19}$$

Based on the imposed condition

$$\tilde{\mathbb{F}}_i(t) \stackrel{!}{=} p_i^{(t)} \qquad (t \geq 0;\ i = 1,...,m),$$

we can define T_i via Equation (3.19) in a way that $\tilde{\tau}_i$, by application of Equation (3.17), matches the PD term structure of asset i,

$$\mathbb{P}[\tilde{\tau}_i \leq t] = p_i^{(t)}, \tag{3.20}$$

such that the first passage time distributions of the time transformed Brownian motions exactly replicate the marginal default time distributions based on the credit curves of the firms.

So far we did not comment on the determination of the thresholds c_i because Equation (3.19) leaves some freedom regarding the definition of c_i. However, it can make sense to fix c_i w.r.t. an *attachment point* $t_0 > 0$ for T_i such that

$$T_i(t_0) = t_0.$$

This can be achieved by setting

$$c_i = N^{-1}\left[\frac{p_i^{(t_0)}}{2}\right]\sqrt{t_0} \tag{3.21}$$

hereby solving Equation (3.19) for the prescribed time point t_0. Figure 3.15 shows time transformations T_i for three different rating classes. We always set $t_0 = 5$ years, in line with the maturity of the CSO. The condition $T_i(5) = 5$ can be clearly seen in Figure 3.15.

In addition, the Figure 3.15 suggests an interpretation for the relation between the default risk of a firm represented by its credit curve and the corresponding time transformation defined in Equation (3.19):

---

[9]For a meaningful determination of c_i, see Equation (3.21).

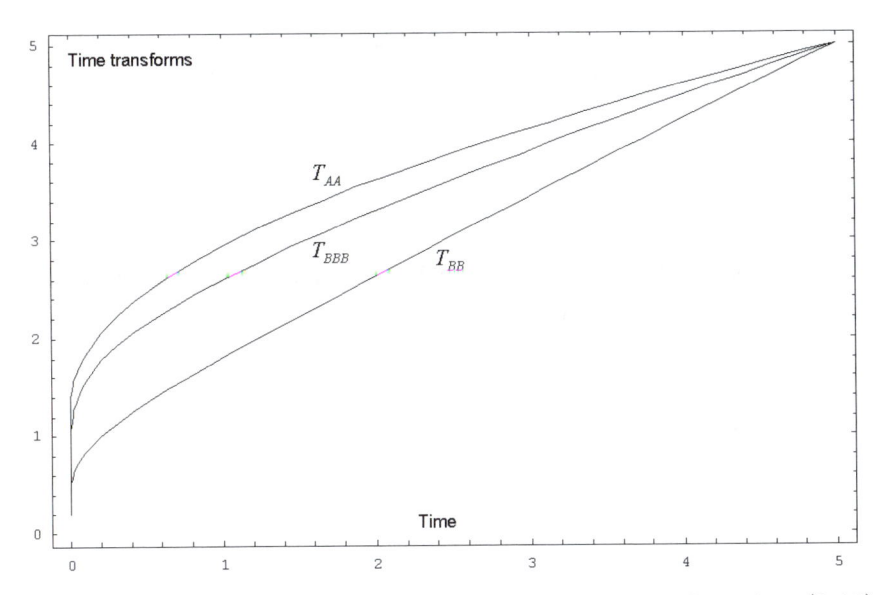

Collateralized Debt and Synthetic Obligations 219

FIGURE 3.15: Time transformations according to Equation (3.19) for three different rating classes

Looking at Figure 3.15, we find that 'normal time' (x-axis) is mapped to a 'transformed time' (y-axis) in a way that at normal time $t_0 = 5$ all transformed times have caught-up again ($T_i(5) = 5$). Hereby, *the transformed time of firms with high credit quality passes along a longer path than it is the case for firms with low credit quality*. In other words, names with higher credit quality take more time to proceed to the *time scale fixpoint* ($T_i(5) = 5$), whereas names with lower credit quality proceed more directly to the time scale fixpoint.

It can be shown (see [97]) that under the assumption of continuity of the term structure (distribution) functions $\mathbb{F}_i$ *the time-transformed Brownian motions have a representation as stochastic integrals with time-varying but deterministic volatilities* where integration is done w.r.t. new Brownian motions. These volatilities can be interpreted as *default speed* parameters: the higher the volatility and speed, respectively, the higher the risk to fall below the threshold and default.

As a last point in this section, we refer to [97] for a calibration of the the time-changed Brownian motion based APP model to given joint default probabilities w.r.t. a fixed time horizon. The calibration is based on an analytical representation of JDPs via modified Bessel functions as described in a paper by ZHOU [117]; see Proposition 2 in [97]. It is natural that in an APP model as described before, the *radial symmetry*[10] of Brownian motion leads to an application of Bessel functions.

These remarks conclude our brief discussion of first passage time models and approaches for their calibration. Following an approcach like the APP model above, one can model the asset side of a CSO. The liability side model and its link to the asset side (following Figure 3.8) then follows the usual routine as described in Section 3.3.3.

3.3.6 Analytic, Semi-Analytic, and Comonotonic CDO Evaluation Approaches

For some structured credit products, *comonotonic*, *analytic*, or *semi-analytic* modeling techniques can be applied. A reference for the following exposition on such techniques are the two papers [27] and [28]. Whether analytic or semi-analytic techniques can be applied or not *depends on the structure of the transaction and the underlying asset pool*. In the sequel, we explore various of these techniques and provide some examples for their application. In certain subsegments of the CDO market like *index tranches* (see Section 3.4.4), related techniques evolved to market standard in the last couple of years. Another important application of (semi-)analytic techniques are CDO rating tools as well as 'quick-and-dirty' evaluation tools where people are interested in getting a rough pre-selection of CDO deals before they start a deeper analysis of pre-selected transactions. CDOs of CDOs are examples where pre-selection procedures can save a good amount of time.

3.3.6.1 Analytic CDO Evaluation

The most typical example where an analytic approach is as good as any Monte-Carlo simulation approach is the case of synthetic (balance sheet, capital relief or risk transfer motivated) transactions, referenced to a large, more or less homogeneous, pool of reference assets, e.g., a large portfolio of retail loans, a highly diversified portfolio of SME loans, or a large portfolio of residential mortgage backed loans underlying an

---

[10] As an example for radial symmetry of Brownian motion, recall the well-known fact, leading to so many applications of Brownian motion in the context of harmonic functions, that Brownian motion started in the origin of a ball hits the sphere of the ball with a distribution equal to the Lebesgue surface measure.

RMBS.[11] In such cases, the structure as well as the reference pool qualify for (semi-)analytic evaluation techniques. More precisely, for the sequel we consider (as an example) a transaction satisfying the following conditions:

1. The underlying reference pool is highly diversified and can be (approximately) represented by a *uniform* reference portfolio with *infinite granularity*. We come back to this condition later in the text; see also Appendix 6.7.

2. Amortization of notes on the liability side follows *sequentially top-down* in decreasing order of seniority (highest seniority tranche first, second highest seniority tranche next, and so on). The asset pool is *static* (non-managed) and has a *bullet*[12] *exposure profile*[13] until maturity.

3. Losses are *written-off sequentially bottom-up* in increasing order of seniority (equity tranche bears the first loss, next higher tranche bears the second loss, and so on).

4. CDO notes are referenced to the underlying pool of assets (e.g., by credit linked notes as discussed in Section 2.7). Interest payments on notes will be paid by the originator in an amount of

$$\text{interest}_T = \text{volume}_T \times [\text{LIBOR} + \text{spread}_T] \qquad (3.22)$$

where T denotes a tranche at the liability side of the transaction.

The only events related to credit risk in the just-described structure are loss write-offs (see Condition 3). Condition 4 describes a very simple *interest stream:* default on the interest stream will only happen if for some reason the *originator* stops to be in line with contractually agreed obligations, a situation that typically is very unlikely. In other words, the interest stream bears only the credit risk of the originator, not the credit risk of the reference pool, because reductions of interest payments due to a melting tranche volume are contractually taken into account and indicated to investors by means of Condition 4.

---
[11] Residential Mortgage Backed Securities.
[12] No repayments until maturity.
[13] However, by application of 'WAL-adjustments' ('WAL' stands for 'weighted average life') the results of this section also can be applied to amortizing asset pools.

For our illustration in the sequel, we assume without loss of generality

$$LGD = 100\%$$

uniformly fixed for all assets in the reference pool. This condition simplifies some of the calculations in the sequel. However, by scaling and substitutions in the following formulas, any other fixed LGD can be implemented, replacing any portfolio gross loss $L^{(t)}$ by a realized net loss $L^{(t)} \times$ LGD; see the example at the end of this section.

For transactions as the one described above, the risk of tranches can be quantified by a *closed-form analytic* approach.

Condition 1 allows to replace the original reference pool by a *uniform* or *homogeneous* portfolio with *infinite granularity* admitting a *uniform credit curve* $(p^{(t)})_{t\geq 0}$ and a *uniform CWI correlation* ϱ. Let us assume that the considered CDO matures at time t. Then, the cumulative portfolio loss $L^{(t)}$ at time horizon t is given by the conditional PD of the assets in the uniform portfolio,

$$L^{(t)} = g(p^{(t)}, \varrho; Y) = N\left[\frac{N^{-1}[p^{(t)}] - \sqrt{\varrho}\, Y}{\sqrt{1-\varrho}}\right] \quad (3.23)$$

where $Y \sim N(0,1)$;

see also Equation (2.12). The derivation of this representation is well-known, due to VASICEK [111], and can be found in a more general setting, e.g., in [25], pages 87-94.

For the sake of convenience, Appendix 6.7 provides some facts around Equation (3.23).

The expression at the right-hand side in Equation (3.23) is the loss variable of a portfolio with infinitely many obligors (the so-called 'limit case') where all obligors have a PD of $p^{(t)}$ and are pairwise correlated with a CWI correlation ϱ. The variable Y has an interpretation as a proxy for a *macro-economic factor* driving the loss of the portfolio. Because $g(p^{(t)}, \varrho; Y)$ corresponds to a portfolio of infinitely many assets, *idiosyncratic or firm-specific credit risk has been completely removed by diversification* in the formula for $L^{(t)}$ so that the randomness of Y is the sole source of the riskiness of the portfolio loss $g(p^{(t)}, \varrho; Y)$.

In the sequel, we will exploit the absolute continuity of the portfolio loss variable $L^{(t)}$ by relying on its density

$$f_{p^{(t)},\varrho}(x) = \sqrt{\frac{1-\varrho}{\varrho}} \times \qquad (3.24)$$

$$\times \exp\left(\frac{1}{2}\left(N^{-1}[x]\right)^2 - \frac{1}{2\varrho}\left(N^{-1}[p^{(t)}] - \sqrt{1-\varrho}\,N^{-1}[x]\right)^2\right)$$

see [25], page 91, and Appendix 6.7. By construction, we get back the credit curve $(p^{(t)})_{t\geq 0}$ via taking expectations,

$$\begin{aligned} p^{(t)} &= \mathbb{E}[g(p^{(t)},\varrho;Y)] \\ &= \int_{-\infty}^{\infty} N\left[\frac{N^{-1}[p^{(t)}] - \sqrt{\varrho}\,y}{\sqrt{1-\varrho}}\right] dN(y) \\ &= \int_{0}^{1} x f_{p^{(t)},\varrho}(x)\,dx, \end{aligned}$$

using the simplifying assumption that the LGD equals 100%.

Going back to Section 3.1.2, we find a notation we want to adopt again in this section. Any tranching of the portfolio consisting of N tranches can be represented by a *partition* of the unit interval

$$[\alpha_0,\alpha_1) \cup [\alpha_1,\alpha_2) \cup \cdots \cup [\alpha_{N-1},\alpha_N]$$

for a strictly increasing sequence of α's, where $\alpha_0 = 0$ and $\alpha_N = 1$. For a tranche T_n with attachment point α_n and detachment point α_{n+1} where $i \in \{0, ..., N-1\}$, we defined in Equation (3.1) the *loss profile* function of tranche T_n as

$$\Lambda_n(L^{(t)}) = \Lambda_{\alpha_n,\alpha_{n+1}}(L^{(t)}) = \qquad (3.25)$$

$$= \frac{1}{\alpha_{n+1} - \alpha_n} \min\left[\alpha_{n+1} - \alpha_n, \max(0, L^{(t)} - \alpha_n)\right]$$

representing the loss, as a function of the overall portfolio loss of $L^{(t)}$, allocated to tranche T_n. The tranche loss profile function is illustrated in Figure 3.3. We are now ready for the following simple proposition.

3.3.1 Proposition *For a CDO with maturity t satisfying the conditions listed at the beginning of this section, the expected loss of tranche T_n, normalized to the tranche size, can be calculated by*

$$EL_{T_n} = \mathbb{E}[\Lambda_n(L^{(t)})] = \int_0^1 \Lambda_n(x) f_{p^{(t)},\varrho}(x)\, dx$$

where Λ_n is the function defined in (3.25).

Proof. The assertion of the proposition is obvious. □

Proposition 3.3.1 offers a closed-form expression for the expected loss of a tranche, given the assumptions on the structure of the transaction as well as on the underlying reference portfolio stated at the beginning of this section are fulfilled.

3.3.2 Proposition *Under the same conditions as in Proposition 3.3.1, the probability PD_{T_n} that tranche T_n is hit by a loss equals*

$$PD_{T_n} = 1 - N\left[\frac{1}{\sqrt{\varrho}}\left(N^{-1}[\alpha_n]\sqrt{1-\varrho} - N^{-1}[p^{(t)}]\right)\right].$$

Remember that α_n denotes the attachment point of tranche T_n.

Proof. First of all note that we have

$$\mathbb{P}[g(p^{(t)}, \varrho; Y) \leq x] = \mathbb{P}\left[-Y \leq \frac{N^{-1}[x]\sqrt{1-\varrho} - N^{-1}[p^{(t)}]}{\sqrt{\varrho}}\right]$$

for all $x \in [0,1]$ according to Equation (3.23); see also Appendix 6.7. Taking $Y \sim N(0,1)$ into account and considering

$$PD_{T_n} = \mathbb{P}[g(p^{(t)}, \varrho; Y) > \alpha_n] = 1 - \mathbb{P}[g(p^{(t)}, \varrho; Y) \leq \alpha_n],$$

the proof of the proposition immediately follows. □.

Proposition 3.3.2 shows how to calculate the PD of a CDO tranche and Proposition 3.3.1 helps calculating its expected loss (EL). The loss given default (LGD) can be defined as in Equation (3.9) as

$$LGD_{T_n} = \frac{EL_{T_n}}{PD_{T_n}} \qquad (3.26)$$

TABLE 3.3: Analytic approach applied to our sample CSO tranching from Section 3.3.1, calculated with an assumed ϱ of 6% and 8%, respectively

Analytic vs MCS	Equity	Class C	Class B	Class A	Super Senior	Total
Vol. [USD]	50,000,000	30,000,000	30,000,000	60,000,000	830,000,000	1,000,000,000
Vol. [%]	5%	3%	3%	6%	83%	100%
Rating	NR	BB	A	AA	AAA	BBB to BB
Maturity	March 2011	March 2011	March 2011	March 2011	March 2011	March 2011
Spreads	Excess Spread	3.50%	1.00%	0.09%	0.10%	2.63%
Gaussian copula Monte Carlo simulation							
PD (cumul.)	99.39%	27.49%	5.88%	1.21%	0.05%	
EL (cumul.) [%]	71.65%	16.08%	3.33%	0.41%	0.00%	
EL (cumul.) [USD]	35,826,083	4,824,083	998,333	243,333	9,167	41,901,000
LGD	72.09%	58.49%	56.56%	33.38%	2.21%	
Analytic approximation (ρ = 6%)							
PD (cumul.)	100.00%	28.65%	4.46%	0.49%	0.003%	
EL (cumul.) [%]	74.39%	13.48%	1.84%	0.10%	0.000%	41,850,000
EL (cumul.) [USD]	37,195,000	4,044,000	552,000	60,000	332	
LGD	74.39%	47.05%	41.26%	20.41%	1.29%	
Analytic approximation (ρ = 8%)							
PD (cumul.)	100.00%	29.62%	6.57%	1.18%	0.03%	
EL (cumul.) [%]	72.02%	15.65%	3.18%	0.31%	0.00%	41,850,000
EL (cumul.) [USD]	36,010,000	4,695,000	954,000	186,000	3,469	
LGD	72.02%	52.84%	48.40%	26.27%	1.67%	

for any tranche T_n, $n = 0, ..., N - 1$. In this way, the three main components of basic risk analysis (PD, EL and LGD) are specified. Another interesting application of Proposition 3.3.2 is the derivation of a model-based *implied rating* of a CDO tranche where standard PD term structures can be compared with PDs for CDO tranches in order to arrive at some *shadow rating* for the tranche.

Let us make an example illustrating Propositions 3.3.1 and 3.3.2. The CSO modeled in Section 3.3.3 does not qualify for an analytic CDO model, especially not in case of an excess spread trap triggered by portfolio losses, but let us nevertheless use the tranching of the CSO and the summary statistics of the underlying reference portfolio of 100 CDS to create an *illustrative* example for this section. As we will see, if the loss trigger re-directing excess spread (see Section 3.3.4) is switched off, the analytic approximation of the cash flow CSO is not too bad. However, *as soon as non-linear credit enhancement features (like loss triggers) are included in order to change the waterfall in dependence on the occurrence of certain events, the analytic approach is no longer feasible*. As an alternative, *semi-analytic approximations* as introduced in Section 3.3.6.2 can be applied.

Figure 3.16 shows the uniform credit curve of the reference portfolio based on the weighted average credit curve of the 100 names in the reference portfolio underlying the CSO from Section 3.3.1; see also Ap-

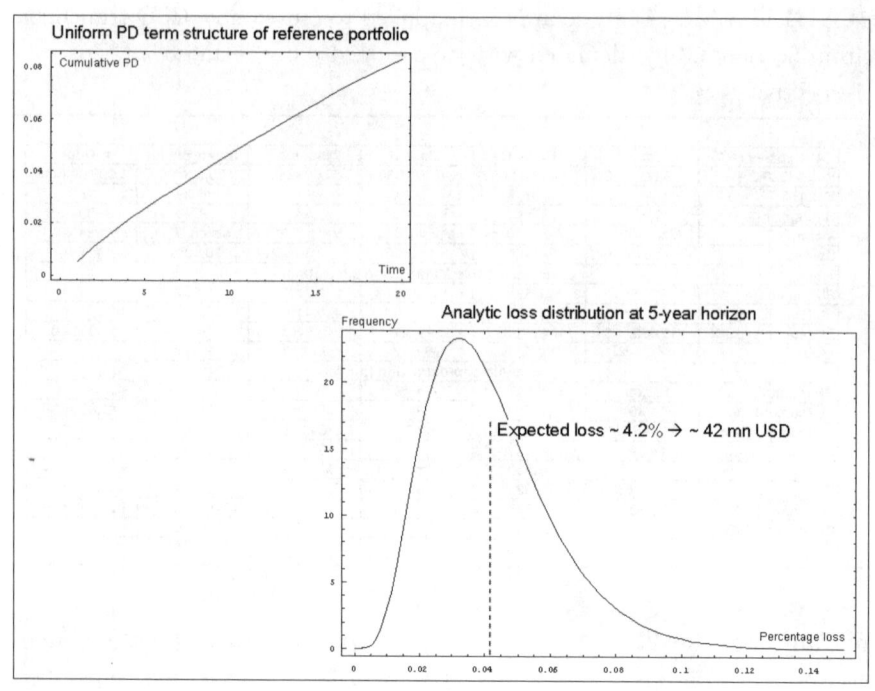

FIGURE 3.16

FIGURE 3.16: Analytic approximation of the reference portfolio from the CSO example in Section 3.3.1, calculated with $\varrho = 6\%$; see Appendix 6.7 and Appendix 6.8 for some background on analytic approximations

pendices 6.7, 6.8, and 6.9. At the lower left-hand side in Figure 3.16 a plot of the *analytically derived* loss distribution according to the density in Equation (3.24) is shown. The expected loss, by construction, is consistent with the calculated EL from Equation (3.2). As in Section 3.3.1, we assumed a uniform fixed LGD of 50%. To incorporate such an LGD into the formulas derived in the context of Propositions 3.3.1 and 3.3.2, a simple substitution has to be made: for instance, the density (3.24) of the portfolio loss then looks like (see Appendix 6.7)

$$f_{p^{(t)},\varrho}(x) = \frac{1}{\text{LGD}} \times \sqrt{\frac{1-\varrho}{\varrho}} \times$$

$$\times \exp\left(\frac{1}{2}\left(N^{-1}[x/\text{LGD}]\right)^2 - \frac{1}{2\varrho}\left(N^{-1}[p^{(t)}] - \sqrt{1-\varrho}\,N^{-1}[x/\text{LGD}]\right)^2\right).$$

Table 3.3 shows the analytically derived tranche PDs, ELs, and LGDs for an assumed ϱ of 6% and 8%, respectively. Figures 3.17 and 3.18

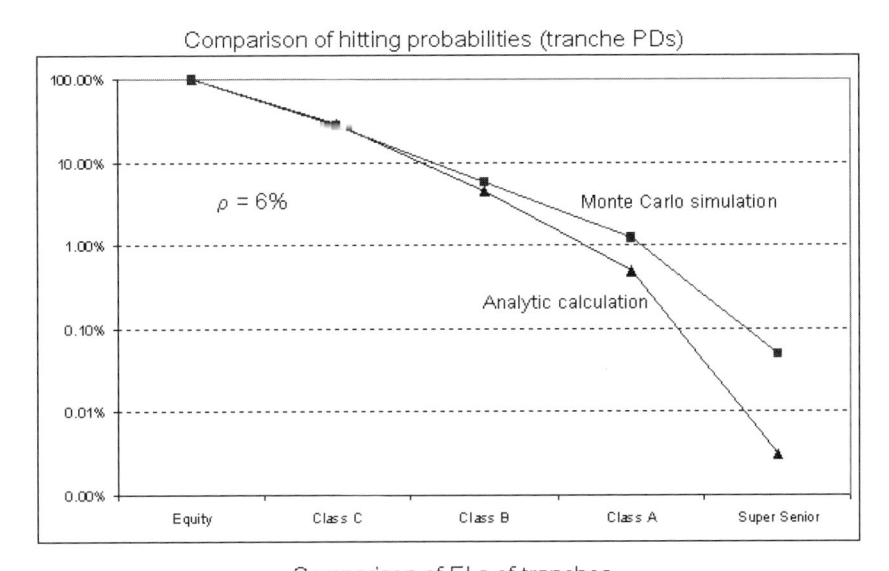

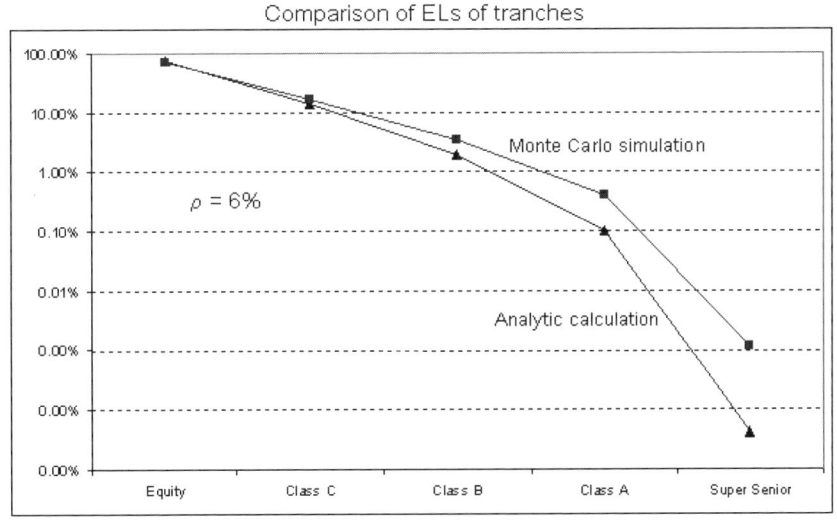

FIGURE 3.17: Comparison of analytic and full Monte Carlo evaluation of the CSO example from Section 3.3.1, calculated with an assumed ϱ of **6%**

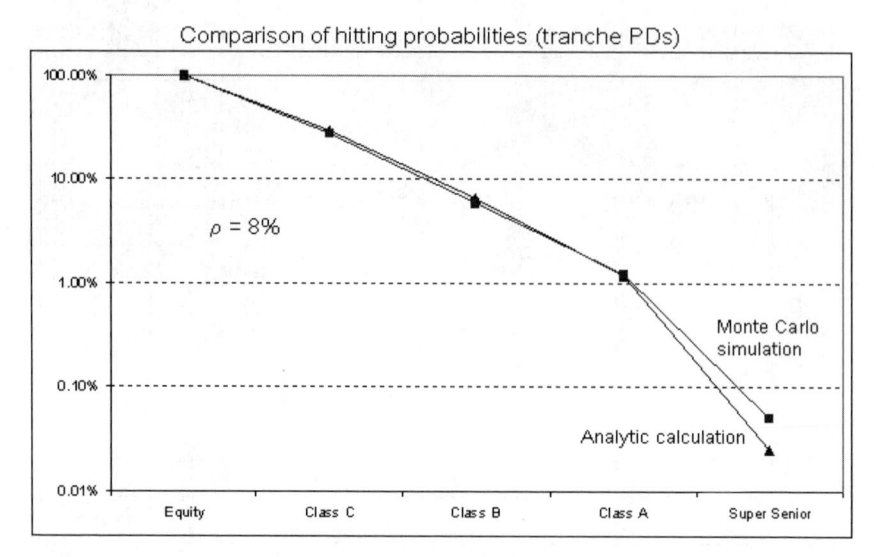

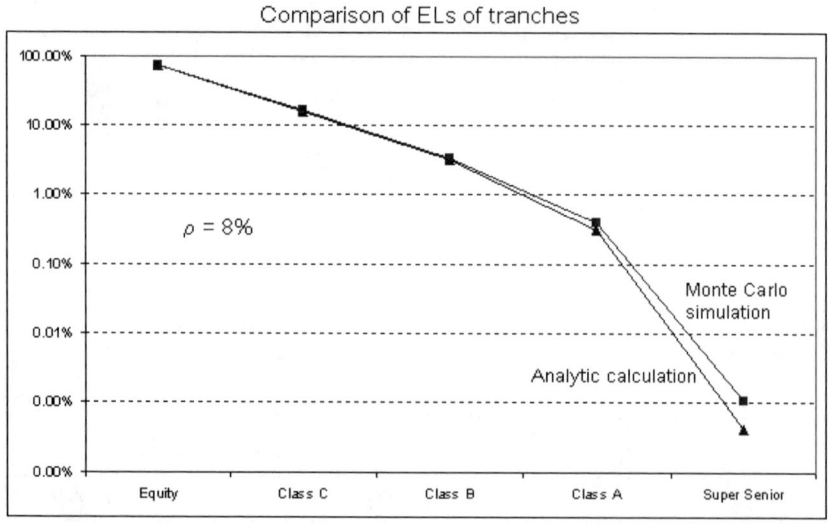

FIGURE 3.18: Comparison of analytic and full Monte Carlo evaluation of the CSO example from Section 3.3.1, calculated with an assumed ϱ of **8%**

graphically illustrate the results by comparing them with tranche PDs, ELs, and LGDs derived by a full Monte Carlo simulation approach as elaborated in Section 3.3.3. In the sequel, we want to make some comments on the results.

Equity tranche:
In analytic approximations, the hitting probability (PD) of the equity tranche typically equals 100% because the loss distribution, represented by the density $f_{p^{(t)},\varrho}(x)$, is absolutely continuous such that

$$\mathbb{P}[L^{(t)} = 0] \;=\; 0.$$

Rule-of-thumb for LGDs:
We already mentioned in Section 3.3.3 that the LGD of a tranche is lower for thicker tranches and higher for thinner tranches. This rule-of-thumb is reflected by the results in Table 3.3, although we clearly recognize that tranche LGDs are also driven by subordination levels of the CDO and by the risk profile of the underlying reference portfolio. The LGD always is driven by a superposition of all of these three drivers.

Comparison with Monte Carlo simulation:
The super senior tranche hardly bears any loss or is hit by a loss at all. A mathematical reason for this observation is the comparably moderate CWI correlation of $\varrho = 6\%$ and $\varrho = 8\%$, which generates fat but not overly fat tails. If in our analytic approximation we would further increase ϱ, we would find that super senior risk would increase (still being low though) but the risk of other tranches would deviate more from the full Monte Carlo simulation results. Altogether, the analytic approximation fits not too bad but also not really well. Either we achieve a good fit in the mezzanine parts of the loss distribution and understate senior risk or we have a better fit in the tail but a worse fit in the mezzanine parts of the loss distribution. Reasons for mismatches are the *difference between multi-factor and one-factor models and the difference between exact granularity ($m = 100$) and an infinite granularity approach as reflected by the density (3.24)*. However, as mentioned before, *if non-linear cash flow features are included in the waterfall, analytic approximations will yield completely misleading results*. We then better use *semi-analytic* techniques.

A quick check on the plausibility of the calculated tranche ELs in Table 3.3 can be done by calculating[14]

$$\sum_{n=1}^{5} \text{volume}_{T_n}[\%] \times \text{EL}_{T_n}[\%]$$

because if our calculations are correct this will give us back the overall portfolio EL. Indeed, doing the calculation yields the portfolio's cumulative EL of 4.2%, hereby confirming the test as 'passed'.

3.3.6.2 Semi-Analytic CDO Evaluation

A much more powerful tool for CDO evaluation than just analytic calculations is the *semi-analytic* approach. The assumption we need to make in this section is the following:

- The underlying reference pool is highly diversified and can be approximately represented by a homogeneous reference pool in the same way as required by Condition 1 in Section 3.3.6.1.

So in contrast to the analytic approach we here only make assumptions regarding the *uniform approximability* of the reference portfolio and do not impose any restriction on the type of the CDO transaction. More technically speaking, the analytic approach as explained in the last section relies on the cumulative loss distribution of the underlying reference portfolio at the time horizon specified by the term of the transaction, whereas the semi-analytic approach, as we will see below, considers all payment periods of the transaction. This makes the approach much more broadly applicable.

The semi-analytic approach works as follows. Instead of considering the default times τ_i of single obligors as in previous sections, e.g., as in Section 3.3.3, we now consider the *fraction of obligors with a default time within the considered payment period*. To make this precise, denote by τ_i the default time of obligor i and by $L_{PF(m)}^{(j)}$ the cumulative loss for a portfolio of m obligors in quarter j, considered over quarterly payment periods $j = 1, ..., T$ where T denotes[15] the maturity of the considered CDO. As an extension to the CSO modeled in

---

[14]Note that some of the figures are rounded.

[15]Note that we also use the capital letter T as a notion for tranches; however, given the respective context, confusions regarding the use of T can be ruled-out.

Section 3.3.3, we now want to incorporate the possibility of including amortizing loans into the reference portfolio. For this, we need to make assumptions regarding the exposure profile of reference assets. The exposure outstanding on loan/asset i in period/quarter j will be denoted by $E_i^{(j)}$. We assume the following natural conditions, considering an increasing number m of obligors in the portfolio:

1. The exposures in the portfolio do not increase over time, i.e., $E_i^{(j-1)} \geq E_i^{(j)}$ for all $i = 1, ..., m$, $j = 2, ..., T$, and $m \in \mathbb{N}$, $m \uparrow \infty$.

2. The total exposure of the portfolio at time j,

$$E_{PF(m)}^{(j)} = \sum_{i=1}^{m} E_i^{(j)},$$

   converges to a limit relative to the portfolio's start exposure,

$$\lim_{m \to \infty} \frac{E_{PF(m)}^{(j)}}{E_{PF(m)}^{(1)}} = w^{(j)}$$

   for every fixed payment period $j = 1, ..., T$. Due to Condition 1, $w^{(j)} \in [0, 1]$.

3. With increasing number of obligors, the total exposure of the portfolio strictly increases to infinity for all fixed payment periods j, i.e., $E_{PF(m)}^{(j)} \uparrow \infty$ for $m \uparrow \infty$ for every $j = 1, ..., T$.

4. For $m \uparrow \infty$, exposure weights shrink very rapidly,

$$\sum_{m=1}^{\infty} \left(\frac{E_m^{(j)}}{E_{PF(m)}^{(j)}} \right)^2 < \infty$$

   for every payment period $j = 1, ..., T$.

These conditions are sufficient but not necessary for establishing the results in the sequel. More relaxed conditions can easily be formulated. In this section, we are only interested in showing the basic principle.

In order to shed some light on the conditions stated above, we make a few examples. For instance, Condition 2 is satisfied in case of exposures admitting *uniform* (amortization) *profiles*,

$$\exists\, 1 = w^{(1)} \geq w^{(2)} \geq \cdots \geq w^{(T)} \geq 0 \quad \forall\, i, j : \frac{E_i^{(j)}}{E_i^{(1)}} = w^{(j)}. \quad (3.27)$$

The most trivial case where Condition (3.27) is fulfilled is the case of typical CSOs where on the asset side protection is sold for m single reference names, typically in form of a 5-year *bullet exposure* profile of m (equal amount) CDS, and on the liability side protection is bought on the (diversified) pool of CDS in form of tranched securities with a suitable leverage regarding spreads on tranche volumes. Then, $w^{(j)} = 1$ for all payment periods j. We modeled a comparable transaction (but with cash flow enhancements) in Section 3.3.3.

An example for exposures fulfilling Conditions 3 and 4 is the case where the exposures are captured in a uniform band,

$$0 < a \leq E_i^{(j)} \leq b < \infty \quad \forall\, i,j.$$

Then, Condition 3 can be confirmed by

$$E_{PF(m)}^{(j)} = \sum_{i=1}^{m} E_i^{(j)} \geq m \times a \uparrow \infty \quad \text{for } m \uparrow \infty,$$

and Condition 4 is fulfilled due to

$$\sum_{m=1}^{\infty} \left(\frac{E_m^{(j)}}{E_{PF(m)}^{(j)}} \right)^2 \leq \sum_{m=1}^{\infty} \frac{b^2}{m^2 a^2} = \frac{b^2}{a^2} \sum_{m=1}^{\infty} \frac{1}{m^2} < \infty.$$

In general, *Conditions 1-4 are not really restrictive and can be confirmed in many cases of transactions of practical relevance.*

We fix Conditions 1-4 for the rest of this section. As in Section 3.3.6.1, we assume for the moment an LGD of 100% (zero collateral) in order to keep formulas slightly better handable, but, as we have seen in the example at the end of Section 3.3.6.1, it is straightforward to drop this simplifying assumption for the more general case of other LGDs. The percentage cumulative portfolio loss for an m-obligor portfolio w.r.t. the time horizon T (maturity of the deal) equals

$$L_{PF(m)}^{(T)} = \sum_{j=1}^{T} \frac{E_{PF(m)}^{(j)}}{E_{PF(m)}^{(1)}} X_{PF(m)}^{(j)} \tag{3.28}$$

$$\text{with} \quad X_{PF(m)}^{(j)} = \sum_{i=1}^{m} w_i^{(j)} \mathbf{1}_{\{j-1 \leq 4\tau_i < j\}}$$

where $w_i^{(j)}$ denotes the exposure weight for obligor i in period j,

$$w_i^{(j)} = \frac{E_i^{(j)}}{E_{PF(m)}^{(j)}} \qquad (i=1,...,m;\ j=1,...,T).$$

Note that here we write '$4\tau_i$' instead of 'τ_i' because we consider quarterly payment periods w.r.t. a time variable t counting in years. A generalization to other payment frequencies is straightforward.

The semi-analytic approach relies on two propositions and a subsequent corollary. A comparable but less general analysis can be found in [25], where Conditions 3 and 4 are essentially 'cloned' from Assumption 2.5.2 in [25].

3.3.3 Proposition *For a homogeneous portfolio with a uniform credit curve* $(p^{(t)})_{t\geq 0}$ *and a uniform CWI correlation* ϱ, *the probability that obligor i defaults in the time period $[s,t)$ with $s<t$ conditional on $Y=y$ is given by*

$$\mathbb{P}[s \leq \tau_i < t \mid Y = y] =$$

$$= N\left[\frac{N^{-1}[p^{(t)}] - \sqrt{\varrho}\, y}{\sqrt{1-\varrho}}\right] - N\left[\frac{N^{-1}[p^{(s)}] - \sqrt{\varrho}\, y}{\sqrt{1-\varrho}}\right].$$

Proof. From Equation (3.23) we conclude

$$\mathbb{P}[\tau_i < t \mid Y = y] = N\left[\frac{N^{-1}[p^{(t)}] - \sqrt{\varrho}\, y}{\sqrt{1-\varrho}}\right].$$

This immediately implies the assertion of the proposition. □

3.3.4 Proposition *Under the conditions of this section, for a homogeneous portfolio with a credit curve $(p^{(t)})_{t\geq 0}$ and uniform CWI correlation ϱ we obtain*

$$\mathbb{P}\left[\lim_{m\to\infty}\left[X_{PF(m)}^{(j)} - \mathbb{P}[(j-1)/4 \leq \tau_1 < j/4 \mid Y]\right] = 0\right] = 1,$$

where $Y \sim N(0,1)$. Recall that in this section, we consider quarterly payment periods, explaining the '/4' in Equations. The index '1' in 'τ_1' can be set because, due to our uniformity assumptions, obligors and, as a consequence, default times are exchangeable; see Proposition 3.3.3.

Proof. The proof is a straightforward modification of the argument provided in [25], pages 88-89, but for the convenience of the reader we provide the argument. The usual 'trick' to prove such results is to condition on the factor Y. We write $\mathbb{P}_y = \mathbb{P}[\,\cdot\, | \, Y = y]$ for the conditional probability measures. Fix $y \in \mathbb{R}$. Then, the random variables

$$Z_i^{(j)} = E_i^{(j)} \mathbf{1}_{\{j-1 \leq 4\tau_i < j\}} - \mathbb{E}[E_i^{(j)} \mathbf{1}_{\{j-1 \leq 4\tau_i < j\}} \, | \, Y]$$

$$(i = 1, ..., m; \; j = 1, ..., T)$$

are i.i.d. w.r.t. $\mathbb{P}_y$ and centered. For k denoting the number of names in the portfolio, the sequence $(E_{PF(k)}^{(j)})_{k=1,2,...}$ is strictly increasing to infinity due to Condition 3 for any (fixed) payment period $j = 1, ..., T$. Moreover, we have

$$\sum_{k=1}^{\infty} \frac{1}{(E_{PF(k)}^{(j)})^2} \mathbb{E}_y[(Z_k^{(j)})^2] \leq \sum_{k=1}^{\infty} \frac{1}{(E_{PF(k)}^{(j)})^2} 4(E_k^{(j)})^2$$

$$= 4 \sum_{k=1}^{\infty} \left(\frac{E_k^{(j)}}{E_{PF(k)}^{(j)}} \right)^2 < \infty.$$

due to Condition 4. Here, $\mathbb{E}_y[\cdot]$ denotes expectation w.r.t. the probability measure $\mathbb{P}_y$. A version of the strong law of large numbers based on Kronecker's Lemma (see, e.g., [16]) then implies that

$$\lim_{m \to \infty} \frac{1}{E_{PF(m)}^{(j)}} \sum_{i=1}^{m} Z_i^{(j)} = 0 \quad \mathbb{P}_y\text{-almost surely.}$$

From this we conlude for every $y \in \mathbb{R}$

$$\mathbb{P}[\lim_{m \to \infty} (X_{PF(m)}^{(j)} - \mathbb{E}[X_{PF(m)}^{(j)} \, | \, Y]) = 0 \, | \, Y = y] = 1.$$

Then, to prove almost sure convergence is straightforward by writing

$$\mathbb{P}[\lim_{m \to \infty} (X_{PF(m)}^{(j)} - \mathbb{E}[X_{PF(m)}^{(j)} \, | \, Y]) = 0] =$$

$$= \int \mathbb{P}[\lim_{m \to \infty} (X_{PF(m)}^{(j)} - \mathbb{E}[X_{PF(m)}^{(j)} \, | \, Y]) = 0 \, | \, Y = y] dN(y) = 1.$$

Proposition 3.3.3 implies that the conditional expectation $\mathbb{E}[X^{(j)}_{PF(m)} \mid Y]$ for $Y = y$ can be written as

$$\mathbb{E}[X^{(j)}_{PF(m)} \mid Y = y] = \frac{1}{E^{(j)}_{PF(m)}} \sum_{i=1}^{m} E_i^{(j)} \, \mathbb{E}[\mathbf{1}_{\{j-1 \leq 4\tau_i < j\}} \mid Y = y]$$

$$= \mathbb{P}\left[\frac{j-1}{4} \leq \tau_1 < \frac{j}{4} \,\Big|\, Y = y \right]$$

because due to our uniformity assumptions, obligors and default times are exchangeable. This completes the proof of the proposition. □

Note that the idea underlying the proof of Proposition 3.3.4 does not rely on $Y \sim N(0, 1)$. The same argument can be used to establish an analogous convergence result for other than normal distributions: *semi-analytic techniques can be applied to other than Gaussian copulas (e.g., Student-t copulas) too*. We finally come to the following conclusion.

3.3.5 Corollary *Under the conditions stated in this section, we have*

$$\mathbb{P}\left[\lim_{m \to \infty} \left(L^{(T)}_{PF(m)} - \sum_{j=1}^{T} w^{(j)} g_j(Y) \right) = 0 \right] = 1,$$

where the functions $g_j(\cdot)$ are defined on $\mathbb{R}$ by

$$g_j(y) = N\left[\frac{N^{-1}[p^{(j/4)}] - \sqrt{\varrho}\, y}{\sqrt{1 - \varrho}} \right] - N\left[\frac{N^{-1}[p^{((j-1)/4)}] - \sqrt{\varrho}\, y}{\sqrt{1 - \varrho}} \right].$$

Numbers $w^{(j)}$ refer to the limit exposure weights from Condition 2.

Proof. The assertion follows without difficulties from the previous two propositions and Condition 2. □

Corollary 3.3.5 provides the tool indicated at the beginning of this section: instead of simulating single-name default times, the *fraction of default times located in a considered payment period* is modeled, which is a much more efficient way of default simulation then modeling default times name by name. Corollary 3.3.5 proves that in the limit $m \to \infty$, this approach becomes exact.

The semi-analytic technique can be further developed and refined in practical applications in several directions, for example, stochastic recoveries could be implemented into the framework quite easily; see [25], pages 86-89, for an approach for a single-period model, which can be extended to a multi-period approach in a straightforward manner.

From a CDO modeling point of view, the tool developed in this section is quite powerful, because it allows for the implementation of all relevant cash flow elements, e.g., redirection of cash flows due to realized losses (as in Section 3.3.4) or other 'triggers' affecting the performance of CDO notes. This flexibility is a consequence of considering every single payment period (see Corollary 3.3.5), such that all of the specialties of a waterfall of a transaction can be implemented in an accurate way. A common example where semi-analytic techniques can be applied are residential mortgage backed securities (RMBS), where underlying reference pools[16] are typically perfectly suited for an approximation by a uniform portfolio, supporting the 'infinite granularity' assumption.

We pass on another example along the lines of the example from the previous section and the example in Section 3.3.3 and refer to [27] for an illustration of the semi-analytic approach. However, the *comonotonic* approach as described in the next section is closely related to the semi-analytic approach and in the following section we present two more examples for illustration purposes to keep the discussion 'applied' and not purely theoretical.

3.3.6.3 Comonotonic CDO Evaluation

A last simulation technique we want to explore, closely related to analytic and semi-analytic modeling approaches, is the *comonotonic copula* approach; see [28] as a reference. In a convenient way, the comonotonic copula approach can be combined with (semi-)analytic techniques in order to arrive at a *comonotonic approximation*.

Let us recall some notation before we get started. Consider a default times vector $\tau = (\tau_1, ..., \tau_m)$ for a portfolio of m credit-risky assets as usual. The vector τ determines the portfolio's *default quote path* $(L^{(t)})_{0 \leq t \leq T}$ until the maturity T of the transaction,

$$L^{(t)} = \frac{1}{m} \sum_{k=1}^{m} \mathbf{1}_{\{\tau_k \leq t\}}. \tag{3.29}$$

---
[16]One can find RMBS transactions in the market including several 10,000 loans in the reference portfolio.

The intertemporal dependence of jumps $L^{(t)} - L^{(s)}$ for times $s < t$, sampled at a discrete time grid

$$0 = t_0 < t_1 < \cdots < t_{q-1} < t_q = T$$

is hidden in the joint distribution of the default quote path evaluated at these times. According to SKLAR's theorem (see Theorems 2.5.6 and 2.5.7), the joint distribution of the time-discrete default quote path can be written by means of some copula function

$$C : [0,1]^q \to [0,1], \ (u_1, ..., u_q) \mapsto C(u_1, ..., u_q)$$

in the following form:

$$\mathbb{P}[L^{(t_1)} \leq x_1, ..., L^{(t_q)} \leq x_q] = \qquad (3.30)$$
$$= C(\mathbb{P}[L^{(t_1)} \leq x_1], ..., \mathbb{P}[L^{(t_q)} \leq x_q])$$

Note that the points x_i always can be written as k_i/m with $k_i \in \{1, ..., m\}$ and that in contrast to the case of continuous distributions the copula C is not uniquely determined by the joint distribution function of the default quote path.

For the following definition, recall that the *comonotonic* or *upper Frechet* copula is given by

$$C_\diamond(u_1, ..., u_m) = \min\{u_1, ..., u_m\}$$

for numbers $(u_1, ..., u_m \in [0,1])$; see Section 2.5.5.5.

3.3.6 Definition *A default quote path is said to have a comonotonic copula representation if it can be written in the form (3.30) with $C = C_\diamond$ being the comonotonic copula.*

Comonotonicity arises very naturally in the context of default quote paths. To see this, choose any credit portfolio model and determine the probabilities

$$p_{t,k} = \mathbb{P}[L^{(t)} = k/m] \qquad (3.31)$$

for a portfolio of m names where $L^{(t)}$ is defined in Equation (3.29). In Section 3.3.6.1 we find examples where we can define the probabilities (3.31) numerically, but in more complicated model situations, e.g., in the multi-factor setup described in Appendix 6.9, these likelihoods have to be determined by simulation techniques. Essential to us in this

238   *Structured Credit Portfolio Analysis, Baskets & CDOs*

section is only that these likelihoods are known. We will throughout this section assume that this is the case.

Given the probabilities (3.31) for a discrete time grid $t = t_1, ..., t_q$,

$$(p_{t,k})_{t=t_1,...,t_q;\ k=0,1,2,...,m} \in \mathbb{R}^{q \times (m+1)},$$

we write $[z] = \max\{k \in \mathbb{N}_0 : k \leq z\}$ for the greatest non-negative integer below $z \geq 0$ and define for any time $t_j > 0$ a distribution function

$$G_{t_j}(x) = \sum_{k=0}^{[mx]} p_{t_j,k} \qquad (x \in [0,1]). \qquad (3.32)$$

The distribution functions G_{t_j} are step functions due to the finite granularity m of the portfolio.

We now compare the original default quote path

$$\boldsymbol{L} = (L^{(t_1)}, ..., L^{(t_q)}) = \left(\frac{1}{m}\sum_{k=1}^{m} 1_{\{\tau_k \leq t_1\}}, ..., \frac{1}{m}\sum_{k=1}^{m} 1_{\{\tau_k \leq t_q\}}\right),$$

determined by the chosen portfolio model applied in Equation (3.29) with the default quote path

$$\tilde{\boldsymbol{L}} = (\tilde{L}^{(t_1)}, ..., \tilde{L}^{(t_q)}) = \left(G_{t_1}^{-1}(Z), ..., G_{t_q}^{-1}(Z)\right)$$

defined by the distribution functions (3.32) where $Z \sim U[0,1]$ is a random variable uniformly distributed in $[0,1]$ and

$$G_{t_j}^{-1}(z) = \inf\{q \geq 0 \mid G_{t_j}(q) \geq z\}$$

denotes the generalized inverse of G_{t_j}.

3.3.7 Proposition

1. The one-dimensional marginal distributions of $\boldsymbol{L}$ and $\tilde{\boldsymbol{L}}$ coincide.

2. $\tilde{\boldsymbol{L}}$ is comonotonic.

Proof. For the marginal distributions consider for $Z \sim U[0,1]$

$$\begin{aligned}
\mathbb{P}\left[G_{t_j}^{-1}(Z) \leq \frac{k}{m}\right] &= \mathbb{P}\left[Z \leq G_{t_j}\left(\frac{k}{m}\right)\right]\\
&= G_{t_j}\left(\frac{k}{m}\right)\\
&= \sum_{i=0}^{k} p_{t_j,i}\\
&= \mathbb{P}\left[L^{(t_j)} \leq \frac{k}{m}\right]
\end{aligned}$$

due to Equations (3.31) and (3.32). This proves the first part of the proposition. Comonotonicity of $\tilde{\boldsymbol{L}}$ can be seen as follows:

$$\begin{aligned}
\mathbb{P}[G_{t_1}^{-1}(Z) \leq x_{t_1}, ..., G_{t_q}^{-1}(Z) \leq x_{t_q}] &= \mathbb{P}[Z \leq G_t(x_t); \; t = t_1, ..., t_q]\\
&= \min\{G_t(x_t) : t = t_1, ..., t_q\}\\
&= C_\diamond(G_{t_1}(x_{t_1}), ..., G_{t_q}(x_{t_q}))
\end{aligned}$$

again due to the uniform distribution of Z in $[0,1]$. As in Section 2.5.5.5 and Definition 3.3.6, $C_\diamond$ denotes the comonotonic copula. □

The coincidence of the marginal distributions of $\boldsymbol{L}$ and $\tilde{\boldsymbol{L}}$ can be illustrated by a simple *simulation scheme* shown in Figure 3.19, here for a yearly time grid $t_1 = 1, ..., t_4 = 4$. Comonotonicity of $\tilde{\boldsymbol{L}}$ refers to the simulation of just one single random variable $Z \sim U[0,1]$. Partitioning the unit interval in appropriate pieces in order to reflect the definition (3.32) of the discrete distribution functions G_{t_j} yields, by construction, marginal default quote distributions exactly matching the original default quote distributions of $L^{(t_j)}$ at any time horizon t_j. Therefore, Proposition 3.3.7 indeed confirms our statement at the beginning of this section that *comonotonicity arises naturally in the context of portfolio default quote paths (independent of the chosen portfolio model)*. In a moment we will find that actually much more is true.

In previous sections, especially in Section 3.3.4, we discovered the importance of an accurate modeling of the *timing of defaults*. We now show that replacing the original copula C in Equation (3.30) by the comonotonic copula $C_\diamond$ not only preserves the original marginal default quote distributions (see Proposition 3.3.7) but also preserves the *timing of defaults* as induced by the original portfolio model. More explicitly, we show that the comonotonic default quote path $\tilde{\boldsymbol{L}}$ generated by means

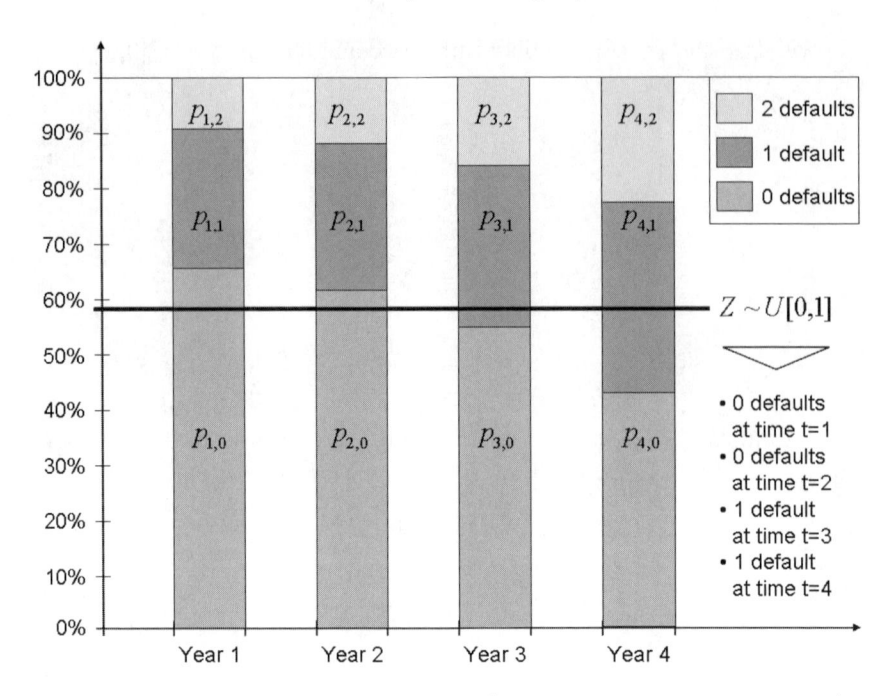

FIGURE 3.19: Simulation scheme ('**comonotonic approach**') of default quote paths based on comonotonicity; see [28]

of a simulation scheme as illustrated in Figure 3.19 and the original default quote path $\boldsymbol{L}$ are perfectly consistent in distribution from an 'nth to default' timing perspective.

To work this out, we need some more notation. Slightly deviating from our notation of *order statistics* in Section 2.6.4, we now denote by $\tau_{n:m}$ the time until the nth default out of m obligors w.r.t. the original default quote path $\boldsymbol{L}$,

$$\tau_{n:m} = \inf\{t \geq 0 \; : \; L^{(t)} \geq n/m\}.$$

In the comonotonic setup we write $\tilde{\tau}_{n:m}$ for the time of the nth default corresponding to the comonotonic default quote path $\tilde{\boldsymbol{L}}$,

$$\tilde{\tau}_{n:m} = \inf\{t \geq 0 \; : \; G_t^{-1}(Z) \geq n/m\} \qquad (Z \sim U[0,1]).$$

The next proposition shows that the distributions of $\tau_{n:m}$ and $\tilde{\tau}_{n:m}$ coincide, which gives us a very powerful simulation tool.

Collateralized Debt and Synthetic Obligations

3.3.8 Proposition *The distributions of the nth default time for default quote paths L and $\tilde{L}$ agree,*

$$\mathbb{P}[\tau_{n:m} \leq t] = \mathbb{P}[\tilde{\tau}_{n:m} \leq t] \quad \text{for all } t \geq 0.$$

Proof. We start with the comonotonic path $\tilde{L}$ and define

$$q_{t,n:m} = \sum_{k=0}^{n-1} p_{t,k} = \sum_{k=0}^{n-1} \mathbb{P}[L^{(t)} = k/m] = \mathbb{P}\left[L^{(t)} \leq \frac{n-1}{m}\right]$$

for $t \geq 0$ and $n = 1, ..., m$.

Note that $t \mapsto q_{t,n:m}$ is a decreasing function; see also Figure 3.19. Based on Proposition 3.3.7 and Figure 3.19, observing n out of m defaults at time t in a comonotonic simulation is equivalent to

$$q_{t,n:m} = G_t\left(\frac{n-1}{m}\right) < Z \leq G_t\left(\frac{n}{m}\right) = q_{t,(n+1):m}. \quad (3.33)$$

Therefore, the time until the nth default must equal the first time t such that $Z > q_{t,n:m}$ for $Z \sim U[0,1]$,

$$\tilde{\tau}_{n:m} = \inf\{t \geq 0 \,:\, Z > q_{t,n:m}\} = \inf\{t \geq 0 \,:\, G_t^{-1}(Z) \geq n/m\}.$$

We now come to the path L. In case of $\tau_{n:m} \leq t$ at least n out of m defaults must have occurred before or at time t. The likelihood for n or more defaults in the time interval $[0,t]$ is given by

$$\mathbb{P}[\#\text{defaults} \in \{n, n+1, ..., m\} \text{ until time } t] =$$

$$= \sum_{i=n}^{m} \mathbb{P}[L^{(t)} = i/m] = \sum_{i=n}^{m} p_{t,i} = 1 - q_{t,n:m}.$$

Therefore, we finally obtain

$$\mathbb{P}[\tau_{n:m} \leq t] = \sum_{i=n}^{m} p_{t,i} = 1 - q_{t,n:m} = \mathbb{P}[Z > q_{t,n:m}] = \mathbb{P}[\tilde{\tau}_{n:m} \leq t]$$

which is the conclusion we need. $\square$

What can we do with Proposition 3.3.8? Now, it says that the distribution of the nth default time can be simulated by a comonotonic approach according to Figure 3.19, *no matter what underlying portfolio model we decided to work with*. There is a very natural condition, which makes this insight broadly applicable:

If all assets in the portfolio have the same amount of exposure at risk, then there is no loss of information arising from a comonotonic approach because based on Proposition 3.3.8 the comonotonic approach yields full information regarding the time grid of default occurrences.

In case of such lucky situations we speak of *exchangeable exposures*, essentially meaning that all assets cause the same loss amount in case of default. This guarantees that we do not have to be bothered with the question *'which particular asset is defaulting at which particular time'* because all assets generate the same amount of loss in any way. Fortunately, the condition of 'exchangeable exposures' can be assumed to be fulfilled in typical collateralized synthetic obligations and default baskets; see the examples in Chapters 2 and 3. Therefore, the comonotonic approach is perfectly suited for such instruments. In 'classical' CDOs, e.g., loan securitizations, the situation is different. Here, the exposure distribution most often is inhomogeneous, although most offering circulars explicitly *do not allow for significant single obligor concentrations* (in terms of lending amount allocated to a single name). Therefore, the comonotonic approach can serve as an *approximation tool* even in such cases; see also the example at the end of this section.

Major benefits we find in the comonotonic setup are

- Simulation efficiency

- Variance reduction

The order of magnitude of both benefits depends on the originally chosen model but even if the original model is a simple one-factor model the benefits are still measurable; see the corresponding remarks in [28].

For illustration purposes we now look at two examples. The first example is a duo basket simulation (in the spirit of Chapter 2) demonstrating the variance reduction and simulation efficiency benefits inherent in the comontonic copula approach. The second example deals with the CSO model from in Section 3.3.1. In this example, we calculate the probabilities $p_{t,k}$ from Equation (3.31) and make a few comments regarding *comonotonic approximations*.

Collateralized Debt and Synthetic Obligations 243

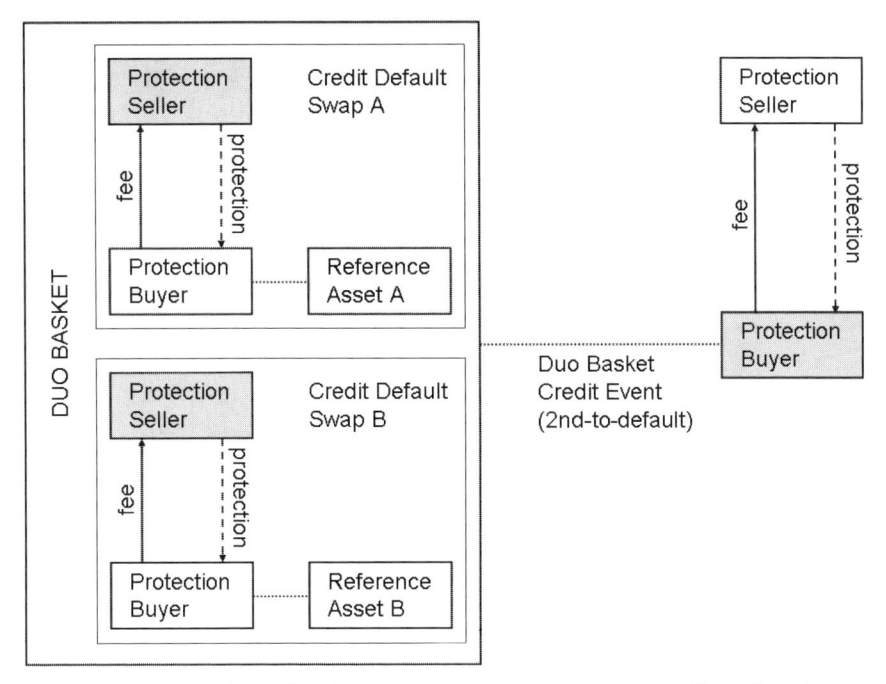

FIGURE 3.20: Duo basket transaction; see [28] as well as Section 2.1

For our first example,[17] we consider a 'duo basket' (see also Section 2.1), this time consisting of the following two assets:

- Instrument A: 5-year CDS written w.r.t. a name with a one-year default probability of $p_A = 8$ bps with a volume of 10 mn EUR

- Instrument B: 5-year CDS written w.r.t. a name with a one-year default probability of $p_B = 36$ bps with a volume of 20 mn EUR

with an assumed CWI correlation of 15% in a Gaussian copula.

The duo basket transaction we have in mind is sketched in Figure 3.20. We sell protection on names A and B but at the same time buy protection against a basket credit event defined by the second-to-default of the basket. Our position is shaded in grey in Figure 3.20. In other words, we are (synthetically) exposed to the default risk of two names but protected against the *joint* default of the two names. Based on the moderate asset correlation of names A and B we can expect a

---
[17]Taken from [28].

moderate premium in our protection buying agreement (at the right-hand side in Figure 3.20), whereas in our protection selling agreement (at the left-hand side in Figure 3.20) we can expect to collect two *single-name* fees driven by the default risk of assets A ($p_A = 0.0008$) and B ($p_B = 0.0036$) in the two CDS we have written for names A and B. For reasons of simplicity, we assume a fixed LGD of 60% for both names and a maturity of the basket in $T = 5$ years. Figure 3.21 shows the comparison of a full Monte Carlo simulation of correlated default times with a comonotonic evaluation of the two assets in the duo basket.

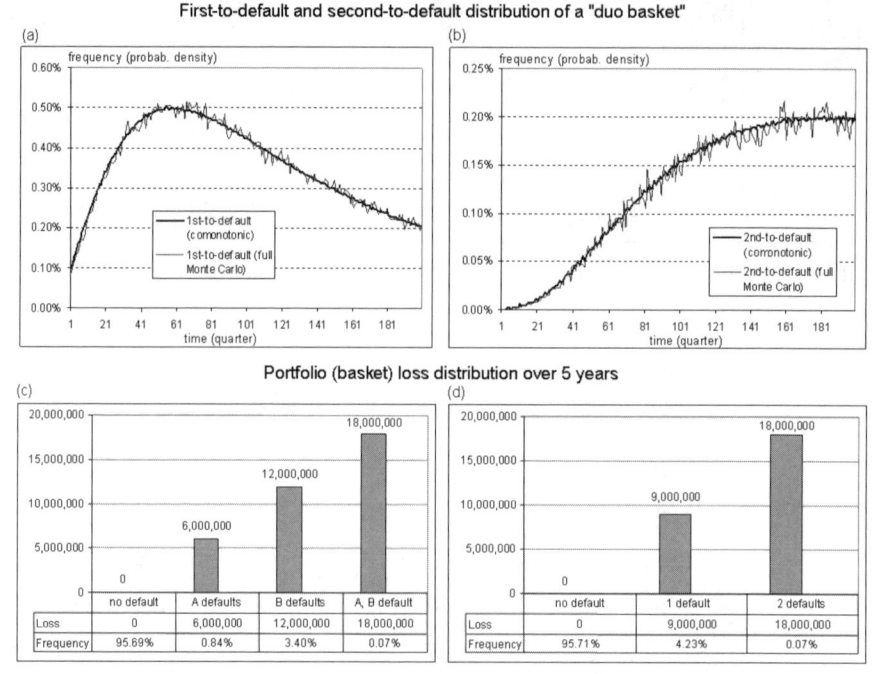

FIGURE 3.21: Comparison of full Monte Carlo with comonotonic simulation approaches: efficiency, variance reduction, and loss of information in case of non-exchangeable exposures, illustrated by means of the duo basket transaction from Figure 3.20; see also [28]

In Figure 3.21, (a) and (b) clearly illustrate the *variance reduction* effect arising from comonotonicity. Note that the full simulation approach is based on *200,000 scenarios of three variables* (one systematic factor and two idiosyncratic effects), whereas the comonotonic approach is based on *50,000 scenarios of just one uniformly distributed random variable* as usual in the comonotonic copula approach.

In Figure 3.21, (c) and (d) illustrate the effect of *non-exchangeable exposures*. The comonotonic approach works fine as long as for both names we have the same amount of exposure at risk. As soon as different obligors cause different losses in case of default, the comonotonic approach *loses the information 'who' is defaulting* although the nth to default distributions still coincide due to Proposition 3.3.8. In case of non-exchangeable exposures it is impossible, based on one single random variable, to correctly address the *multinomial character of non-exchangeable exposures* by means of a comonotonic approach. Therefore, for the comonotonic approach we have to make an assumption regarding the amount that is lost in case of the first and second default. Because in the comonotonic setup we do not have this information at our disposal we decide for equal 'exposures at default' (see Figure 3.21 (d)) for both assets. Alternatively, we could have assumed that the exposure of the asset with higher default probability always constitutes the first loss. No matter what assumptions we make, we have to live with a lack of information in the comonotonic approach and can only try to overcome this information gap by subjective assumptions leading to an *approximation* of the truth.

As a second example, let us again consider the CSO modeled in Section 3.3.3. First of all, we focus on the default quote probabilities from Equation (3.31)

$$p_{t,k} = \mathbb{P}[L^{(t)} = k/m] \qquad (t \geq 0; \ k = 0, 1, 2, ..., m).$$

For the sake of an illustration, we calculate these probabilities out of the Monte Carlo simulation of the CSO in order to see how they look like. Figure 3.22 shows the corresponding picture for $k = 0, 2, 3, 5$.

Now, in *cash flow* transactions as in case of the CSO from Section 3.3.1, the condition of *exchangeable exposures* is not sufficient to guarantee full compliance of the comonotonic with the full Monte Carlo simulation, assuming a correct parameterization. The reason is that *although the reference assets are exchangeable from an exposure point of view they are not exhangeable regarding the spread payment contribution to the interest stream;* see Figure 3.11. We have two related effects making the comonotonic approach a little more tricky to apply:

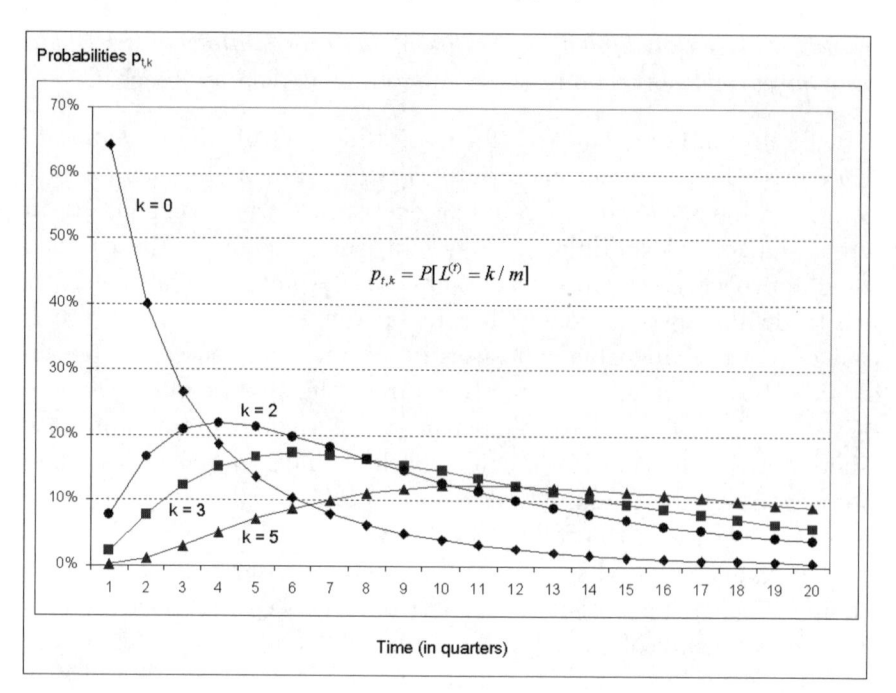

FIGURE 3.22: Illustration of the probabilities $p_{t,k}$ from Equation (3.31); calculated for the reference portfolio underlying the CSO from Section 3.3.1; see also Appendix 6.9 for a portfolio description

- Spreads as illustrated in Figure 3.11 have a clear tendency to strongly increase with decreasing credit quality
- Names with lower credit quality default earlier on average than names with better credit quality

The superposition of both effects has consequences for the *interest stream* of the CSO: we can expect a clear trend that spread payments from CDS premium legs on average will decline over the lifetime of the transaction. So for an application of the comonotonic approach where we know *when* defaults occur but not *which* names are defaulting we have to find a proxy for the expected decline of the stream of spread payments feeding the interest stream of the CSO. An easy way to work on this problem is the following approach.

Given the PD term structures $(p_R^{(t)})_{t\geq 0}$ assigned to rating classes R, we can define (cumulative) survival probabilities

$$1 - p_R^{(t)}$$

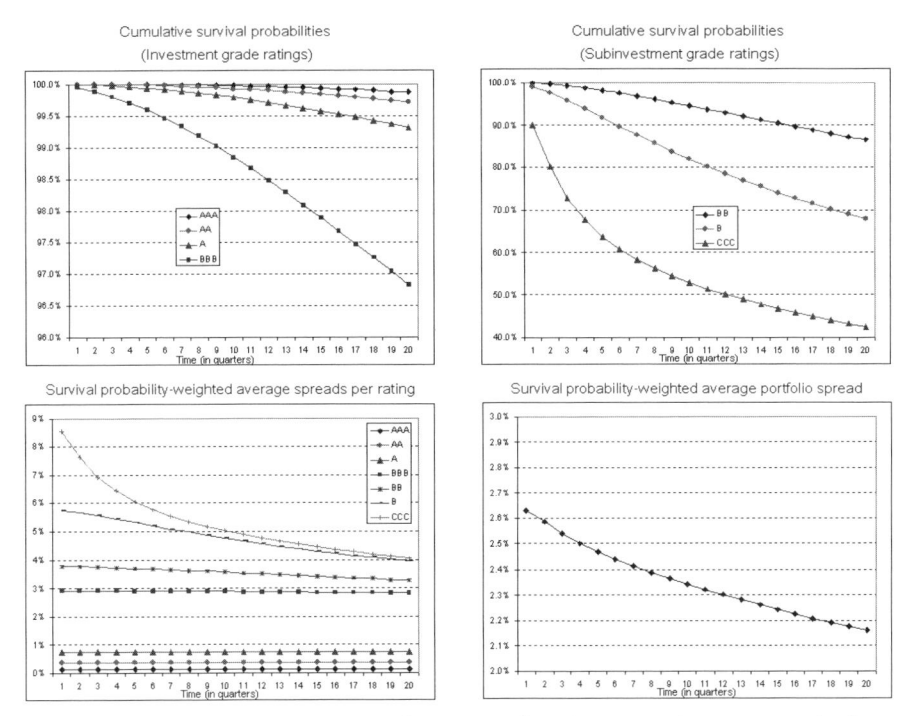

FIGURE 3.23: Illustration of a proxy for the spread payment decline applied in a comonotonic approximation

for every rating class R w.r.t. a given time horizon t. Figure 3.23 graphically illustrates the survival probabilities over time for the 7 performing rating clases in our example. Next, we can use Figure 3.11 to calculate the average spread for all names in the 100-names reference portfolio carrying a given rating R. Figure 3.23 at the lower left-hand side shows the so generated survival probability-weighted average spread curves per rating class. In a last step, we aggregate these curves into a single portfolio average spread curve, taking the rating distribution of the reference portfolio into account; see Figure 6.4. Figure 3.23 at the lower right-hand side shows the results of this exercise. On average we can expect to lose almost 20% of spread income based on early defaults of heavy spread payment contributors in the reference portfolio.

Calculations like the determination of an average spread curve for the portfolio change the comonotonic approach from an *exact* modeling approach to a comonotonic *approximation*. Another approximation we can make is a replacement of the 'true' probabilities $p_{t,k}$ from (3.31) by

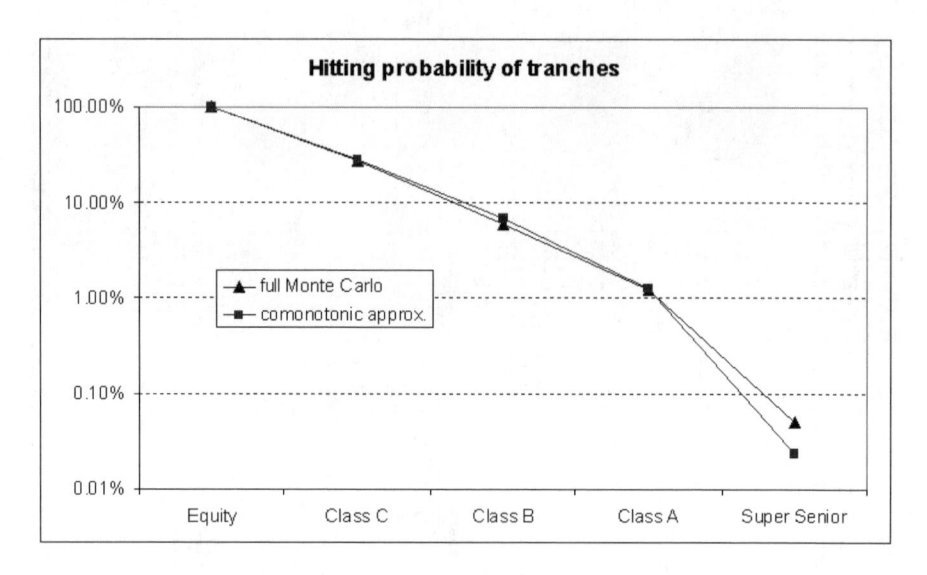

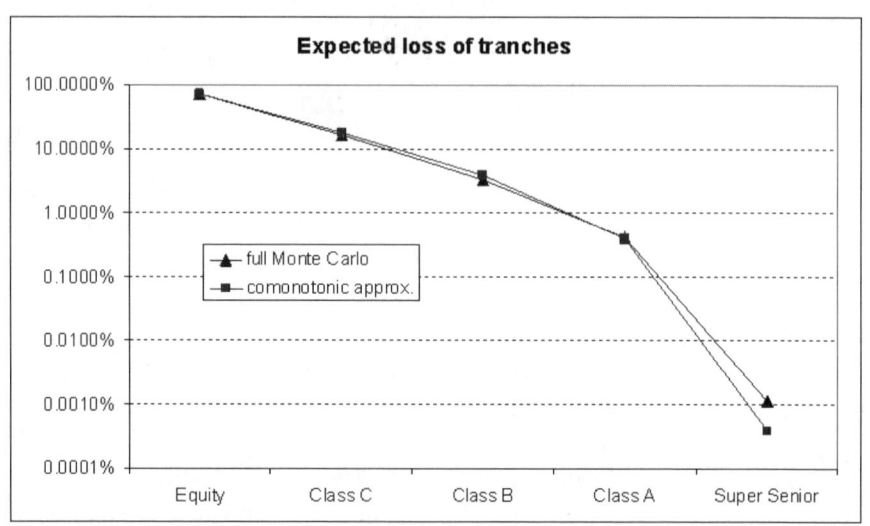

FIGURE 3.24: Comparison of tranche PDs and ELs derived by full Monte Carlo simulation versus comonotonic approximation in CSO evaluation; see Section 3.3.3 for the full Monte Carlo CSO evaluation

LGD of tranches

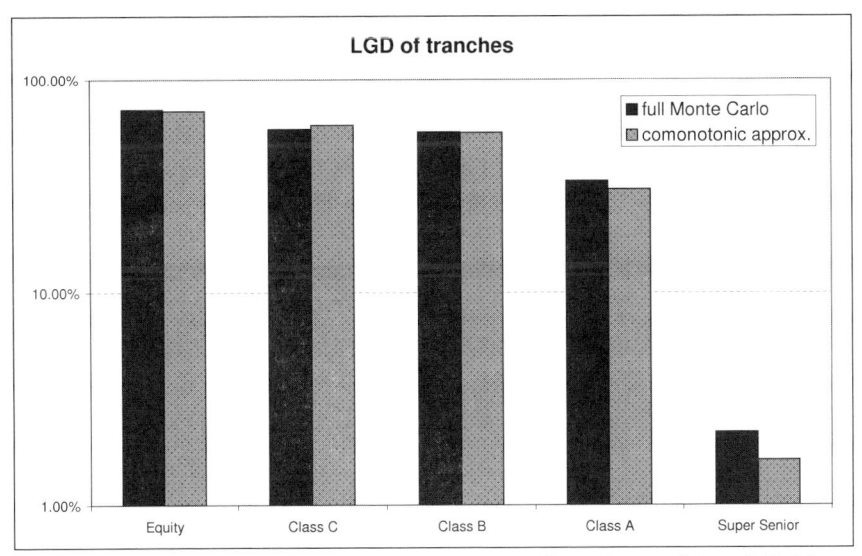

FIGURE 3.25: Comparison of tranche LGDs derived by full Monte Carlo simulation versus comonotonic approximation in CSO evaluation; see Section 3.3.3 for the full Monte Carlo CSO evaluation

easier-to-derive proxies like

$$p_{t,k} = \binom{m}{k} \int_{-\infty}^{\infty} g(\overline{p}^{(t)}, \varrho; y)^k \left(1 - g(\overline{p}^{(t)}, \varrho; y)\right)^{m-k} dN(y) \qquad (3.34)$$

where $\varrho = 6.5\%$ is chosen as a uniform CWI correlation and $(\overline{p}^{(t)})_{t \geq 0}$ is a uniform credit curve representing the average PD term structure of the reference portfolio; see Appendix 6.8 for an extensive discussion on such *analytic approximations*. If we then exercise a comonotonic simulation based on (3.34) and Figure 3.23, we obtain Figures 3.24 and 3.25 for the base case simulation (no triggers in place), which compares to the Gaussian case in Table 3.1. We find that, except for the super senior tranche where we see a few basispoints deviation, the comonotonic approximation yields quite satisfying results regarding the tranche risk statistics; see also [27] and [28].

A major advantage compared to analytic approximations as explained in Section 3.3.6.1 is the *flexibility* in the semi-analytic and comonotonic approach to model every single payment period such that all relevant cash flow elements can be implemented. For instance, with the semi-analytic approach we can very well model the excess cash trap situation elaborated in Section 3.3.4, whereas the analytic approach does not incorporate sufficient flexibility to model such deviations from plain vanilla tranched structures. Nevertheless, as we have seen in the context of Figure 3.23, proxies for the average of *non-exchangeable* input parameters like heterogeneous spread distributions etc. have to be constructed in order to apply the comontonic approach in a meaningful way as an approximation tool.

However, there are situations in which only aggregated information is available, for instance, in CDO *presale reports*. In such cases, the comonotonic approach can be quickly implemented in an efficient way to provide information on the performance of a considered CDO.

3.4 Single-Tranche CDOs (STCDOs)

In this section, we focus on so-called *single-tranche CDOs* (STCDOs). A main subclass of STCDOs are *index tranches* referenced to an index (or portfolio) of liquid CDS. This gives us the opportunity to briefly explain the mechanics of iTraxx Europe and CDX.NA.IG (see Section 3.4.2, which are the two most actively traded CDS indices in the market today. Index tranches in their role as important examples for STCDOs will also lead us to the topic of *implied correlation*, a concept comparable to *implied volatility* in the *Black & Scholes* option pricing framework. Index tranches are traded w.r.t. special quoting standards arising from the need to *delta hedge* mark-to-market fluctuations of credit positions in the considered index and in certain parts of the (tranched) capital structure of the index. The mechanism of delta hedging will be explained in the sequel.

3.4.1 Basics of Single-Tranche CDOs

STCDOs constitute a class of structured credit products exhibiting growing market share and importance. The idea behind STCDOs is best explained by means of a description of the typical three-step negotiation process in a STCDO deal.

Assume an investor wants to engage in a short or long position in a special class of credit risks, for instance, in investment grade European names. In addition, the investor has certain restrictions regarding the worst rating he is allowed to invest in. He wants the transaction to be unfunded so that he has no funding costs. In order to get such a *tailor-made* transaction implemented, the investor contacts the CDO trading desk of an investment bank in order to start negotiations on a STCDO suiting his needs and meeting the imposed conditions. The negotiation process typically has three steps.

1. Step. **Selection of reference names:**
   As mentioned above, the investor has individual preferences regarding the underlying reference names of the transaction. The investment bank he talks to picks up the wishes of their client and offers a range of European investment grade names the investor can choose from. Criteria playing a role are typically the regional and industrial diversity of the reference pool, possibly supplemented by certain 'include' or 'not include' votes of the client regarding specific names he has a special opinion or investment strategy on.

2. Step. **Choice of capital structure:**
   In a second step, the investor negotiates an attachment and detachment point for the STCDO with the investment bank he talks to. The attachment point determines the amount of subordinated capital, which absorbs potential losses in the reference pool before losses eat into the tranche of the investor. The amount of subordinated capital strongly impacts the hitting probability and, as a consequence, the rating of the STCDO. The detachment point then determines the thickness of the tranche and, as we know from previous sections, has a significant influence on the LGD of the tranche. It often is the case that the investor has certain preferences or contrains regarding the risk he is allowed to or wants to invest in. Attachment and detachment points are major triggers of the risk profile of the tranche, besides the risk profile of the reference portfolio.

3. Step. **Determination of the maturity of the STCDO:**
   The last step is to agree in a maturity of the tranche. Typical maturities include 3, 5, 7, and 10 years, but potentially every maturity is thinkable. The maturity combined with the risk profile of the reference names determines the so-called risky duration of the reference pool, a notion we will explain later in the text.

In our example, the investor wants an unfunded transaction, which is a pure derivative- or basket-type transaction. However, in Section 2.7, we explained how to structure a CLN like a basket swap but with eliminated counterparty risk. In some cases, such a funded analogon of a synthetic STCDO is preferred by investors due to various reasons. STCDOs can be found funded and unfunded in the market. The flexibility for the investor in a STCDO inherent in the structuring process described above is the reason why STCDOs and related instruments are also called *bespoke CDOs*.

In our example, the investment bank who arranges the deal upon request of the investor, has to deal with the following challenge. In a standard CDO transaction, *the whole capital structure of the CDO is in the market so that losses can be routed through the capital structure 'bottom-up' through tranches* as discussed in many places in this book. For a STCDO, the situation is different. The transaction is tailor-made for meeting one particular investor's preferences, *covering only one part of the overall capital structure*, namely the tranche between attachment and detachment points of the STCDO. Other parts of the capital structure remain with the investment bank who arranges the deal. This situation generates a natural demand for hedging. In Section 3.4.4, we will explain how mark-to-market fluctuations of index tranches are *delta hedged*, which is a good example for hedging strategies in the context of STCDOs. The need for hedging in STCDOs is one of the reasons why reference names in STCDOs typically are *highly liquid* names; see also our remarks in Section 3.4.2.

Another discussion going on in the context of STCDOs among market participants is the allowance for a certain *management of the reference names* in the underlying pool. In case of standardized index tranches, the situation is clearly defined, but in deals where the underlying reference pool deviates from a standardized index, the decision if a STCDO should be *managed* or not leads to interesting, sometimes controversial, discussions. There are managed STCDOs in the market and it can be expected that we will see more of it in the next years. As always in managed CDO structures (even in standard CLOs with replenishment), managing of reference names has to be in line with clearly specified eligibility and management criteria. In addition, it is natural to let the investor in a STCDO participate in the management of the pool, again in compliance with strict rules and guidelines. Managed STCDOs offer an even greater flexibility than STCDOs exhibit anyway.

Value drivers for STCDO are spreads, interest rates, recovery rates, model assumptions, and so on; see also our remarks in Sections 3.4.4 as well as Figures 3.37, 3.38 and 3.39, and so on. STCDOs can be traded, and in case of standardized STCDOs like index tranches on liquid CDS indices (see Section 3.4.4), the secondary market for certain STCDOs works very well.

We close our general discussion on STCDOs here and turn our attention to *index tranches*, which constitute an important subclass of STCDOs. Index tranches provide a certain standardization of STCDOs but still keep a fair amount of flexibility for investors and investment banks. Standardization boosts liquidity so that investors and banks might prefer slight restrictions of flexibility rewarded by a significantly higher liquidity of the structured credit instrument.

3.4.2 CDS Indices as Reference Pool for STCDOs

If one wants to boost active trading of STCDOs, two major conditions help enormously to achieve this goal:

- Underlying names in the reference portfolio of the STCDO should exhibit *highest possible liquidity*

- The CDO tranches should be part of *highly standardized* and not too complicated CDO structures

The combination of both conditions guarantees a reasonable level of transparency and comparability of trading products in the same way as standardization elements in ISDA[18] agreements helped a lot in developing a liquid market of credit derivatives; see our comments below.

Both conditions mentioned above are met in case of so-called *CDS index tranches*. Underlying reference names in index tranches are typically taken from a *CDS index*. For the convenience of the reader, Figure 3.26 recalls the basic definition of a CDS contract in which one party (the *protection buyer*) pays a periodic fee (the *premium*; typically on a quarterly basis) to the other party (the *protection seller*) who has the obligation to compensate the protection buyer for credit losses on the underlying reference entity in case of a *credit event*[19];

---
[18] www.isda.org
[19] Credit events typically include bankruptcy, failure to pay, and restructuring.

254 *Structured Credit Portfolio Analysis, Baskets & CDOs*

```
                    Periodic fee (premium)
 ┌──────────┐ ─────────────────────────────▶ ┌──────────┐
 │Protection│                                 │Protection│
 │  buyer   │                                 │  seller  │
 └──────────┘ ◀ ─ ─ ─ ─ ─ ─ ─ ─ ─ ─ ─ ─ ─ ─ ─└──────────┘
                     Credit protection
```

Credit event triggered by
- bankruptcy
- failure to pay
- repudiation
- material debt restructuring

Reference name

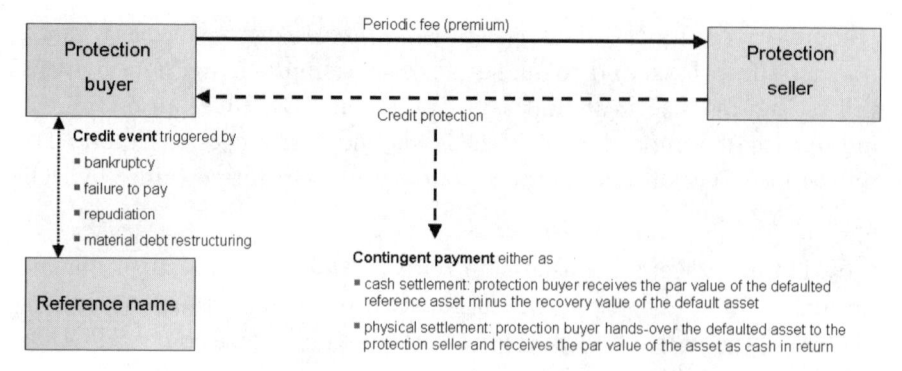

Contingent payment either as
- cash settlement: protection buyer receives the par value of the defaulted reference asset minus the recovery value of the default asset
- physical settlement: protection buyer hands-over the defaulted asset to the protection seller and receives the par value of the asset as cash in return

FIGURE 3.26: Single-name credit default swaps (CDS)

see also previous sections in this book. CDS contracts are typically standardized via ISDA definitions ruling the mechanism of common types of credit derivatives. Standardization and the fact that CDS are *unfunded/synthetic* so that no principal payments have to be made created a liquid CDS trading market referring to a specified universe of credit risk bearing names with different ratings risk profiles, ranging from investment to subinvestment grade.

The basic idea of CDS index tranches is to collect liquid CDS into a pool ('the index') and use the pool as underlying reference portfolio for a CSO. Hereby, index contracts typically consider 'bankruptcy' and 'failure to pay' as credit events triggering a default. Restructuring and other credit events sometimes used in credit derivatives are not necessarily considered[20]. In case of a credit event, the defaulted asset is removed from the portfolio or index and the basket CDS contract continues but now with a tranche notional reduced by defaulted exposures. Figure 3.27 illustrates the mechanism for the index CDX.NA.IG described below. It shows tranches with specified attachment and detachment points. This reflects part of the *standardization* of index tranches. For instance, index tranches on CDX.NA.IG are tranched according to the following strikes:

---

[20]Note that iTraxx follows more or less the standard CDS terms (e.g., credit events include bankruptcy, failure to pay, and restructuring, whereas CDX.NA.IG, as it is the case for all CDS indices in the United States, does not consider restructuring as a credit event. In other words, European indices typically trade w.r.t. the same credit events as the underlying single-name CDS, whereas in the United States there can be differences, sometimes leading to certain spread discounts.

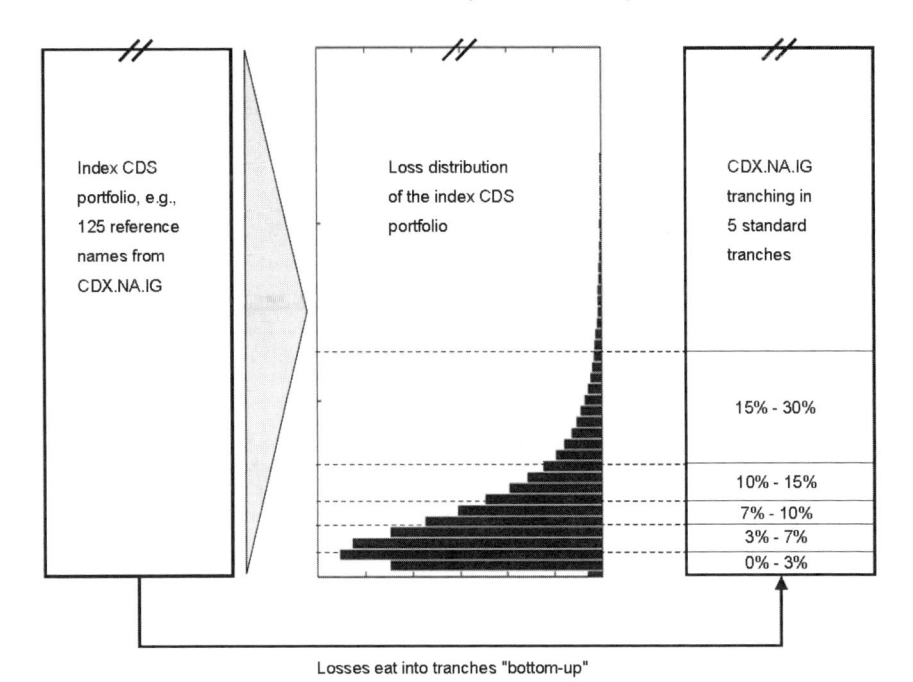

Collateralized Debt and Synthetic Obligations

FIGURE 3.27: CSO index tranches on the index CDX.NA.IG

- 0% to 3%
- 3% to 7%
- 7% to 10%
- 10% to 15%
- 15% to 30%

Other indices like iTraxx Europe (see below) have different but comparably standardized tranche levels for corresponding index tranches. As we will discuss later, index tranches provide a certain flexibility for tailored short and long positions in credit risk. Index tranches also initiated an increased awareness of *correlations* (see, e.g., Figure 3.37 and the corresponding discussion) as the most basic but already far-reaching concept of credit risk dependence. Market participants interested in index tranches include a whole range of investors including banks of all sizes, pension funds, and insurance companies. CDX index tranches significantly contribute to the steadily growing market of STCDOs as mentioned in Section 3.4.

TABLE 3.4: Dow Jones CDX Indices; see www.djindexes.com and [38]

Region	Index Name	Index Notation	No. of Names
North America	Investment Grade	CDX.NA.IG	125
	Investment Grade - High Volatility	CDX.NA.IG.HVOL	30
	High Yield	CDX.NA.HY	100
	High Yield - BB	CDX.NA.HY.BB	depends
	High Yield - B	CDX.NA.HY.BB	depends
	Crossover	CDX.NA.XO	35
Emerging Markets	Emerging Markets	CDX.EM	14
	Emerging Markets - Diversified	CDX.EM.DIVERSIFIED	40

Let us now look at CDS indices before we move on to tranched indices. We already mentioned CDX.NA.IG and iTraxx Europe as well-known examples. As a historical remark in the context of the iTraxx index family note that two earlier index families (Trac-x and iBoxx), initially brought to the market in 2003, were merged and consolidated in 2004 and are now under the label of *Dow Jones*.[21] The new entity created out of the merger of Trac-x and iBoxx is the *International Index Company* (IIC).[22] The process of index creation and the process of index maintenance including substitution of names is accompanied by a consortium of global investment banks who are active players in the CDS market. Substitutions and replacements steer the breakdown of index portfolios in industry classes, countries, and risk categories, hereby following strict rules and guidelines. Main criteria of 'inclusion eligibility' of a CDS into an index are (besides others) *liquidity* and *trading activity*, which is closely tracked by the market making derivative dealers/investment banks. Indices are typically reviewed and modified periodically (the so-called 'roll', typically every 6 months) in order to keep them updated regarding market developments.

Dow Jones and iTraxx CDS indices are provided w.r.t. different regions and subclasses. To give an example, Table 3.4 provides an overview on the Dow Jones CDX index family; see also [38]. For instance, the table shows that CDX.NA.IG contains 125 investment grade names in the index CDX.NA.IG, which actually is the most actively traded index of the CDX index. The 30 most volatile names out of the 125 names in CDX.NA.IG are collected into a subindex CDX.NA.IG.HVOL where 'HVOL' stands for 'high volatility'. An-

---
[21] See www.djindexes.com
[22] See www.intindexco.com

Collateralized Debt and Synthetic Obligations 257

other CDX index containing more risky names is the high yield index CDX.NA.HY from which all BB- and B-rated names are grouped into subindices CDX.NA.HY.BB and CDX.NA.HY.B, respectively. The number of names in these indices, therefore, always depends on the respective number of BB- and B-rated names in CDX.NA.HY. The two CDX emerging market indices are given by CDX.EM, which contains 14 sovereign entities, and CDX.EM.DIVERSIFIED which contains 40 names consisting of 28-32 sovereigns and 8-12 corporate entities. The first index in Table 3.4, CDX.NA.IG, is in its 5th series in the market and is the most actively traded index in the CDX family. It consists of a portfolio of 125 names with equal weighting such that each name contributes 0.8% to the index. Table 6.4 in Appendix 6.10 shows a list of all of the 125 names included in the index.

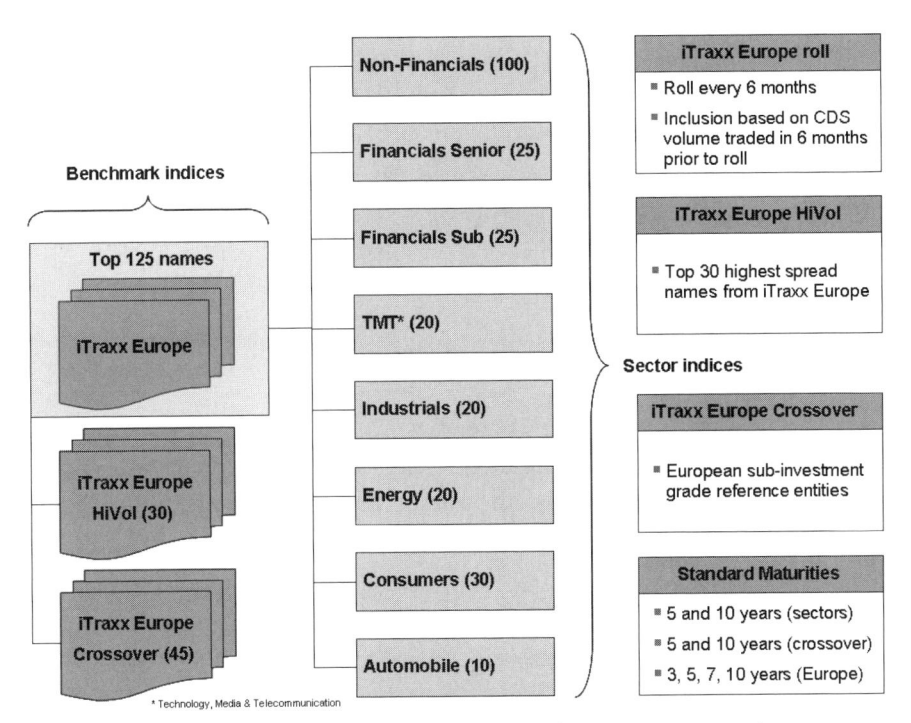

FIGURE 3.28: The iTraxx index family (source: iTraxx documentation from the IIC website www.intindexco.com)

A similar summary is provided in Figure 3.28 for the iTraxx index family (numbers in brackets refer to the number of names in the respective (sub)index). Here, the most actively traded index is the *iTraxx Europe*, which will be explained below. Analogously to CDX.NA.IG, every name in iTraxx Europe is equally weighted by 0.8%.

There are several products in the market derived from iTraxx indices. We focus on *iTraxx based index tranches* because these products are the most relevant for the topic of this book.

From now on, we focus exclusively on iTraxx Europe as an example for a common CDS index. Table 6.5 in Appendix 6.10 lists the 125 names in the 5th series of the iTraxx Europe CDS index. It contains names from a blend of different industries whereof

- Banking and finance
- Utilities
- Telecommunications
- Automobiles
- Food and beverage

constitute the top-5 most heavy industries in the index. The index is also diversified w.r.t. various European countries whereof

- United Kingdom
- Germany
- France

are the countries contributing most. The iTraxx Europe functions as a *master index* for certain sub- and sector indices as indicated in Figure 3.28.

In the same way as CDX.NA.IG, iTraxx Europe serves as reference portfolio for a series of actively traded and standardized CSO tranches with the following tranche boundaries:

- 0% to 3%
- 3% to 6%

- 6% to 9%
- 9% to 12%
- 12% to 22%

Loss allocation (the *default leg* of the deal) to these tranches and premium payments (the *premium leg* of the transaction) are straightforward and follow the usual pattern of plain-vanilla CSO products:

- Let us assume party A buys protection from party B on an iTraxx single tranche with attachment point α and detachment point β.

- In every payment period, the protection buyer A pays a premium of X to the protection seller B. If during the lifetime of the transaction none of the reference entities in iTraxx Europe defaults, the premium is paid on the full notional agreed in the contract in every payment period until the maturity of the transaction.

- However, if sufficiently many reference names in iTraxx Europe default such that the total realized loss (net of recoveries) exceeds the attachment point α of the considered tranche, then the realized loss is allocated to the protection seller B up to a maximum amount defined by the detachment point β of the tranche.

- Defaulted names are removed from the reference portfolio for that deal. If losses eat into the considered tranche, premium payments from A to B are applied to the reduced tranche outstanding.

Overall we find that the tranche loss profile function

$$\Lambda_{\alpha,\beta}(L^{(t)}) = \frac{1}{\beta - \alpha} \min\left[\beta - \alpha, \max(0, L^{(t)} - \alpha)\right] \qquad (3.35)$$

from Equations (3.1) and (3.25) can be applied in this context because STCDOs, indended to be standardized and simple, are just based on tranched loss distributions and do not involve complicated cash flow waterfalls as cash flow CDOs.

3.4.3 ITraxx Europe Untranched

In the following sections, we want to describe the basic evaluation principles underlying iTraxx Europe based index tranches. For this

260   *Structured Credit Portfolio Analysis, Baskets & CDOs*

purpose, we include several numerical examples based on real spread quotations from the first week of April in 2006. However, for the sake of a more accessible presentation we rely on several simplifying assumptions, which will be explicitly mentioned to keep readers aware of certain complications like market and quoting conventions as well as maturity mismatches and other effects making it difficult to replicate market prices.

In the sequel, *the word 'index' always refers to iTraxx Europe* (Series 5). We start our exposition by focussing on the index itself before we become more specific and consider iTraxx Europe based STCDOs.

CDS 5-year mid-spreads for iTraxx names (1st week of April 2006)

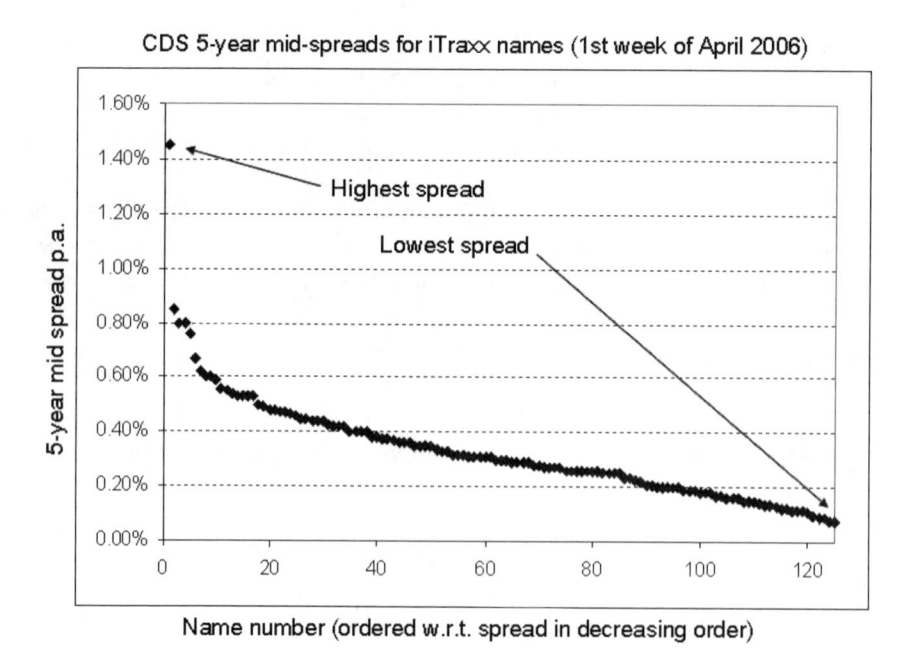

FIGURE 3.29: CDS spreads for the iTraxx Europe portfolio Series 5 (quotations from the 1st week of April)

Figure 3.29 shows 5-year mid-spreads[23] (mean of bid and ask) of the 125 names constituting the index as of the first week in April 2006.

---
[23] Because the index contains the most actively traded European CDS, the difference between bid and ask spreads is very low.

Collateralized Debt and Synthetic Obligations 261

The index itself can be considered like an *index tranche with attachment point 0 and detachment point 1*. In Section 3.4.3 we discussed the mechanism of *premium* and *default legs* of index tranches. In the same way, the index itself has a premium and a default leg, where premium payments are based on a *quoted index spread* S_{Index}. In our example and data sample, the index spread equals (roundabout)

$$S_{\text{Index}} = 32 \text{ bps}. \qquad (3.36)$$

Note at this point that we make the simplification of ignoring bid/ask spreads for CDS spreads as well as for the index spread. However, for highly liquid products, as it is the case for iTraxx Europe, these simplifications do not make a large difference to 'exact' calculations.

How do investment banks arrive at a quotation of index spreads, given the single-name CDS spreads as shown in Figure 3.29? The answer to this question is not free from complications but in the sequel we will discuss an approach on how to obtain such index spreads. Without further reflections, one could come up with the suggestion to take the arithmetic mean of single-name spreads

$$S_{mean} = \frac{1}{m} \sum_{i=1}^{m} S_i$$

as a natural candidate for the index spread, where m denotes the number of names in the index ($m = 125$ in case of iTraxx Europe) and S_i denotes the spread of name i. However, we will see in a moment that a certain risk-related weighting scheme will do a better job than just the unweighted mean value of single-name CDS spreads.

Let us make a **generic example**, which one can easily understand in order to illustrate the problem. Assume we are given two CDS,

- CDS A on name A pays an annual spread of $S_A = 100$ bps
- CDS B on name B pays an annual spread of $S_B = 50$ bps

We further assume that the default legs of both CDS are determined by a *recovery rate* of 40%. Consider now a (duo) index consisting of the two CDS with equal weights of 50%. The premium and default legs of the index work exactly in the same way as described above for index tranches and the indices itself:

- If no defaults occur, the index pays the index spread applied to the 100% notional of the index.

- If one default occurs, the index pays the index spread applied to 50% of the outstanding notional. The defaulted CDS is removed from the index and an amount of

$$(1 - [\text{recovery quote}]) \times [\text{notional}]$$

has to be paid on the default leg by the protection seller.

- If two defaults occur, the index has a payout of zero, both CDS are removed and the default leg asks for a payment of realized losses on both CDS.

Figure 3.30 summarizes in its upper half the payout w.r.t. different default scenarios for the index in case the index spread is equals to S_{mean}. In case of no defaults, the payout equals an index spread of $S_{mean} = (100+50)/2 = 75$ bps. In case of one default, the index spread is applied to the non-defaulted notional of the index such that the payout equals half of the index spread ($= 75/2 \approx 38$ bps) in percentage notation. In case of two defaults, the payout of the index is zero because no notional exposure is left over.

Figure 3.30 in its lower part shows the payout of a *portfolio* consisting of 50% of CDS A and 50% of CDS B. The payout looks different to the payout of the index in cases where we have only one default in the index or portfolio, respectively. For instance, if A defaults, the portfolio loses half of its notional. To the remaining notional, the spread of CDS B is applied, which results in a percentage payout of $S_B/2 = 25$ bps. Analogously, if B defaults, the percentage payout is $S_A/2 = 50$ bps.

Summarizing, we find in Figure 3.30 that the portfolio payout not exactly *replicates* the payout profile of the index product, although the underlying CDS are the same. However, what we would expect at least from a 'fair' index spread $\tilde{S}_{\text{Index}}$[24] (note: not S_{mean}) is that it generates the same *expected present value* for the index and the portfolio payout. In order to get this right, we continue our analysis as follows.

---

[24] In this 2-asset excursion, we write $\tilde{S}_{\text{Index}}$ in order to avoid any confusion with S_{Index} from Equation (3.36).

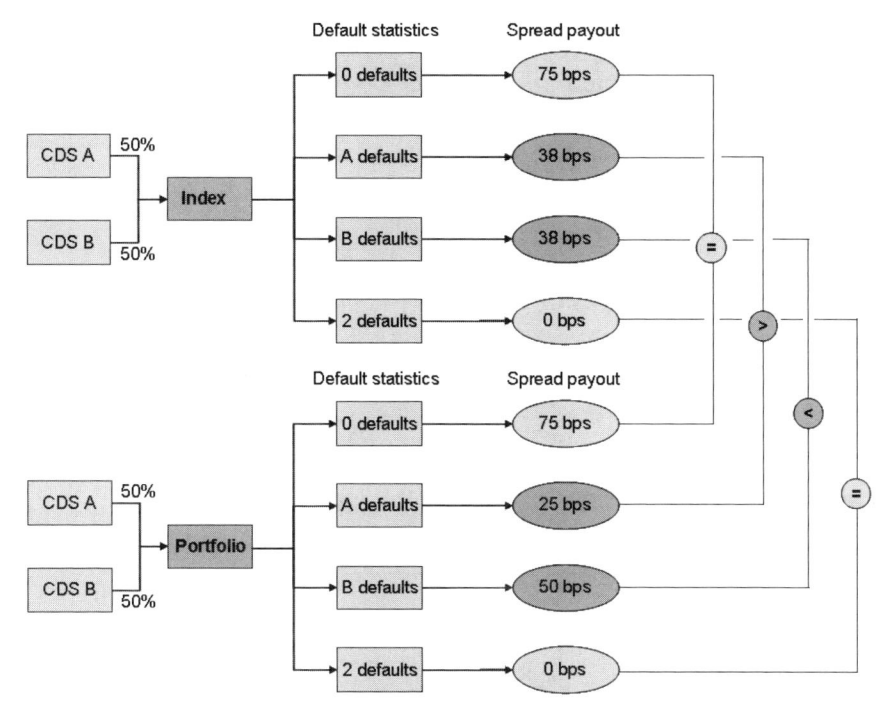

Collateralized Debt and Synthetic Obligations 263

FIGURE 3.30: Illustration of the difference between an index with arithmetic mean index spread and a portfolio consisting of two equally weighted CDS contracts

In our example, the *premium legs* of the CDS have present values[25]

$$\text{PV}_{PL}(A) = \frac{(1-p_A)S_A}{1+r}$$

$$\text{PV}_{PL}(B) = \frac{(1-p_B)S_B}{1+r}$$

where PVs are quoted in percentage of outstanding notionals, p_A and p_B denote the *risk-neutral*[26] default probabilities of names A and B, and r denotes the risk-free interest rate. We come back later in this section to the derivation of risk-neutral default probabilities. The *default legs* (in percentage of notional amounts) of the two swaps are

$$\text{PV}_{DL}(A) = \frac{p_A(1-R_A)}{1+r}$$

$$\text{PV}_{DL}(B) = \frac{p_B(1-R_B)}{1+r}$$

where R_A and R_B denote the expected *recovery quotes* of CDS A and CDS B. If the CDS contracts are fairly priced, their PVs at time $t=0$ should be zero such that we can impose that

$$\text{PV}_{PL}(A) = \text{PV}_{DL}(A) \quad \text{and} \quad \text{PV}_{PL}(B) = \text{PV}_{DL}(B).$$

This yields the equations

$$\frac{(1-p_A)S_A}{1+r} - \frac{p_A(1-R_A)}{1+r} = 0$$

$$\frac{(1-p_B)S_B}{1+r} - \frac{p_B(1-R_B)}{1+r} = 0$$

which finally leads to

$$S_A = \frac{p_A(1-R_A)}{1-p_A} \quad \text{and} \quad S_B = \frac{p_B(1-R_B)}{1-p_B} \qquad (3.37)$$

---

[25] The simplifying assumption we make here is that we can focus on a discrete time grid of $t=0$ and $t=1$ only such that the annual premium is paid in arrears in case the underlying name did not default.

[26] The nature of risk-neutral PDs can be explained as follows. Risk-neutrality expresses the risk aversion of market participants, which rely in, e.g., default-adjusted cash flows on typically substantially larger default probabilities than implied by really measured historical default rates. Because risk-neutral default probabilities are implicitly driving market pricing, a standard assumption is that risk-neutral PDs can be bootstrapped from market prices, e.g., term structures of credit spreads.

Collateralized Debt and Synthetic Obligations   265

This confirms the well-known fact *that theoretically spreads are determined by the risk-neutral default probability of the referenced name and the expected recovery rate in the swap*. If we combine both CDS into an index, the fair index spread $\tilde{S}_{\text{Index}}$ must be chosen[27] in a way making the PVs of premium and default legs equal. This can be written as

$$\tilde{S}_{\text{Index}} \times \frac{\text{PV}_{PL}(A) + \text{PV}_{PL}(B)}{2} = \frac{\text{PV}_{DL}(A) + \text{PV}_{DL}(B)}{2} \quad (3.38)$$

because the index spread appears linear in our simplified equations. Factoring terms in Equation (3.38) out, we finally obtain

$$\tilde{S}_{\text{Index}} = \frac{S_A(1-p_A) + S_B(1-p_B)}{(1-p_A) + (1-p_B)} \quad (3.39)$$

such that the fair index spread is a (risk-neutral) *survival probability weighted average of the single-name CDS spreads* and not just the arithmetic mean as in the example in Figure 3.30.

For illustration purposes, we calculate the fair index spread in our example from Figure 3.30. For the CDS A and B we have

$$p_A = \frac{S_A}{S_A + 1 - R_A} = 164 \text{ bps}, \quad p_B = \frac{S_B}{S_B + 1 - R_B} = 83 \text{ bps}$$

for the risk-neutral default probabilities implied by spreads and recovery rates, and

$$1 - p_A = 98.36\%, \quad 1 - p_B = 99.17\%$$

for the corresponding risk-neutral survival probabilities. Recall that in our introduction of the example above we assumed a recovery rate of

$$R_A = R_B = 40\%$$

for both CDS contracts. Altogether we obtain

$$\tilde{S}_{\text{Index}} = 74.9 \text{ bps} \quad \text{compared to} \quad S_{mean} = 75.0 \text{ bps}.$$

The difference between the survival probability weighted (fair) spread and the arithmetic mean spread seems negligible at first sight. However, note that in our simplified example we considered *one period* only. With

---
[27]In the same way as we saw it in case of the single-name CDS.

increasing time, survival probabilities will decline and the difference between $\tilde{S}_{\text{Index}}$ and S_{mean} will become more significant. In Figure 3.28 we find different maturities for iTraxx products. Over a 5- or 10-year horizon, default probabilities (risk-neutral or historical;[28] see also the PD term structure calibrations in Chapter 2) can be substantially high so that survival probabilities can significantly decrease over time. Then, the effect of using the 'correct' weighting for the derivation of index spreads becomes more material.

After this little excursion with a simplified index consisting of two names only, we go back to our iTraxx example.

The index spread of $S_{\text{Index}} = 32$ bps (p.a.) for iTraxx Europe from (3.36) is based on a term of 5 years, index spreads for different terms look differently. For instance, the corresponding index spread for a term of 10 years is given by 55 bps. In the sequel, we introduce the multi-period extension of Formulas (3.38) and (3.39), but before we need to calibrate multi-year (spread-implied) risk-neutral default probabilities.

A common and simple model for multi-year PDs, which can be applied to spread-implied PDs, is given by the distribution function $\mathbb{F}_\lambda$ of an *exponentially distributed* random variable,

$$\mathbb{F}_\lambda[t] = 1 - \exp(-\lambda t) \qquad (t \geq 0) \qquad (3.40)$$

where $\lambda > 0$ in this context for obvious reasons is called a *default intensity*. Recall that the exponential distribution with parameter λ has the density

$$f_\lambda(t) = \lambda \exp(-\lambda t) \qquad (t \geq 0)$$

and an expected value of $1/\lambda$ as well as a variance of $1/\lambda^2$. The exponential distribution is well-known in the natural sciences as well as in finance as a distribution for a *memoryless*[29] *waiting time*, which in our context will be interpreted as the *default time* of a credit-risky asset; see also Appendix 6.5 for another reason why the exponential distribution can be considered as a 'natural choice' for a distribution under certain circumstances.

---

[28]The attribute 'historical' here refers to PDs, which have been 'historically observed' in real life.
[29]The rationale for this attribute is the equation $\mathbb{P}[\tau \leq s+t \mid \tau \geq s] = \mathbb{P}[0 \leq \tau \leq t]$ for any exponentially distributed random variable τ.

Collateralized Debt and Synthetic Obligations 267

Note that in line with Proposition 2.5.2 the exponential term structure corresponds to the most simple case of a *constant hazard rate function*; see Section 2.5.2. Given the spreads from CDS in the index, we can use Equation (3.37) to derive a risk-neutral one-year default probability, apply Equation (3.40) for $t=1$ to obtain a (risk-neutral) intensity λ, and then use again Equation (3.40) with $t>0$ to derive a risk-neutral PD term structure implied by just one single spread value applied in Equation (3.37).

A slightly more refined version of (3.37) to derive a spread-implied one-year PD is the following. For a spread S for a CDS in the index, we can solve the quarterly cash flow equation

$$\sum_{t=1}^{20} \frac{S/4}{(1+r/4)^t} e^{-\lambda t/4} \stackrel{!}{=} \sum_{t=1}^{20} \frac{1-R}{(1+r/4)^t} \left(e^{-\lambda(t-1)/4} - e^{-\lambda t/4}\right) \quad (3.41)$$

for $\lambda > 0$ to find a corresponding risk-neutral PD. The underlying model still is Equation (3.40), but term effects are now taken into account to some extent. By construction, (3.40) determines for an intensity λ determined by Equation (3.41) the corresponding term structure of risk-neutral default probabilities.

Equation (3.41) can easily be understood: at the left-hand side we find the premium leg PV of the considered CDS represented by the sum of discounted spread payments weighted by the time-congruent survival probabilities. At the right-hand side we see the discounted cash flows at the default leg of the CDS, weighted by the respective marginal default probabilities. Both sides must be equal in case of a fair spread S.

Note that the spreads we have for the CDS in the index are one-year spreads quoted for a 5-year term. In fact, *it is common market practice to not only look at one particular point on the spread curve (as we did in our illustrative examples for the sake of simplicity) but to bootstrap a term structure of risk-neutral default probabilities from a whole term structure of credit spreads*; see, e.g., SCHMIDT and WARD [102]. Then, the above-mentioned constant hazard rate function in the exponential waiting time approach has to be replaced by a step function or an even more sophisticated (non-constant) function.

Another way to generate a proxy[30] for the risk-neutral PD term

---
[30] More or less based on kind of a 'misuse' of a well-known truly time-dynamic model

structure of a CDS name based on just one spread value is the following. Fix a CDS name i in the index and assume its *historical* (rating based) PD term structure $(p_i^{(t)})_{t\geq 0}$ to be given for that name. Examples are the HCTMC and NHCTMC term structures introduced in Chapter 2. According to Equations (2.50) and (2.51) we can write $p_i^{(t)}$ as

$$p_i^{(t)} = N\left[\frac{\ln(\tilde{c}_i^{(t)}/\mathrm{CWI}_i^{(0)}) - (\mu_i - \tfrac{1}{2}\sigma_i^2)t}{\sigma_i\sqrt{t}}\right] \quad (3.42)$$

where $\mathrm{CWI}_i^{(t)}$ denotes the CWI of name i at a fixed horizon t and μ_i and σ_i denote the corresponding mean return and volatility, respectively. The *risk-neutral* version of this Equation can be obtained by replacing the return μ_i by the risk-free rate r, which gives us

$$p_{i;\mathrm{rn}}^{(t)} = N\left[\frac{\ln(\tilde{c}_i^{(t)}/\mathrm{CWI}_i^{(0)}) - (r - \tfrac{1}{2}\sigma_i^2)t}{\sigma_i\sqrt{t}}\right] \quad (3.43)$$

as the risk-neutral default probability of the CDS; see also [25], Section 6.4. Equation (3.42) can be re-written as

$$N^{-1}[p_i^{(t)}] + \frac{(\mu_i - \tfrac{1}{2}\sigma_i^2)t}{\sigma_i\sqrt{t}} = \frac{\ln(\tilde{c}_i^{(t)}/\mathrm{CWI}_i^{(0)})}{\sigma_i\sqrt{t}} \quad (3.44)$$

and substitution of (3.44) into (3.43) finally yields

$$p_{i;\mathrm{rn}}^{(t)} = N\left[N^{-1}[p_i^{(t)}] + \frac{\mu_i - r}{\sigma_i}\sqrt{t}\right]. \quad (3.45)$$

Equation (3.45) is a very convenient representation for the risk-neutral PD term structure of name i because the term $(\mu_i - r)/\sigma_i$ does not depend on the time parameter t such that the time dependence of $(p_{i;\mathrm{rn}}^{(t)})_{t\geq 0}$ is driven exclusively by the well-known time dependence of the historical PD term structure $(p_i^{(t)})_{t\geq 0}$ and the term $\sqrt{t}$ in Equation (3.45). All we need to do in order to obtain the whole term structure is one attachment point giving us the value of $(\mu_i - r)/\sigma_i$. For this, one spread value is sufficient for applying Formula (3.37) in order to obtain $p_{i;\mathrm{rn}}^{(1)}$. Then,

$$\frac{\mu_i - r}{\sigma_i} = N^{-1}[p_{i;\mathrm{rn}}^{(1)}] - N^{-1}[p_i^{(1)}]$$

---

in the context of the CWI model with fixed time horizons, where the time dynamics is ignored.

which can then be inserted in Equation (3.45) to get a proxy for the whole risk-neutral PD term structure. Figure 3.31 summarizes the calculation for a particular example.

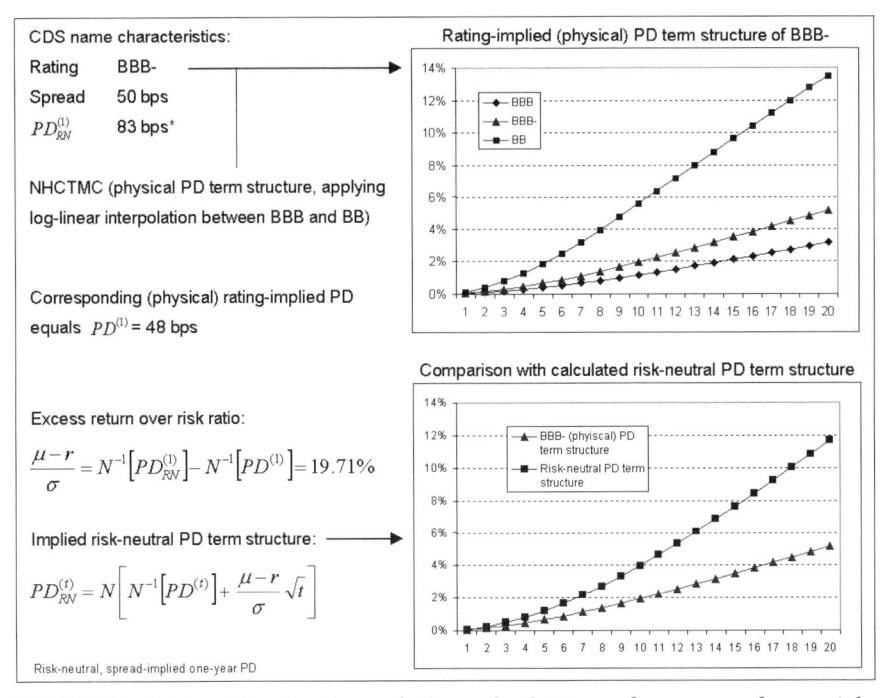

FIGURE 3.31: Illustration of the calculation of a proxy for a risk-neutral PD term structure based on Equation (3.45); ratings-based PD term structures are NHCTMC-based curves from Section 2.11

Given a risk-neutral PD term structure $(p_{i;\text{rn}}^{(t)})_{t\geq 0}$ for each CDS name i in the index, we can calculate corresponding survival probabilities

$$1 - p_{i;\text{rn}}^{(t)}$$

for each name and each point in time. If for reasons of simplicity we choose quarterly payment periods and ignore any payment time mismatches etc., we can generalize Equation (3.38) in a straightforward manner to obtain the fair index spread S_{Index} in the multi-period setup as the solution of the equation

$$\sum_{i=1}^{m} S_{\text{Index}} \sum_{t=1}^{T} \frac{1-p_{i;\text{rn}}^{(t/4)}}{(1+r/4)^t} \overset{!}{=} \sum_{i=1}^{m} S_i \sum_{t=1}^{T} \frac{1-p_{i;\text{rn}}^{(t/4)}}{(1+r/4)^t} \qquad (3.46)$$

where T denotes the term length, which is equal to 5 years in our index example. Again, at the left-hand side we find the *premium leg* PV of the index and at the right-hand side we have the sum of premium leg PVs for single-name CDS, which, in fact, must agree with the *sum of default legs* for single-name CDS, based on our discussion before. Arranging terms in (3.46), we get the multi-period analogue of Equation (3.39) as

$$S_{\text{Index}} = \sum_{i=1}^{m} \frac{S_i \sum_{t=1}^{T} \frac{1-p_{i;\text{rn}}^{(t/4)}}{(1+r/4)^t}}{\sum_{k=1}^{m} \sum_{t=1}^{T} \frac{1-p_{k;\text{rn}}^{(t/4)}}{(1+r/4)^t}} \qquad (3.47)$$

which can be used to calculate the fair (multi-period based) index spread. The term

$$\sum_{t=1}^{T} \frac{1-p_{i;\text{rn}}^{(t/4)}}{(1+r/4)^t} \qquad (3.48)$$

in Formula (3.47) is sometimes called the *risky duration* of name i. From this perspective, S_{Index} in (3.47) is an average spread weighted by the risky duration of names in the index. Figure 3.32 illustrates the dependence of risky durations on spreads. The point of clouds arises from a blend of two major effects: the risky duration depends on the term structure of risk-neutral default probabilities, which depends on spreads as well as on the agency ratings of the CDS names and the corresponding ratings-based PD term structures; see Figure 3.31. As expected, higher spreads correspond (ignoring other spread components like liquidity) to higher risk and, therefore, to a lower risky duration.

If we apply formula (3.47) and calculate the fair spread S_{Index} of the index we get

$$S_{\text{Index}} = 32.9\text{bps}.$$

The difference of almost 1 bps to the quoted index spread is due to our simplifying assumptions, mainly

- Using a proxy for the risk-neutral PD term structure based on one spread value per name only instead of bootstrapping it from the term structure of spreads

Collateralized Debt and Synthetic Obligations 271

Risky duration versus CDS Spread

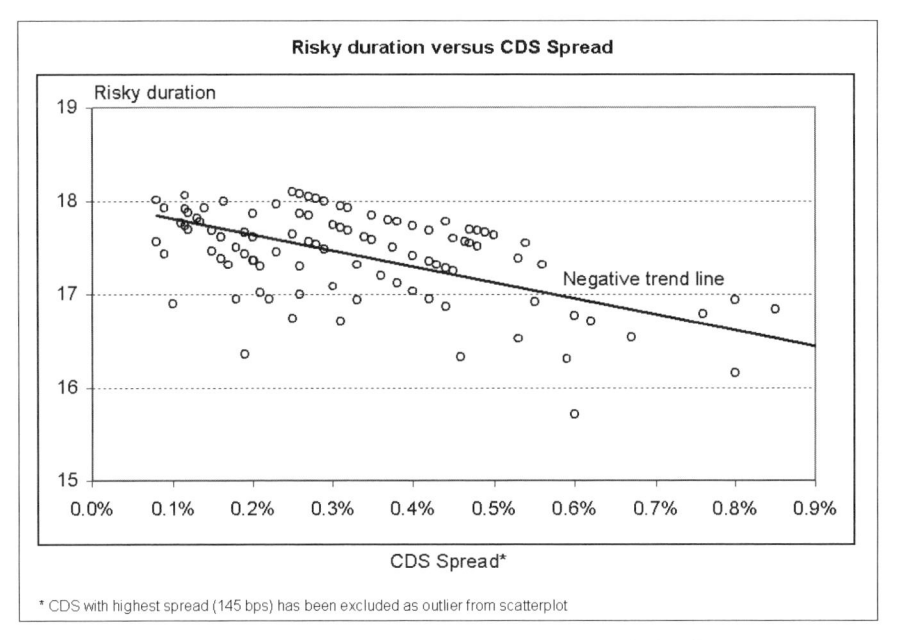

* CDS with highest spread (145 bps) has been excluded as outlier from scatterplot

FIGURE 3.32: Scatterplot of risky duration of CDS names in the index against spreads; see Equation (3.48)

- Ignoring maturity mismatches and payment date incongruencies

However, we can perfectly live with a deviation of less than a basispoint in our example. Figure 3.33 shows the historic development of index spreads for 5-year iTraxx Europe over 1.25 years, starting at the beginning of 2005.

3.4.4 ITraxx Europe Index Tranches: Pricing, Delta Hedging, and Implied Correlations

We now focus on the tranched index. We already mentioned earlier in this section that iTraxx Europe has a standardized tranching where for the sake of an easier reference we attach letters to tranches:

- Tranche FLP: 0% to 3%
- Tranche D: 3% to 6%
- Tranche C: 6% to 9%
- Tranche B: 9% to 12%

272   *Structured Credit Portfolio Analysis, Baskets & CDOs*

Historic index spreads iTraxx Europe

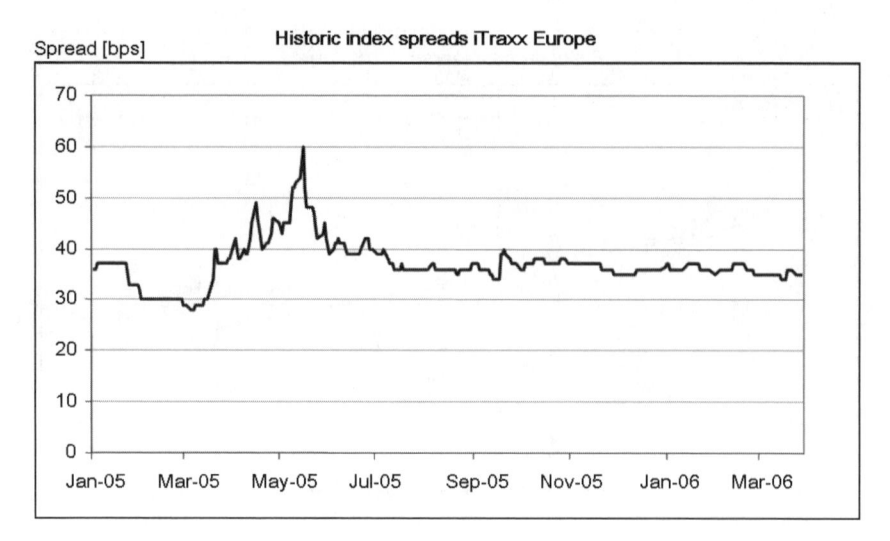

FIGURE 3.33: Historic index spreads of 5-year iTraxx Europe (interpolated; from January-05 to March-06); note that in April 2006 the index (mid) spreads dropped down to a level between 31 and 33 bps; data of this kind can be obtained from any investment bank's CDS trading desk or from Bloomberg or other sources of CDS/index spread data

- Tranche A: 12% to 22%

Every tranche is traded for a specified spread corresponding to a certain maturity like 5 or 10 years. Figure 3.34 illustrates tranche spreads by means of historic spread levels for 5-year iTraxx Europe. The equity tranche (FLP, as we call it here) is an exception regarding quoting: it has a fixed running spread of 500 bps and involves an upfront payment (quoted in percentage of referenced notional), which varies with the market as one can see in Figure 3.34.

To explain the basic principle of the functionality of iTraxx Europe based STCDOs, we have to look at some illustrative but realistic data. Table 3.5 shows (rounded, smoothed) quotes of STCDOs referenced to iTraxx Europe Series 5. Comparable quotes could have been obtained from investment bank's correlation trading desks in the first week of April 2006. We are now going to explain quotes in Table 3.5 step by step, hereby demonstrating the concept of *implied correlations* as well as the basic *hedging scheme* common in index tranche trading.

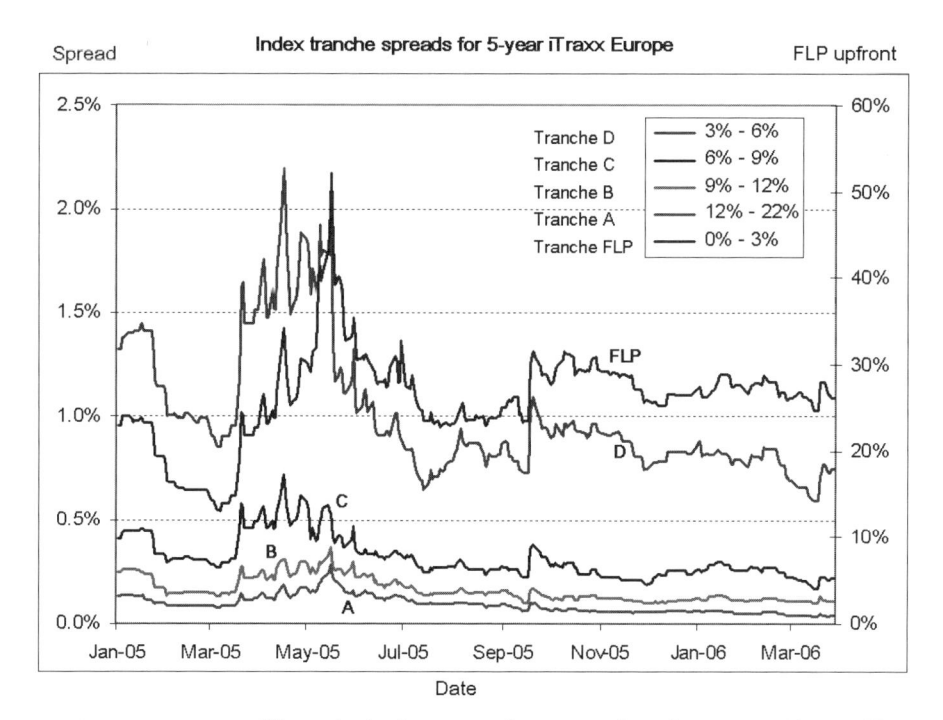

Collateralized Debt and Synthetic Obligations 273

FIGURE 3.34: Historic index tranche spreads of 5-year iTraxx Europe (interpolated; from January-05 to March-06); regarding data sourcing see the caption of Figure 3.33

Let us start with the second and third upper columns. So far we introduced the lower and upper boundary of tranches as *attachment* and *detachment* points. However, considering the tranche loss profile function (3.1)

$$\Lambda_{\alpha,\beta}(L^{(t)}) = \frac{1}{\beta - \alpha} \min\left[\beta - \alpha, \underbrace{\max[0, L^{(t)} - \alpha]}_{\text{long call option}}\right] \qquad (3.49)$$

illustrated in Figure 3.3 we find that it looks like the capped and normalized payout of a *long call option* written w.r.t. the cumulative loss of the underlying reference portfolio with a strike at the attachment point α of the tranche. Based on this analogy, the detachment point of a tranche is the strike of the next more senior tranche so that the notion of *lower* and *upper* strike can be motivated by this association.

The fourth upper column is headlined as 'Upfront Payment'. Such a payment takes place at the beginning of a STCDO contract and is com-

TABLE 3.5: STCDO quotes p.a. (rounded/smoothed for illustration purposes) based on (5-year) iTraxx Europe; time window: first week of April 2006

Tranche	Lower Strike	Upper Strike	Upfront Payment	Spread	Spread leverage
A	12%	22%	0%	0.05%	0.2
B	9%	12%	0%	0.10%	0.3
C	6%	9%	0%	0.20%	0.6
D	3%	6%	0%	0.65%	2.0
FLP	0%	3%	22%	5.00%	29.4

Tranche	Tranche PV01 [bps]	Traded Delta	Compound Correlation	Base Correlation	Index Spread (S_{ix})
A	-2.0	0.4	25%	54%	
B	-4.0	0.8	18%	43%	
C	-7.5	1.5	14%	36%	0.32%
D	-25.0	5.0	8%	28%	
FLP	-125.0	25.0	18%	18%	

mon for equity tranches only. The buyer of protection on the tranche 'FLP' has, in addition to a *running spread* (see column 'Spread'), to make a single payment of (in our example)

$$22\% \times [\text{equity tranche notional}]$$

to the protection seller. This compensates the risk-taking party for the effective first loss risk position. The running spread (500 bps) for the equity tranche then is comparably low. However, one could combine the upfront payment and the running spread to a *fair equity tranche running spread*, although it does not make much sense to follow other than market quotation standards (which is 'upfront + running').

The column 'Spread' shows (except for equity where the upfront payment has to be taken into account) *fair* STCDO spreads. For each tranche T, the column *spread leverage* is determined by

$$[\text{spread leverage}](T) = \frac{[\text{fair spread}](T)}{S_{\text{Index}}}$$

which provides[31] a measure for the non-linear risk-adjusted premium allocation to tranche long positions.

The *tranche delta* column 'Traded Delta' shows so-called *hedge ratios*, which are determined as follows.

---
[31] For the equity tranche we take '500 bps + [upfront payment]/5' as a proxy for the fair equity spread, hereby ignoring discounting effects.

- The first type of deltas people typically look at is the *spread delta* of a tranche T defined by

$$\Delta_{spread}(T) = \frac{\partial S_T}{\partial S_{\text{Index}}} \approx \frac{\Delta S_T[\text{bps}]}{\Delta S_{\text{Index}}[\text{bps}]}.$$

where T is the considered tranche, S_T denotes the fair spread of the tranche and S_{Index} is the fair index spread. The spread delta provides information about the impact of 1 bps index spread move[32] on fair tranche spreads.

- Next, we look at *PV deltas* defined by

$$\Delta_{PV}(T) = \frac{\partial PV_T}{\partial S_{\text{Index}}} \approx \frac{\Delta PV_T[\% \text{ of notional}]}{\Delta S_{\text{Index}}[\text{bps}]}$$

which captures tranche PV changes w.r.t. changes in the underlying index spread. The PV deltas are often denoted by 'PV01', addressing present value or mark-to-market changes given a 1 bps change of the underlying, such that one shortly speaks about the 'tranche PV01'. In Table 3.5 we included a column showing the tranche PV01 values. For instance, a 1 bps spread widening in the index spread induces a loss of 125 bps for FLP long positions. In other words, a 10 mn EUR equity/FLP long position suffers a mark-to-market loss of 125,000 EUR in case of a 1 bps spread widening of the underlying index.

- Finally, the *hedge ratios* shown in column 'Traded Delta' in Table 3.5 are defined as

$$\delta_T = \frac{\Delta_{PV}(T)}{\Delta_{PV}(\text{Index})} \approx \frac{\Delta_{PV}(T)[\% \text{ of notional}]}{\Delta_{PV}(\text{Index})[\% \text{ of notional}]}$$

where $\Delta_{PV}(\text{Index})$ denotes the PV delta of the whole index, which corresponds to a tranche PV delta with attachment point 0 and detachment point 1. In our example we have

$$\Delta_{PV}(\text{Index}) = \text{PV01}(\text{Index}) = -5 \text{ bps}$$

such that, e.g., a spread widening by 1 bps of the index induces a mark-to-market loss on a 100 mn EUR index long position of

---
[32] Some banks quote note at bps but 0.1 bps level.

50,000 EUR.

The hedge ratio sometimes is also called the *leverage ratio* of a tranche. Hedge ratios compare the mark-to-market change of a tranche with the mark-to-market change of the underlying index given a 1 bps change of the index spread, or, in PV01 language, the 'tranche PV01 in relation to the index PV01'. As explained before, there are traded and theoretical deltas or hedge ratios. We will come back to that when we explain the meaning of *delta exchange prices/quotes*.

In our example, for an FLP long position of 10 mn EUR we can read off a hedge ratio of 25 from Table 3.5 such that we need a short position in size of

$$25 \times 10 \text{ mn EUR} = 250 \text{ mn EUR}$$

in the total index in order to neutralize our equity long position w.r.t. spread moves in the underlying index.

Figure 3.35 provides another example and an illustration for the mechanics of delta hedging.

Delta hedging has to be done *dynamically* in order to 'follow' the index spread to a certain extent. Therefore, quoted deltas have to be adapted and updated over time accordingly. The same principle actually holds in the more general context of STCDOs: the two parties of such a transaction have to adjust the deltas periodically to keep the hedge effective. Hereby, a common problem is that delta hedging can be expensive due to *negative carry* effects. For instance, in Figure 3.35, the tranche long position pays a spread of 65 bps, whereas the short position (hedge) in the index requires a spread payment of

$$5 \times S_{\text{Index}} = 5 \times 32 \text{ bps} = 160 \text{ bps}$$

which leads to an expensive negative carry of $160 - 65 = 95$ bps. In general, mark-to-market valuations can vary substantially over time and often the smoothed P&L in one book is bought for the price of a highly volatile other book, e.g., the hedging book of the trading unit. In general, delta hedging not always produces a *perfect hedge*, especially, in cases where a bespoke STCDO is hedged via an index etc. Banks have invented a whole bunch of strategies to reduce hedge costs in different ways, sometimes for the price of an imperfect hedge. In addition, note

Collateralized Debt and Synthetic Obligations

Delta hedging for the 3% - 6% tranche

Loss distribution of the index

iTraxx Europe standard tranches:
- 12% - 22%
- 9% - 12%
- 6% - 9%
- 3% - 6%
- 0% - 3%

Index PV01 = -5

- Long position in tranche
- Notional of 10 mn EUR
- Tranche PV01 = -25
- Tranche hedge ratio = -25 / -5 = 5 = tranche delta
- Hedge: index short position in size of 5 x 10 mn EUR = 50 mn EUR
- Scenario: index spread + 1 bps
- Corresponding position combination:
  - tranche: - 0.0025 x 10 mn EUR
  - index: + 0.0005 x 50 mn EUR
- MtM loss based on spread widening neutralized for tranche long position

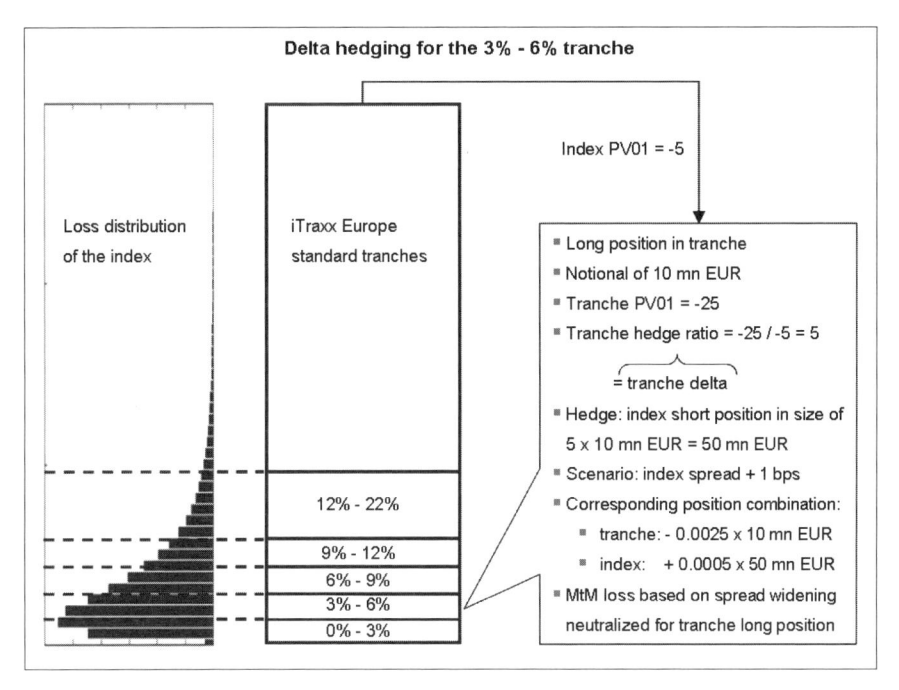

FIGURE 3.35: Example for delta hedging of an i-Traxx based STCDO (data from Table 3.5)

that *delta hedging can be based on an index or individually on single-name CDS underlying the STCDO*. The latter-mentioned is especially useful in situations where one party in the deal has a special opinion on particular names in the portfolio. The overall principle of hedging against single names in a STCDO transaction is basically the same as if one hedges against the overall underlying index. The only difference is that then the PV01 of the tranche has to be divided by the PV01 of the CDS name instead of the PV01 of the overall index.

Here is another remark one should have in mind when considering delta hedging, namely, *spread neutrality does not mean default risk neutrality*. For instance, a long position of 10 mn EUR in the equity index tranche plus a short position in the index in size of 250 mn EUR (see our discussion above) neutralizes mark-to-market fluctuations but is not neutral regarding defaults. To see this, consider the percentage FLP loss arising from one default in iTraxx Europe. We get

$$0.8\% \times 60\% \times \frac{1}{3\%} \;=\; 16\%$$

for the percentage loss, based on a uniform obligor notional contribution of 80 bps, a recovery rate of 40%, and the iTraxx typical FLP size of 3%. In contrast, the percentage loss in the delta adjusted index arising from one default equals

$$0.8\% \times 60\% \times 25 = 12\%$$

because the index is an index tranche with lower strike 0 and upper strike 1. So indeed we find that the impact of defaults on the two parts of the transaction is different, default risk is not neutralized. Another important remark in this context is that *delta hedging does not address value changes due to changed correlations or recovery rates*. For these risk drivers, delta hedging is not the appropriate answer.

Another aspect in the context of delta hedging is that between sophisticated players who both have strong opinions regarding hedge ratios the actual negotiation of deltas has some potential to delay deals. The reason for this is that *deltas are calculated based on proprietary internal CDO and portfolio models and calibrations* such that it is very unlikely that quant teams in two different investment banks end-up at the exactly same delta for a STCDO and and index. A solution out of this is to quote deltas (as in Table 3.5) as 'traded' deltas and to clearly point out that quoted index tranche spreads are *delta exchange* spreads. Delta exchange STCDO transactions are good examples for increased efficiency and reduced costs in the context CDO hedging. Often, investment banks prefer to trade index based STCDOs in combination with an index trade hedging the open part of the capital structure. Doing both deals with one single counterparty increases efficiency and reduces transaction costs. Another reason for delta exchange spread quotes are guidelines of *correlation desks* who are often not allowed to take on any other positions than correlations. Trading desks, therefore, often quote STCDO spreads as *delta exchange spreads*, which means that the quoted price only can be exploited if the client in parallel to the STCDO transaction also engages in an 'off-setting' delta adjusted position on the index with the trading bank. If the client does not accept this, she or he can ask for 'clean' spreads, which do not include delta exchange adjustments, but it can be expected that these spreads count in the costs the investment bank has to hedge the transaction somewhere else down the street.

Let us make an example for a delta exchange index tranche deal; see Figure 3.36. To illustrate the effect, we now have to take bid/ask

Delta exchange STCDO transaction

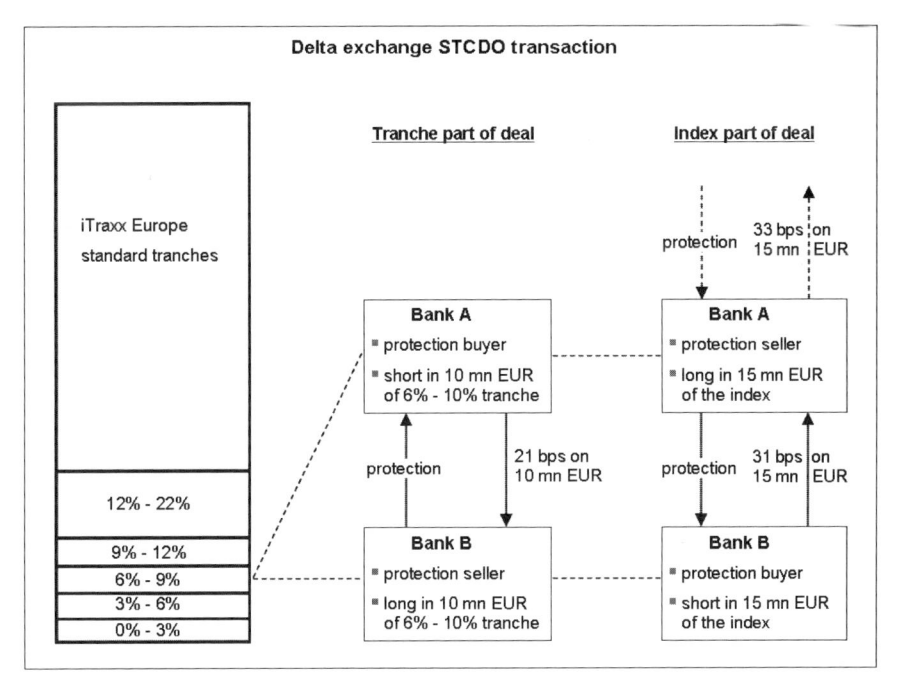

[Figure: diagram showing iTraxx Europe standard tranches (12%-22%, 9%-12%, 6%-9%, 3%-6%, 0%-3%) with Tranche part of deal and Index part of deal.

Tranche part: Bank A (protection buyer, short in 10 mn EUR of 6%-10% tranche) ↔ 21 bps on 10 mn EUR ↔ Bank B (protection seller, long in 10 mn EUR of 6%-10% tranche).

Index part: Bank A (protection seller, long in 15 mn EUR of the index) — 33 bps on 15 mn EUR protection — Bank B (protection buyer, short in 15 mn EUR of the index), 31 bps on 15 mn EUR protection.]

FIGURE 3.36: Example for delta exchange STCDO trading

spreads into account, for instance, we assume a spread bid of 19 bps for the 6%-9% tranche and an ask/offer of 21 bps. The arranger or sponsor of an index tranche often is the protection buyer, so let us assume Bank A wants to buy protection on 10 mn EUR of the 6%-9% tranche. Negotiations with Bank B reveal that Bank B is willing to sell protection for a 21 bps spread, which is a *delta exchange price* involving a second deal where Bank A sells protection to Bank B for the whole index for a quoted spread of 31 bps with a notional amount given by the tranche's traded delta of 1.5 times the 10 mn EUR tranche notional. As a consequence, *Bank B has a mark-to-market neutral deal position*. In order to have a 'flat' position in the index, Bank A has to buy protection in the market against the 15 mn EUR index long position (A is protection seller on the index to bank B) as soon as the delta exchange 'tandem deal' starts. Let us assume that Bank A pays 33 bps for buying protection on 15 mn EUR notional of the index. Given these parameters, we can now calculate the effective spread Bank A pays for protection on the 6%-9%-tranche. Altogether, we obtain

$$21 + 1.5 \times (33 - 31) = 24 \text{ bps}$$

280   *Structured Credit Portfolio Analysis, Baskets & CDOs*

as spread effectively paid due to delta exchange trading. As said before, if Bank A asks for a spread clean of delta exchange effects, one can expect that 24 bps (instead of 21 bps delta exchange spread) is the most likely answer of bank B.

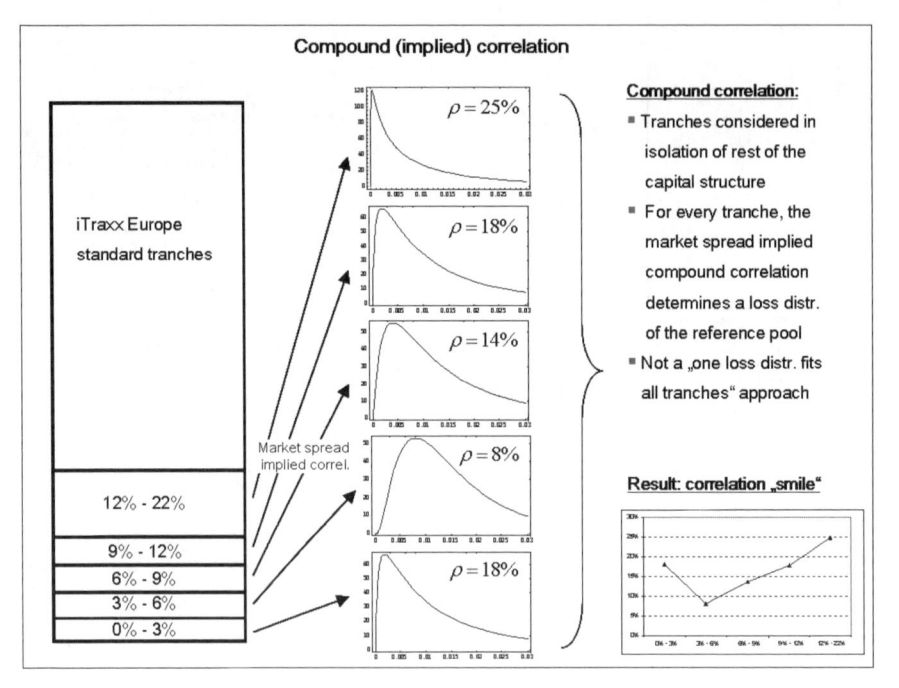

FIGURE 3.37: Illustration of implied compound correlations; loss distributions in the middle are based on a one-factor Gaussian copula model with infinite granularity assumption

A last topic indicated in Table 3.5 we want to discuss is *implied correlation*. Figures 3.37 and 3.38 illustrate the concept. The starting point and main driver of implied correlations are *market spreads attached to CDO tranches*. Market spreads are assumed to capture in full the risk of CDO tranches such that they provide kind of a benchmark for the calibration of CDO models. The standard model in the context of pricing index tranches is the *one-factor Gaussian copula model* as explained in Section 3.3.6.1 and Appendix 6.7. As already mentioned in our discussion above on delta hedging, the actual implementation of the Gaussian copula model and its parameterization is varying among dif-

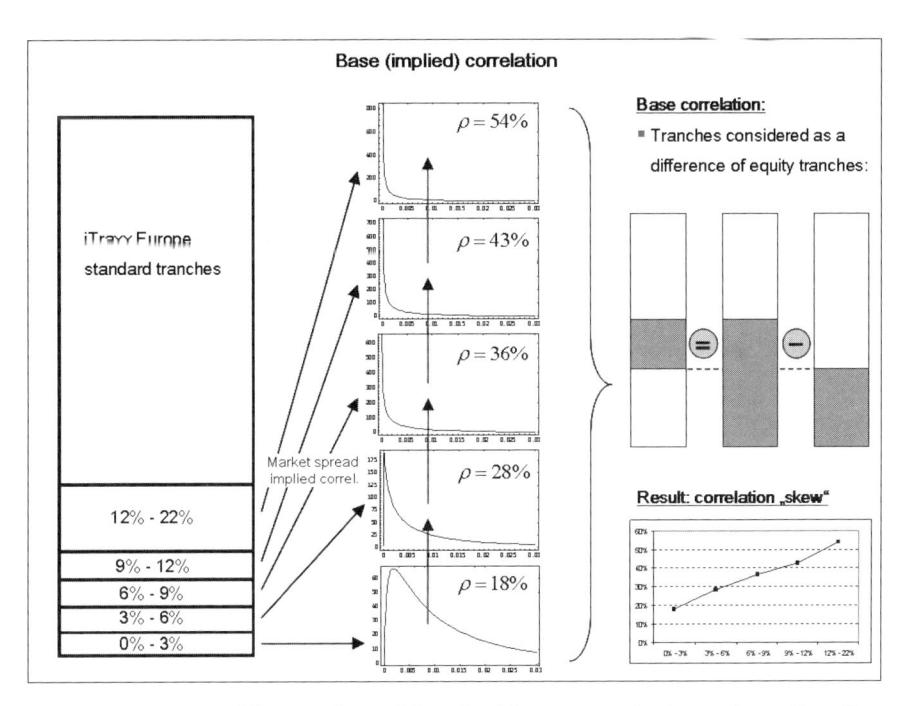

FIGURE 3.38: Illustration of implied base correlations; loss distributions in the middle are based on a one-factor Gaussian copula model with infinite granularity assumption; upward arrows in the middle indicate the bootstrapping scheme explained later in the text

ferent banks and market players, but a broad consensus is given in the same way as option pricing in a base case version (before adjustments) typically is done via the BLACK & SCHOLES option pricing framework. The role of implied volatility is then replaced by the concept of implied correlation. In the sequel, we provide some simple formulas for the calculation of implied correlation and then conclude this section with a few remarks.

Starting point for our calculation is Proposition 3.3.1 for the expected loss of CDO tranches in an analytic[33] framework based on VASICEK's limit distribution; see Appendix 6.7. For the convenience of the reader, we briefly recall the formulas. The loss profile of a tranche $T_{\alpha,\beta}$ with

---

[33]Making an infinite granularity assumption for the 125-pool (market standard).

282   *Structured Credit Portfolio Analysis, Baskets & CDOs*

lower and upper strike α and β can be described by the function

$$\Lambda_{\alpha,\beta}(L^{(t)}) = \frac{1}{\beta - \alpha} \min\left[\beta - \alpha, \max[0, L^{(t)} - \alpha]\right]$$

where $L^{(t)}$ denotes the portfolio loss cumulated over the time interval $[0,t]$; see Figure 3.3. Then, the expected loss of the tranche $T_{\alpha,\beta}$ for the time interval $[0,t]$ is given by

$$\mathrm{EL}^{(t)}_{\alpha,\beta;\varrho} = \mathbb{E}_{\varrho}[\Lambda_{\alpha,\beta}(L^{(t)})] = \qquad (3.50)$$

$$= \int_0^1 \Lambda_{\alpha,\beta}(x[1-R]) f_{p_{\mathrm{rn}}^{(t)},\varrho}(x)\, dx$$

(see Proposition 3.3.1) for a prescribed recovery rate of R in [%]. Note that we explicitly include the correlation parameter in the notation of the tranche EL because later on we determine ϱ implicitly by Equation (3.51). The underlying credit curve $(p_{\mathrm{rn}}^{(t)})_{t\geq 0}$ in (3.3.1) is the uniform term structure of risk-neutral PDs best matching the index. Based on an index spread of

$$S_{\mathrm{Index}} = 32\ \mathrm{bps}$$

according to Table 3.5, we can apply, e.g., Formula (3.41) to obtain a uniform default intensity λ_{Index} for the index and then apply Equation (3.40) to obtain a proxy for the risk-neutral PD term structure of the index. This term structure, the (CWI) correlation ϱ and the assumed recovery rate of 40% determine $\mathrm{EL}^{(t)}_{\alpha,\beta;\varrho}$.

The basic equation leading to *compound correlations* as shown in Table 3.5 implied by market spreads is the equilibrium between premium and default legs

$$\mathrm{PV}_{PL}(T_{\alpha,\beta}) \stackrel{!}{=} \mathrm{PV}_{DL}(T_{\alpha,\beta}) \qquad (3.51)$$

where the PVs are calculated based on Formula (3.50) as follows:

$$\mathrm{PV}_{PL}(T_{\alpha,\beta}) = \frac{S_{\alpha,\beta}}{4} \sum_{t=1}^{20} \frac{1}{(1+r/4)^t} \left(1 - \mathrm{EL}^{(t/4)}_{\alpha,\beta;\varrho}\right) \qquad (3.52)$$

$$\mathrm{PV}_{DL}(T_{\alpha,\beta}) = \sum_{t=1}^{20} \frac{1}{(1+r/4)^t} \left(\mathrm{EL}^{(t/4)}_{\alpha,\beta;\varrho} - \mathrm{EL}^{((t-1)/4)}_{\alpha,\beta;\varrho}\right) \qquad (3.53)$$

where $S_{\alpha,\beta}$ is the spread quoted for tranche $T_{\alpha,\beta}$. In (3.52), the last factor addresses the expected survived notional amount up to time horizon $t/4$, and in (3.53), the last factor represents the marginal expected loss in the time interval $[(t-1)/4, t/4]$. Note that in our presentation we count time in quarters but adjust time for a yearly scale. Also note that for the equity tranche (FLP) the upfront premium (22% in our example) has to be added to the premium leg. Given we believe in the recovery rate of 40% and the risk-neutral PD term structure, the uniform correlation ϱ in Formula (3.3.1) is the only parameter we can use to enforce Equation (3.51). Calibrating for each tranche in Table 3.5 the correlation ϱ making (3.51) true, we obtain what is called the *compound correlation* for each iTraxx Europe based tranche.

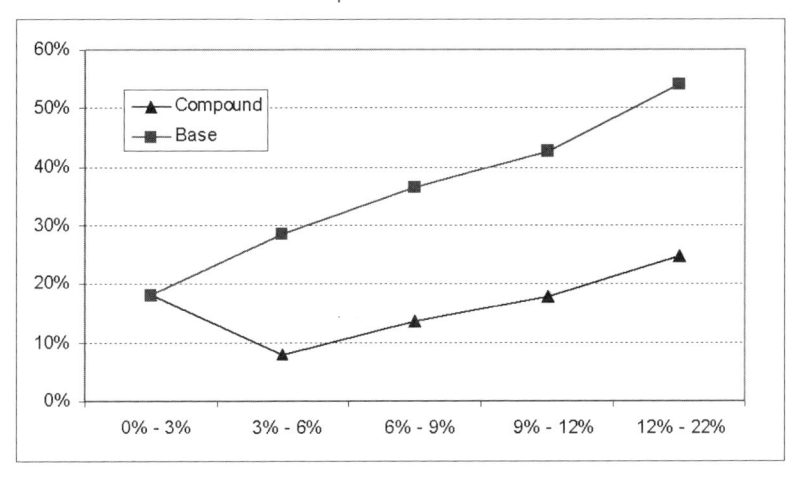

FIGURE 3.39: Correlation smile (compound correlation) and correlation skew (base correlation) according to Table 3.5

Figure 3.39 shows a graph of implied *compound correlations* for each tranche quoted in Table 3.5. The chart also shows a plot of so-called *base correlations*, a concept we explore next before we make some general remarks and comments on implied correlations. Because compound correlations induce a smile-like shape, whereas base correlations are strictly increasing, one speaks of the *correlation smile* in the compound and *correlation skew* in the base correlation case.

In contrast to compound correlations where every tranche is considered in isolation of the rest of the capital structure, *base correlations* arise from the following bootstrapping algorithm; see [95]. For the equity/FLP tranche, the calculation of base correlation follows exactly Equation (3.51) so that compound and base correlations for FLP agree. For tranches senior to FLP, we use the following representation

$$\Lambda_{\alpha,\beta}(L^{(t)}) = \frac{1}{\beta - \alpha}\left(\beta\Lambda_{0,\beta}(L^{(t)}) - \alpha\Lambda_{0,\alpha}(L^{(t)})\right) \quad (3.54)$$

which shows that, up to scaling, the loss profile of a tranche $T_{\alpha,\beta}$ can be written as the loss profile function of an equity tranche $T_{0,\beta}$ with the same detachment point β reduced by the loss profile function of an equity tranche $T_{0,\alpha}$ with detachment point given by the attachment point of the considered tranche $T_{\alpha,\beta}$. Let us now consider the tranche next senior to FLP, namely the tranche from $\alpha = 3\%$ to $\beta = 6\%$. The base correlation of this tranche again is determined by Equation (3.51) but with premium and default legs given by

$$\mathrm{PV}_{PL}(T_{\alpha,\beta}) = \quad (3.55)$$

$$= \frac{S_{\alpha,\beta}}{4} \sum_{t=1}^{20} \frac{1}{(1+r/4)^t}\left(1 - \frac{\beta\mathrm{EL}_{0,\beta;\varrho_D}^{(t/4)} - \alpha\mathrm{EL}_{0,\alpha;\varrho_{FLP}}^{(t/4)}}{\beta - \alpha}\right)$$

$$\mathrm{PV}_{DL}(T_{\alpha,\beta}) = \sum_{t=1}^{20} \frac{1}{(1+r/4)^t}\Delta\mathrm{EL}_t \quad (3.56)$$

where

$$\Delta\mathrm{EL}_t = \left(\frac{\beta\mathrm{EL}_{0,\beta;\varrho_D}^{(t/4)} - \alpha\mathrm{EL}_{0,\alpha;\varrho_{FLP}}^{(t/4)}}{\beta - \alpha} - \frac{\beta\mathrm{EL}_{0,\beta;\varrho_D}^{((t-1)/4)} - \alpha\mathrm{EL}_{0,\alpha;\varrho_{FLP}}^{((t-1)/4)}}{\beta - \alpha}\right)$$

where we applied representation (3.54) but with two different correlation parameters ϱ_{FLP} (for the equity piece, coinciding with the FLP compound correlation) and ϱ_D (for tranche D from 3%-6%). The correlation parameter ϱ_D making Equation (3.51) with premium and default legs given by (3.55) and (3.56) true, is the *base* correlation of the 3%-6% index tranche. For tranche C ($\alpha = 6\%$ and $\beta = 9\%$) we calculate

$$\mathrm{PV}_{PL}(T_{\alpha,\beta}) = \frac{S_{\alpha,\beta}}{4}\sum_{t=1}^{20}\frac{1}{(1+r/4)^t}\left(1 - \frac{\beta\mathrm{EL}_{0,\beta;\varrho_C}^{(t/4)} - \alpha\mathrm{EL}_{0,\alpha;\varrho_D}^{(t/4)}}{\beta - \alpha}\right)$$

$$\mathrm{PV}_{DL}(T_{\alpha,\beta}) = \sum_{t=1}^{20} \frac{1}{(1+r/4)^t} \Delta \mathrm{EL}_t \quad \text{where}$$

$$\Delta \mathrm{EL}_t = \left(\frac{\beta \mathrm{EL}_{0,\beta;\varrho_C}^{(t/4)} - \alpha \mathrm{EL}_{0,\alpha;\varrho_D}^{(t/4)}}{\beta - \alpha} - \frac{\beta \mathrm{EL}_{0,\beta;\varrho_C}^{((t-1)/4)} - \alpha \mathrm{EL}_{0,\alpha;\varrho_D}^{((t-1)/4)}}{\beta - \alpha} \right).$$

The correlation parameter ϱ_C making Equation (3.51) calculated w.r.t. the above-stated premium and default legs true, is the base correlation for the 6%-9% tranche. Continuing in this way, we obtain base correlations for every index tranche, hereby generating the *correlation skew* visible in Figure 3.39.

We now come to a series of comments and remarks. First of all, note that as in our whole exposition on STCDOs we neglected several possible refinements in our analysis like payment date incongruencies and maturity mismatches. Also note that our approach to implied correlation is slightly simplified: banks use different variations of the scheme (like heterogeneous PD distributions etc.) to calculate their view on implied correlations. Next, let us try to understand what we see in Figure 3.39 and why we see it.

Starting with the correlation smile, the question arises why compound correlation decreases when moving from equity to mezzanine and then again increases when moving from mezzanine to senior, hereby generating the smile shape of compound correlation. In the literature, the reader will find many reasonable and also some more speculative reasons and explanations for this phenomenon. A simple but at least to us convincing argument is the following. In many cases, mezzanine tranches have tight spreads due to a strong demand[34] for such tranches in the market, whereas senior and especially super senior tranches typically exhibit wide spreads despite the fact that the risk of a hit by losses is low and the loss severity is close to zero. Senior swappers nevertheless insist in a certain market price for selling protection on such tranches. Since credit curves and recovery rates are fixed in the formulas above, the correlation parameter is the only 'tuning parameter' the implied correlation model can use to match mezzanine and

---

[34] Strong demand for mezzanine tranches can be expected in economic cycle periods as we currently have it where investors do not reach their target returns by investing in senior tranches and, therefore, step-down the credit quality stair to more risky classes like mezzanine tranches, which are still protected by a certain amount of subordinated capital but already pay an attractive yield.

senior tranche market spreads. Therefore, to enforce (equilibrium), ϱ has to be chosen lower for mezzanine and higher for senior tranches. For us, this is a reasonable argument and we stop the discussion here by saying that the correlation smile is a blend of different effects among which the pricing argument regarding mezzanine and senior tranches just made is one of the most intriguing.

The important conclusion from the correlation smile is that obviously *the one-factor Gaussian copula (with infinite, but also with finite granularity assumption) is not capable of producing one single loss distribution making (3.51) true simultaneously for all tranches*. One can make comparable observations also in standard CDOs and CLOs.

Now let us consider base correlations. First of all, base correlation by construction is a function of the upper strike, which defines the tranche size for base tranches. To see why base correlation increases with increasing seniority we provide the following 'heuristic'[35] arguments.

- Base correlation for FLP is given by the compound correlation ϱ_{FLP} for FLP. This is our starting point.

- For the 3%-6% tranche we insert ϱ_{FLP} into the PV formula for the lower base tranche (FLP) but use the spread from the 3%-6% tranche in the PV formula. This spread is much lower than the running spread for equity so that the PV typically becomes negative. The only way to get a balance between premium and default leg (a PV of zero) is to use a higher correlation ϱ_D for the upper base tranche (0%-6%) such that (3.51) is fulfilled. Recall that, as we have seen in many places in this book, the value of the equity piece benefits from increasing correlation. In this way, one could say that equity is *long in correlation*, because increasing the correlation shifts mass of the loss distribution to the tail in a way beneficial for equity note holders.

- For the next higher tranche, we proceed analogously.

Given this chain of arguments, we typically expect base correlation to produce a 'skew'. As a last remark, we should mention that base correlation is an arbitrage-free concept: if one would sell protection

---

[35]Note that implied correlations need not necessarily to behave the way we saw it in our examples, although the results we presented are kind of typical.

on each single index tranche and buy protection in a corresponding notional amount on the whole index then the arising expected losses on the two positions coincide. If this would not be the case, one could make arbitrary gains with a risk-free combination of positions. The reason why this EL-balance holds is given by a simple telescope sum argument applied to the difference of base tranches in combination with the fact that portfolio EL is not affected by correlations.

This concludes our discussion on STCDOs. There would be more to say, e.g., on *interpolation* between standard index tranches (using the correlation skew) to price non-standard index tranches, and so on, but going any further is beyond the scope of this section. In the literature (see Chapter 5), some suggestions for further reading are given. In all that it is important that one keeps in mind what implied correlation can do and what it cannot do. Shortly speaking, *implied correlation is a market quoting standard and not a fully-fledged mathematical theory*. It is used by CDO traders for the exchange of information and opinions on the risk and chances of certain STCDOs. It does not help us to *model* index tranches appropriately, because in contrast to CDO models we considered in previous sections, implied correlation according to current market standards needs *several* loss distributions to price *one* CDO; see also Figure 3.37. In recent research papers one finds different suggestions on how to solve this problem. Clearly, one has to leave the Gaussian copula (large pool) one-factor model if one wants to find a loss distribution for an index replicating spreads of all tranches simultaneously; see references in Chapter 5.

3.5 Tranche Risk Measures

For many applications in credit risk like pricing or capital allocation, *risk measure contributions* are the key to success. For an overview on risk measure contributions in credit risk we refer to [25], Chapter 5. In this section, we want to explore the contribution of individual names in the reference portfolio described in Appendix 6.9 to the risk of tranches in the sample cash flow CSO modeled in Section 3.3.3. We divide this section in two parts. In a first part we describe the calculation of tranche hit and tranche expected loss contributions of single names

3.5.1 Expected Shortfall Contributions

Recall from Chapter 1, Equation (1.10), that the *expected shortfall* (ESF) w.r.t. a given loss threshold[36] q is defined as the conditional expectation

$$\text{ESF}_q = \mathbb{E}[L^{(t)} \mid L^{(t)} > q] \tag{3.57}$$

of the cumulative portfolio loss $L^{(t)}$ up to time t conditional on losses exceeding the threshold q. ESF has the nice additivity property

$$\text{ESF}_q = \mathbb{E}[L^{(t)} \mid L^{(t)} > q] = \sum_{k=1}^{m} \mathbb{E}[L_k^{(t)} \mid L^{(t)} > q], \tag{3.58}$$

inherited from the linearity of conditional expectations. Here, $L_k^{(t)}$ denotes the loss generated by name k in the portfolio. Equation (3.58) indicates that ESF is kind of an ideal risk measure for *capital allocation* purposes; see also [25], Chapter 5, for a discussion on the *coherency* of ESF and related risk measures.

The following calculations are exercised in the context of the CSO modeled in Section 3.3.3. The copula we use in the Monte Carlo simulation is the Gaussian copula w.r.t. the factor model described in Appendix 6.9. Our exposition uses some simplifying assumptions also made in Section 3.3.1, e.g., uniform notional amounts of names (10 mn USD) and a uniform LGD of 50%. However, it is straightforward to generalize our findings to more general portfolio parameterizations and other copula functions.

In Monte Carlo simulations as elaborated in Section 3.3.3, the calculation of ESFs is straightforward. We denote the number of simulated scenarios by n and define the set of all scenarios as

$$I = \{1, ..., n\}.$$

---

[36] Typically a quantile of the loss distribution.

Collateralized Debt and Synthetic Obligations   289

The simulated portfolio loss in scenario i is denoted by $\hat{L}^{(i)}$, where for the sake of an easier notation we drop the time superscript '(t)'. Each loss scenario $\hat{L}^{(i)}$ can be decomposed into a sum[37] of single-name losses, which contribute to $\hat{L}^{(i)}$ in scenario i,

$$\hat{L}^{(i)} = \sum_{k=1}^{m} \hat{L}_k^{(i)}, \qquad (3.59)$$

where $\hat{L}_k^{(i)}$ denotes the realized loss of name k, given by the realization of the name's indicator/Bernoulli variable in scenario i times its LGD and exposure. Given a threshold q, we define

$$I_q = \{i \in I : \hat{L}^{(i)} > q\} \subseteq I$$

as the set of scenarios meeting the ESF condition. The ESF w.r.t. q can then be calculated via

$$\widehat{\mathrm{ESF}}_q = \frac{1}{|I_q|} \sum_{i \in I_q} \hat{L}^{(i)}$$

where $|M|$ denotes the number of elements in a set M. Let us make an example. For the CSO portfolio in Section 3.3.1 we obtain

$$q = q_{99.9\%}(L^{(t)}) = 155{,}000{,}000 \text{ USD}$$

for the 99.9%-quantile of the reference portfolio's loss distribution. In Table 3.1 we find that

- The hitting probability of tranche A with a subordinated capital of 110,000,000 USD equals 121 bps whereas

- The hitting probability of the super senior tranche with a subordinated capital of 170,000,000 USD equals 5 bps

Therefore, we actually expected the 10 bps loss quantile to be located in tranche A and find it confirmed by the 155 mn USD. Estimating ESF_q yields

$$\widehat{\mathrm{ESF}}_q \approx 176{,}000{,}000 \text{ USD}$$

so that ESF_q is located in the super senior tranche.

---
[37] Recall that m in our example is equal to 100.

Next, we calculate from the simulated set of scenarios I_q meeting the ESF condition the *single-name contributions* to ESF_q. Again, this is straightforward in Monte Carlo simulations. We just have to define another scenario set

$$I_q^{(k)} = \{i \in I_q : \hat{L}_k^{(i)} > 0\} = \{i \in I_q : \hat{\tau}_k^{(i)} \leq T\}$$

counting the scenarios where name k contributes to a loss scenario meeting the ESF condition. Here, $\hat{\tau}_k^{(i)}$ denotes the realization of the default time of name k in scenario i and T denotes the maturity of the CSO in years ($T = 5$ in our example). The figure

$$\widehat{\text{ESFC}}_q^{(k)} = \frac{I_q^{(k)}}{I_q} \times 10{,}000{,}000 \times 50\% \text{ USD} \qquad (3.60)$$

gives us an estimation for the *ESF contribution* of name k to ESF_q, given the 10 mn USD notional per name and a recovery rate of 50%. If q is a quantile of the loss distribution, ESF is a *coherent risk measure* (see [25], Chapter 5) such that ESF contributions provide reasonable *risk measure contributions*. Note also that $\widehat{\text{ESFC}}_q^{(k)}$ indeed is the empirical (simulation-based sample) version of

$$\text{ESFC}_q^{(k)} = \mathbb{E}[L_k^{(t)} \mid L^{(t)} > q].$$

From Equation (3.60) we get back $\widehat{\text{ESF}}_q$ by summing up all $\widehat{\text{ESFC}}_q^{(k)}$'s,

$$\sum_{k=1}^{m} \widehat{\text{ESFC}}_q^{(k)} = \widehat{\text{ESF}}_q.$$

Again note that we made our life slightly easier by uniformly weighted names in the reference pool and by a fixed non-random LGD, etc. However, the principle for ESF contributions remains unchanged in more complicated portfolio parameterizations.

Figure 3.40 shows the single-name ESF contributions calculated via Formula (3.60). In addition, the average ESF contribution (in USD) is shown as a horizontal line at $1{,}760{,}000$. We also compare the ESF contributions with the log-PD of the name in the reference portfolio (see the step function in Figure 3.40 referenced to the right y-axis). As expected, we find a certain trend that names with higher PD on average contribute more to ESF_q than names with good credit quality.

Expected shortfall contributions of single-names

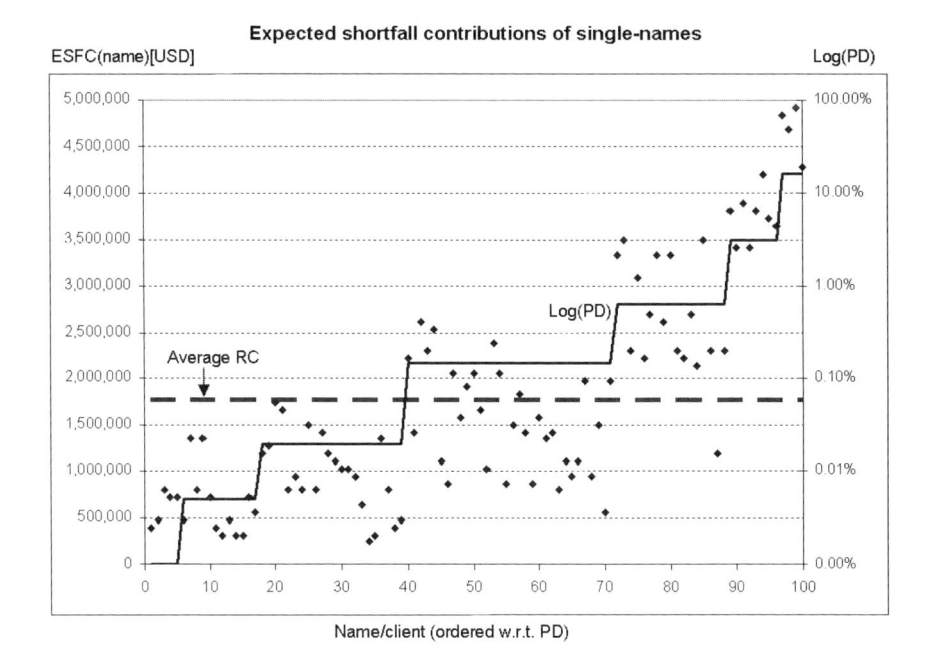

FIGURE 3.40: Single-name ESF contributions to ESF_q according to Formula (3.60) with $q = q_{99.9\%}(L^{(t)})$.

However, the wide spread cloud of points in the chart shows that the PD is not the only driver of risk contributions. ESF contributions are portfolio model-driven quantities, and our portfolio model includes a *multi-factor* model making ESF contributions not predictable based purely on default probabilities; see Appendix 6.9 for a description of the used factor model.

ESF contributions give us valuable information about the *risk a single-name or subportfolio contributes to the overall risk of the portfolio, seen in the light of the considered reference portfolio and evaluated w.r.t. to ESF as risk measure*. It is important to keep in mind that ESF contributions are strongly dependent on the surrounding reference pool: the same name in the context of some other reference portfolio can have a substantially different ESF contribution.

3.5.2 Tranche Hit Contributions of Single Names

We are now ready for the next step. Following the same lines as for ESF contributions, we can calculate the *tranche hit contribution* of a single name w.r.t. a given tranche of our sample CSO. For this purpose, assume a tranche $T_{\alpha,\beta}$ with attachment point α and detachment point β to be given. Denote the total original notional of the pool by $V = V_0$. The tranche boundaries in monetary amounts are then given by αV and βV. Analogously to (3.60), we can now calculate 'hit statistics'

$$\widehat{\text{THC}}_{\alpha V}^{(k)} = \frac{I_{\alpha V}^{(k)}}{\sum_{j=1}^{m} I_{\alpha V}^{(j)}} \qquad (3.61)$$

where THC stands for *tranche hit contribution*. Here, the underlying rationale is that every loss hitting a tranche is what it is due to loss-contributing single-name defaults. In this way, every name contributing to a loss hitting a tranche takes part in the 'shared responsibility' for the tranche shortfall. THCs yield the relative frequency of the involvement of a particular name in losses hitting a particular tranche.

Figure 3.41 shows tranche hit contributions for our CSO example. Here are two comments on the result.

- For lower tranches, especially visible in case of the equity piece, the PD of a name is the main driver of the risk contribution. With increasing amount of subordinated capital, moving from equity to super senior, THCs more and more deviate from an almost purely PD-driven quantity and exhibit fluctuations driven by the dependence structure of the portfolio. Because here we work with a Gaussian copula, correlations drive THCs (besides PDs).

- Another aspect we expect to see in CDO tranche hit contributions is a reflection of the fact that high quality assets tend to default later in time than low quality assets such that tranches with large subordinated capital should get hit (if at all) by high credit quality assets after low quality name defaults already wiped out subordinated tranches. In the chart, we see this effect indicated: for names with high credit quality, THCs are typically higher for more senior tranches than for tranches with low or zero subordinated capital. For low credit quality names, the effect is opposite: tranches with high subordinated capital have

Tranche hit contributions of single names

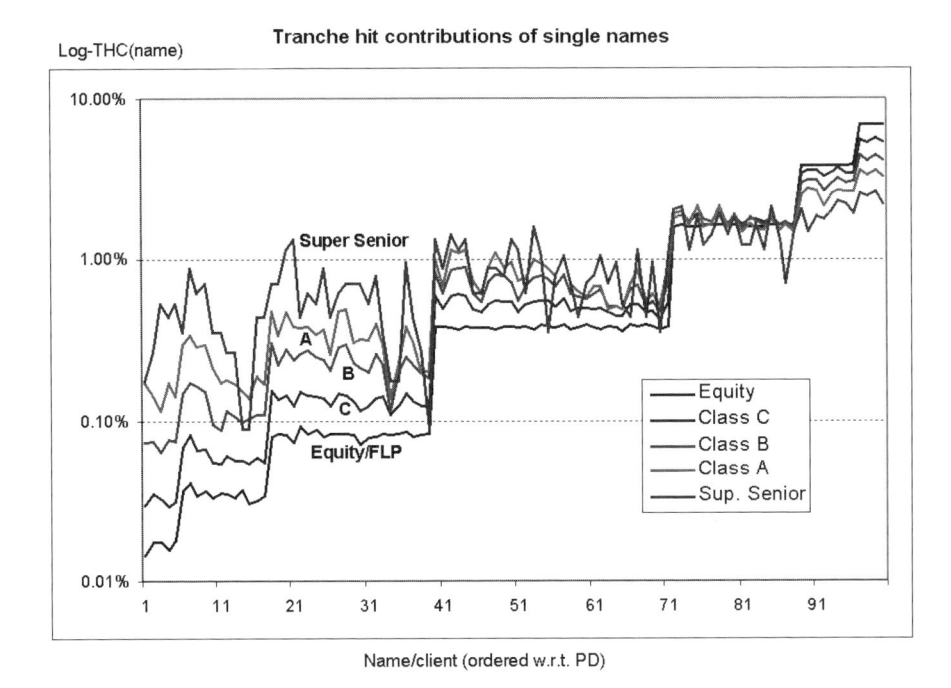

FIGURE 3.41: Tranche hit contributions of single names

lower TCHs, whereas names with high PD have higher junior tranche THCs.

THCs only consider the *hit of tranches*, the tranche loss *severity* is not taken into account. One way to capture also tranche LGDs is to switch to *tranche expected loss constributions* (TELCs) where the *tranche risk measure* and its corresponding contributions are defined in terms of the expected loss of a tranche, see, e.g., Proposition 3.3.1 for an analytic formula for tranche's EL in particular situations.

TELCs can be determined as follows.

For a tranche $T_{\alpha,\beta}$, we measure which names are actually contributing to the loss of a tranche by taking the *timing of defaults* into account. For instance, if we consider a mezzanine tranche surviving the fourth default in the pool but getting a hit if the fifth default occurs, then the first four defaults in the reference pool would not count for the expected loss contribution of single-names to the considered mezzanine tranche, but the fifth defaulting name would be considered as a contributor to the mezzanine tranche's expected loss. Aggregating all default time adjusted tranche loss contributions of single names over all simulated scenarios, yields TELCs such that the tranche's expected loss

$$\mathbb{E}[L_{T_{\alpha,\beta}}] = \sum_{k=1}^{m} \text{TELC}_{\alpha V, \beta V}^{(k)} \qquad (3.62)$$

over the lifetime of the transaction equals the sum of single name TELCs, reflecting a very natural property of risk measure contributions. The concept of TELCs and applications in credit portfolio management are extensively discussed in [24].

3.5.3 Applications: Asset Selection, Cost-to-Securitize

Knowing risk measure contributions of single names to the risk of a particular tranche in a CDO can be used for different purposes. In the context of CDO management and optimization, *asset picking* is one of the major applications we would like to mention.

In Section 3.4, we explained the basic mechanism underlying STCDOs. We mentioned that the selection of reference names by the investor is a major part in the negotiation process of the STCDO transaction. This provides a very natural example for a meaningful application of TELCs in CDO optimization and asset picking. If investors have the technology (they actually do in many times) to simulate the CDO in terms of different combinations of capital structures and underlying reference pools, the investor can even better 'engineer' a STCDO tailor-made for her or his needs, risk appetite, and yield targets.

A closely related example is to route the risk of a single name in a CDO's reference portfolio through the capital structure of the considered CDO. Especially if an investor is skeptical about the risk profile of a single name included in a pool underlying a CDO, the investor can calculate the ESF contribution of this particular name to the tranches relevant for the investment and manage the transaction accordingly.

Another example we want to mention is the concept of a *cost-to-securitize*. The basic idea is easily explained. For private clients (retail or corporate) it is often difficult to come up with estimates for a sound credit risk premium or credit price. However, if for the considered asset class a capital market transaction can be found in the market reflecting the risk of the portfolio and the standing of the issuer in the market,

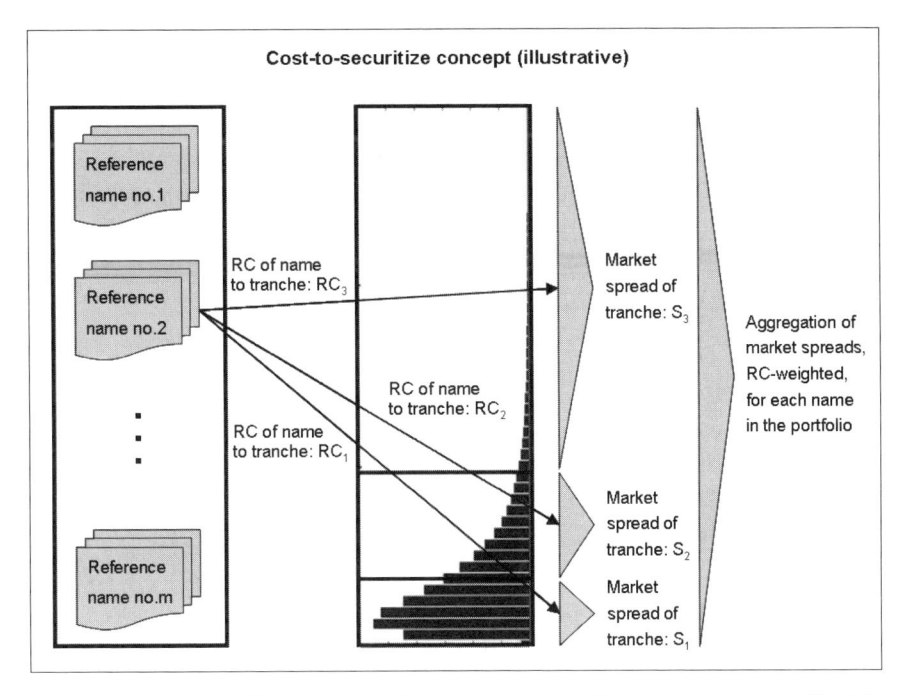

FIGURE 3.42: Illustration of a cost-to-securitize concept, applicable for instance as a proxy for a mark-to-model for illiquid but securitizable credit instruments ('RC' stands more general for 'risk contribution')

then the market spreads attached to securitized tranches reflect the market view on the diversified risk of the pool. Based on the risk measure contribution (e.g., TELCS) of obligors in the reference portfolio to tranches, we can calculate a cost-to-securitize for each obligor in the portfolio. A meaningful application could be to integrate the cost-to-securitize into the pricing framework of the bank to be applied as a *credit risk premium floor* for, e.g., *limit breachers*, given the bank has such a limit framework for single obligors. The philosophy underlying such a pricing floor is that certain risks should be only included in the bank's portfolio if they can be *liquidated* at any time, hereby providing certain flexibility to the portfolio management unit of the bank. For private (corporate) clients, which cannot be marked-to-market, the cost-to-securitize is a useful proxy for a *mark-to-model* based on the assumption of being able to securitize the risks in appropriately pooled and diversified form. Note that, in general, securitizations can be used as tools to enforce a mark-to-model for illiquid credit risks, transforming illiquid (pooled) assets into marketable tranched securities.

Being more explicit, the spread paid on a tranche expresses the market view on the potential of cumulative portfolio loss scenarios $\hat{L}^{(i)}$ ($i = 1, ..., n = \#$scenarios) to eat into a considered tranche. As said in (3.59), each individual portfolio loss scenario $\hat{L}^{(i)}$ is an aggregation of single obligor losses

$$\hat{L}^{(i)} = \sum_{k=1}^{m} \hat{L}_k^{(i)}, \qquad (3.63)$$

such that tranche losses can be routed back to loss contributions of underlying reference names. Our exposition in the sequel is fairly general and can be universally applied so that we denote by RMC[38] any meaningful notion of a risk measure contribution of single reference names to tranche losses, e.g., TELCs could be used as RMCs. Let us now make an example, how the specified RMCs can be used to develop a *cost-to-securitize* concept for every name in the reference pool of the CDO. For the sake of an easy exposition, let us assume that, as in the illustration in Figure 3.42, we have a CLO with 3 tranches: equity, mezzanine, and super senior swap, requiring a market spread of

$$S_{\text{equ}}, \quad S_{\text{mez}}, \quad S_{\text{sen}}.$$

Based on the RMCs of underlying reference names, we can decompose the spread for every tranche in spread components[39] via

$$S_{\text{equ}} = \sum_{k=1}^{m} \widehat{\text{RMC}}_{\text{equ}}^{(k)}[\%] \times S_{\text{equ}}$$

$$S_{\text{mez}} = \sum_{k=1}^{m} \widehat{\text{RMC}}_{\text{mez}}^{(k)}[\%] \times S_{\text{mez}}$$

$$S_{\text{sen}} = \sum_{k=1}^{m} \widehat{\text{RMC}}_{\text{sen}}^{(k)}[\%] \times S_{\text{sen}}$$

where $\text{RMC}_T^{(k)}$ denotes the RMC of name k to tranche T and

$$\sum_{k=1}^{m} \widehat{\text{RMC}}_T^{(k)}[\%] = 1$$

---
[38] Risk measure contribution.
[39] Hereby ignoring spread components like liquidity and complexity premium and assuming that the tranche spread is a function of credit risk only.

Collateralized Debt and Synthetic Obligations

for every tranche T; see also (3.62). Note that again we write $\widehat{\mathrm{RMC}}$ instead of RMC in order to indicate that we derive all relevant quantities from scenarios arising from Monte Carlo simulation. Then, for any reference name k,

$$\begin{aligned}\hat{S}^{(k)} &= \Big(\widehat{\mathrm{RMC}}_{\mathrm{equ}}^{(k)}[\%] \times S_{\mathrm{equ}} \times \mathrm{size}(\mathrm{equ})[\%] \\ &\quad + \widehat{\mathrm{RMC}}_{\mathrm{mez}}^{(k)}[\%] \times S_{\mathrm{mez}} \times \mathrm{size}(\mathrm{mez})[\%] \\ &\quad + \widehat{\mathrm{RMC}}_{\mathrm{sen}}^{(k)}[\%] \times S_{\mathrm{sen}} \times \mathrm{size}(\mathrm{sen})[\%]\Big) \times \frac{V}{V^{(k)}}\end{aligned}$$

is a proxy for the spread required to bring obligor k to the market via the considered securitization. Here, $\mathrm{size}(T)$ denotes the size of a tranche T, V again denotes the total original notional of the reference pool, and $V^{(k)}$ denotes the notional of reference name k. As a brief check that this produces a total spread payment in the reference pool sufficient for financing the total liability spread, we can make the following calculation as a cross-check:

$$\begin{aligned}\frac{1}{V}\sum_{k=1}^{m}\hat{S}^{(k)}V^{(k)} &= S_{\mathrm{equ}} \times \mathrm{size}(\mathrm{equ})[\%] \times \sum_{k=1}^{m}\widehat{\mathrm{RMC}}_{\mathrm{equ}}^{(k)}[\%] \\ &\quad + S_{\mathrm{mez}} \times \mathrm{size}(\mathrm{mez})[\%] \times \sum_{k=1}^{m}\widehat{\mathrm{RMC}}_{\mathrm{mez}}^{(k)}[\%] \\ &\quad + S_{\mathrm{sen}} \times \mathrm{size}(\mathrm{sen})[\%] \times \sum_{k=1}^{m}\widehat{\mathrm{RMC}}_{\mathrm{sen}}^{(k)}[\%] \\ &= \sum_{T \in \{\mathrm{equ,mez,sen}\}} S_T \times \mathrm{size}(\mathrm{T})[\%]\end{aligned}$$

which exactly equals the (weighted) spread at the liability side of the transaction necessary for paying the tranche investors. Note that, as we have seen in Section 3.3.1, there are other costs required in a securitization, for instance upfront costs and administration or maintenance costs. However, these costs are not risk-driven so that they can be allocated equally weighted to clients in the underlying reference portfolio. Altogether, the scheme can be used to determine a *cost-to-securitize* for every obligor in the reference pool.

Cost-to-securitize – illustrative example

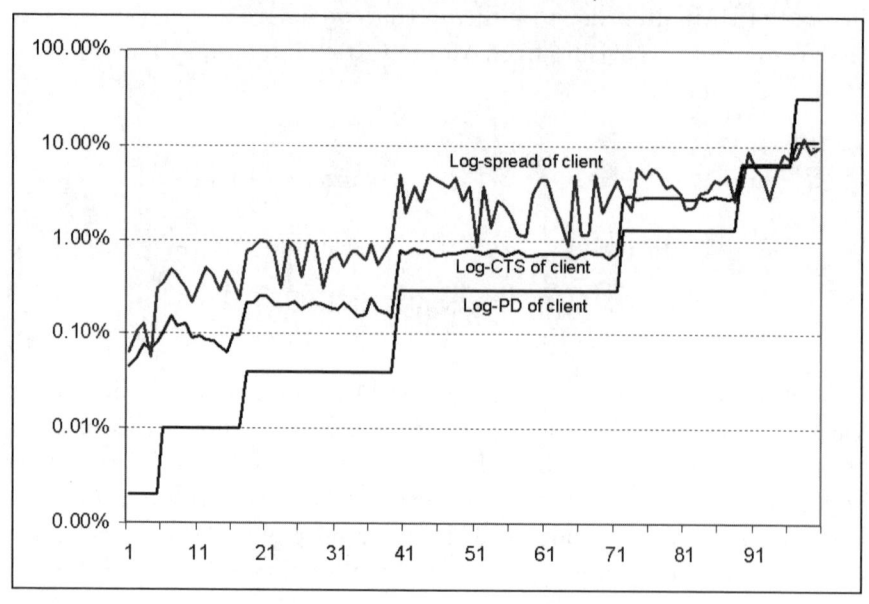

FIGURE 3.43: Cost-to-securitize (CTS) illustration in the context of the sample CSO from Section 3.3.1; for equity, which gets excess spread and no fixed spread, we assumed a required fixed spread of 30% in the calculation

As an illustrative[40] example, Figure 3.43 shows the cost-to-securitize calculated for the 100 reference names from Appendix 6.9 underlying the CSO in Section 3.3.1, where RMCs have been calculated as simple loss amount weighted tranche hit contributions.

Here we stop our discussion on the *cost-to-securitize*. Analogous calculations can be done for other classes of structured credits like baskets, RMBS, CMBS, CLOs, etc., serving as a proxy calculation of a *'mark-to-model'* for illiquid but securitizable credit risks. A more comprehensive exposition of this topic can be found in [24].

---

[40]'Illustrative' because the sample CSO from Section 3.3.1 is referenced to names, which already have a market spread, so that we do not need to calculate a cost-to-securitize, except, we want to find out how much money we can save by transferring the name's credit risk to the market via a CSO in 'pooled diversified form' instead via buying single-name protection.

3.5.4 Remarks on Portfolios of CDOs

We conclude this section with a few indicative remarks on the management of mixed portfolios, consisting of CDOs, loans, bonds, CDS, and so on, some of them are long and some of them are short positions. In such portfolios, it is essential to capture *cross dependencies* in an appropriate way. For instance, in previous sections we saw how one can calculate a PD (hitting probability) and an LGD (tranche loss severity) of a CDO tranche. However, if the reference pool of the CDO contains a name, which also is part of some other credit risk position in our portfolio, e.g., as underlying name in a CDS or as a corporate loan, then we cannot make the simplifying assumption that the CDO is just a bond-type instrument with a PD and an LGD. Instead, we have to think in terms of Monte Carlo simulations and follow a *bottom-up approach*, which simulates every underlying credit-risky name at lowest (single name) level and then, step-by-step, aggregates simulated single name risks bottom-up into different instruments until all risks are at a common highest level; see Figure 3.44. Bottom-up approaches treat dependencies at the *single name* level such that *cross dependencies*, e.g., in *overlapping CDO pools* are taken into account. Indeed, cross dependencies are a typical problem CDO investors have. For instance, when Enron defaulted some years ago, many CDOs were hit because Enron was part of many CDO reference pools. In Section 3.4.4, we discussed index tranches based on iTraxx. Remember, iTraxx and CDX include the 125 most actively traded names, iTraxx in Europe and CDX in the United States, altogether 250 names. This means we can expect to have a few hundred ($\approx 600+$) tradable names in the market, which are all included over and over again in reference pools for correlation products. *If such a name defaults, various transactions suffer.* Therefore, cross dependencies are a major issue CDO investors have to manage.

Another challenge CDO portfolio managers and also so-called *CDOs of CDOs* (CDO-squared) have to deal with is *systematic risk*. Because CDO tranches are referenced to an already *diversified* pool of risks, idiosyncratic risk can expected to be low but systematic risk can expected to be high. Buying many CDO tranches into a book or including them as reference names into a CDO pool basically means to create a *boost of systematic risk*.

In both challenges (cross dependencies as well as systematic risk), *sensitivities* and *risk measures* are market standard tools for the man-

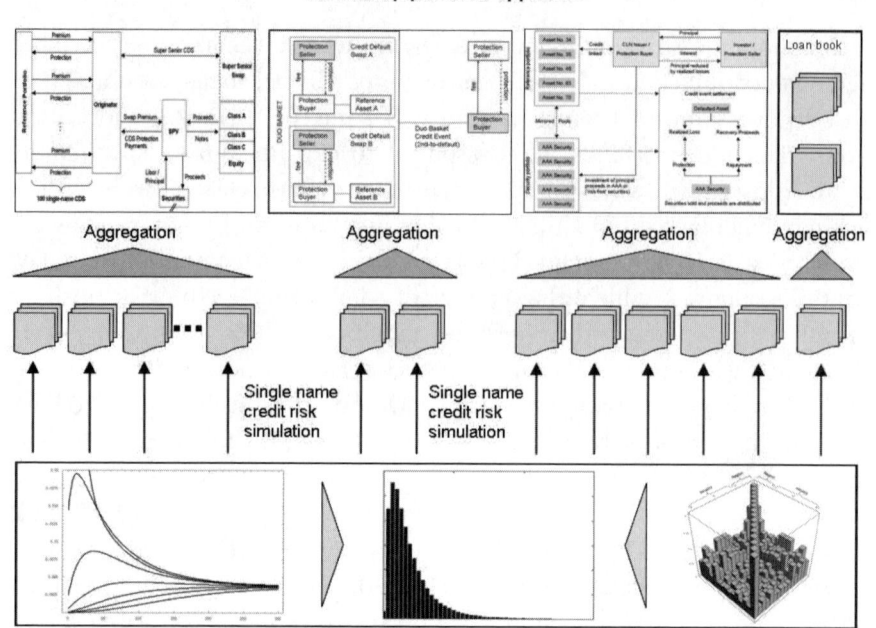

FIGURE 3.44: Illustration of a bottom-up approach for portfolios of CDOs and CDO-related instruments or hybrid portfolios

agement of such portfolios, especially when they are *hybrid* as, e.g., in the example in Section 3.1.3. Common risk measures in this context are the following:

- **Tranche delta w.r.t. portfolio/index spread:**
  We have already extensively discussed deltas in the context of STCDOs; see Section 3.4.4. In general, delta quantifies the PV change of an instrument, e.g., a CDO tranche, given a 1 bps spread widening of the underlying, e.g., a CDS index or a single name CDS. Observations one makes when considering delta in dependence on the position in the capital structure are as follows:

    - For equity tranches, tranche delta decreases in case of a spread widening of the underlying index/portfolio spread.
    - For senior tranches, tranche delta increases if the underlying portfolio spread widens.
    - For mezzanine tranches, tranche delta depends on the particular position of the mezzanine tranche in the capital struc-

ture of the portfolio.

Both observations do not come much as a surprise and are consistent with insights we gained in previous sections: as spread moves wider, the risk profile of the reference portfolio increases. This makes it more likely that senior tranches get hit and suffer a loss. In contrast, from a certain spread level on, the equity tranche is less and less sensitive to spread widenings because the odds to avoid losses are bad anyway.

- **Tranche gamma w.r.t. portfolio/index spread:**
The observations just discussed for tranche deltas have natural consequences on *tranche gamma*, which measures the sensitivity of delta w.r.t. 1 bps widening of the underlying (as always in market risk, gamma is the 2nd-order derivative w.r.t. changes in the underlying spread). Because tranche delta for equity decreases with wider spreads, equity tranche gamma is negative, whereas senior tranche gamma is positive because senior tranche delta increases with wider spreads.

- **Jump-to-default w.r.t. single names:**
The *jump-to-default* as a risk measure is not standardized so far, although certain players in the market actually do jump-to-default analysis. It measures the mark-to-market or mark-to-model impact (PV change) on a tranche if a name in the portfolio defaults. The impact of a single default on CDO tranche value is obvious for all tranches above equity: default means a reduction of subordination to senior tranches, which is a decline in credit enhancement. The jump-to-default as a risk measure is comparable to *scenario analysis*. However, thinking in terms of conditional PDs, the default probabilities of all surviving reference names increase simultaneously if a jump-to-default occurs in the reference pool. This shifts the whole PD term structure of the pool with a materiality driven by the underlying dependence structure of the reference pool.

- **Gap risk:**
Gap risk is another non-standardized risk measure used by market participants. Instead of tracking the impact of *small* spread changes in the underlying, gap risk measures the impact of *large spread moves*, e.g., a 20 bps widening of iTraxx index spread and

its impact on index tranches, or the impact of a 100 bps spread widening of a single reference name.

In mixed portfolios, risk measures and 'Greeks' (delta, gamma) have to be evaluated at single name *and* at aggregated portfolio level. Effects just mentioned can be enlarged or deflated by certain combinations of positions. Depending on the strategy of the particular bank, or, at smaller scales, investment unit, different hedging strategies are necessary to achieve a certain tailor-made portfolio profile. For instance, correlation desks typically have to make sure they are not long or short in other positions than correlation. It is the job of portfolio managers to ensure full control of sensitivities and risk measures. We conclude here, but refer to two references in the bibliography that we recommend to readers interested in diving deeper into one or the other aspect of this discussion, namely, the paper by MOROKOFF [89] and the paper on delta hedging in CDOs of CDOs by BAHETI et al. [14].

Chapter 4

Some Practical Remarks

This last section is a brief collection of basic comments we include for the convenience of the reader, intended for 'newcomers' in the field. Experienced quants should skip the following text passage!

In this book, we gave an overview on modeling techniques for baskets, CDOs, and related credit instruments. As said in Section and the preface to this book, we made a selection of topics and techniques, which we find useful for our own daily work. After reading this book, the reader should have a solid understanding about dependence modeling in credit risk in general and the modeling of structured credit portfolios in particular. However, having a theoretical understanding of modeling techniques is only one side of the coin. The second, maybe even more important side, is *gaining modeling experience*. It is the daily practical work with structures from the market, which makes it possible to migrate from a theoretician to a CDO modeler. Here are some comments and remarks for the way from theory to practice:

- As a starting point, we recommend to become part of the various distribution lists in the market (from rating agencies and investment banks) announcing new CDOs and related structures in weekly or monthly periods.

- Next, it is a very good exercise to go through such periodic emailings and select a few transactions that seem to be interesting and innovative.

- For the look into a presale report or offering circular of a CDO, we found the following 4-step approach very useful:

    1. Step: Read the first few pages of the document, which often give kind of an executive summary about the deal.

    2. Step: Try to get a clear picture about the underlying reference pool: what asset classes are included, what is the

303

rating distribution, are there collateral securities taken into account or what recoveries can be expected, is it a static or managed pool, what are terms and durations, etc. This part of the exercise later on defines the *asset side* of the deal; see Figure 3.8, the left-hand side.

3. Step: Try to find the waterfall or cash flow and loss allocation description in the document. Typically, there is a clear 'bullet point ordered' list of cash flow elements in such documents, e.g., saying how interest and amortizations are paid, what fees are to be deducted from the cash flow, how losses are allocated to tranches (bottom-up provisioning or via principal deficiency ledgers, etc.), and so on. Based on observations in this part of the work, the *liability side* of the transaction can be modeled; see Figure 3.8, the right-hand side.

4. As a last step before VBA or C^{++} is started on the computer in order to model the deal, the whole document should be checked for making sure that one does not miss an important point relevant for modeling the deal.

After these steps, one is ready for implementing the deal with all the typical challenges regarding insufficient data, time pressure from the business side, and limited computational power.

- When modeling deals as a basis for *investment decisions*, it is extremely important that model results are carefully checked. A principle we found useful, although it is resource intensive, is *double programming*, which means that two quants model the same deal independently and at the end compare their calculation or simulation results. If results are not matching, a debugging phase has to be initiated. This phase has a final maturity at the time when both quants have reasonable coincidence in their results.

- Regarding modeling time we think that 1-2 days as a rule of thumb should be the average time an experienced modeler needs to develop an opinion about a CDO. In plain-vanilla CDOs (good cash top-down, bad cash bottom-up, no complicated waterfall, or pure swap-type basket derivative) the modeling exercise should take less than a day. For more complicated cash flow CDOs, modeling can take 1-2 days plus eventually 1 day for testing and verification/validation. Of course, the 'slide battle' (conversion of quant results into a presentation suitable for senior management purposes) is not included in the just-mentioned time horizons.

- As a last comment, we should say that it is extremely helpful to have an exchange of ideas with other quants working in the field. We personally owe a lot to different people who had the right advice for us at the right time. Even in at first sight 'trivial' aspects like using an efficient random number generator, experience made by others can be very helpful.

The good thing about CDOs is that there are new innovative structures coming to the market on an ongoing basis. This means that CDO modeling never completely turns into routine and never will become a boring or tedious exercise.

Chapter 5

Suggestions for Further Reading

In this section, we make a few comments regarding further readings. The collection of references we present is not complete by far. It is a small sample from a large universe of books and papers where the latter-mentioned are often available for download from the internet.

Risk Management in General

For an introduction to risk management, various books and papers are recommendable. However, a combination of the books [37] from CROUHY et al. and [81] from MCNEIL et al. will cover every aspect of risk management one can think of. The first book has an emphasis on management aspects, whereas the second book has a strong focus on quantitative modeling.

Credit Risk Modeling

Let us start with mentioning some new and very promising approaches to credit risk modeling. Often data insufficiencies generate natural model constrains. Helpful in this context are *Bayesian* methods like Markov chain Monte Carlo. In general, models have best chances to make an important contribution to the model world if they are powerful but at the same time not over-engineered, e.g., requiring too much data. A good example for an easy but widely applicable model is VASICEK's analytic portfolio model (see Appendix 6.7), which finally made its way even into the new capital accord [15] and into implied correlation modeling; see Section 3.4.4. Bayesian methods belong to this category of powerful but reasonably handable models. Recent papers in this direction are from GOESSL [55] and MCNEIL and WENDIN [82].

Regarding an introduction to credit risk modeling in general, we have a natural positive bias toward the book [25]. However, there are many other books on credit risk in the market which will do a perfect job in providing introductory guidance to the topic. Examples include DUFFIE and SINGLETON [40], LANDO [71], and ONG [96]. A book which needs a slightly stronger background in math is from

307

BIELIECKI and RUTKOWSKI [22]. Furthermore, it must not necessarily be a book that can be used as introductory material. For instance, the papers by BELKIN et al. [19, 20] provide basic know how on credit risk modeling approaches in the CreditMetrics context, CROSBIE [36], FINGER [47, 48], FREY and MCNEIL [51] are excellent introductions to the topic, and FRYE [54] and HILLEBRAND [57] can be read for a first encounter with stochastic LGDs correlated with PDs. Other papers in this direction include LOTTER et al. [79], AKHAVEIN et al. [6], PITT [99], and ZENG and ZHANG [116] (for research on correlations). An introduction to risk measures (e.g., expected shortfall) can be found in the seminal paper by ARTZNER et al. [11].

PD Term Structures

A landmarking paper in this direction is the work by JARROW, LANDO, and TURNBULL [62]. Further research has been done by various authors, see, e.g., KADAM [66], LANDO [72], SARFARAZ et al. [101], SCHUERMANN and JAFRY [104, 105], TRUECK and OEZTURKMEN [110], just to mention a few examples. A new approach via Markov mixtures, deviating from the basic Markov assumption, has been presented recently by FRYDMAN and SCHUERMANN [53]. The NHCTMC approach we propose in this book is taken from [29].

Asset Value Processes, APPs, and First Passage Times

A classical paper is the already-mentioned work by CROSBIE [36]. Other related approaches are NICKEL et al. [93], ALBANESE et al. [7], AVELLANEDA and ZHU [12], OVERBECK and SCHMIDT [97], PAN [98], ZHOU [117], and the self-contained introduction in [25], Section 3.4.

Copula Functions

A highly readable exposition on copula functions is from EMBRECHTS et al. [42]. Starting point for the copula success story were the initial papers by SKLAR [106, 107]. Other papers we want to mention are BOUYÉ et al. [30], BURTSCHELL et al. [21] (providing a nice overview of the application of different copula functions to CDO tranche evaluations), LINDSKOG [76] (which is a self-contained introduction providing an overview on copula approaches in credit risk), NELSON [91] (as a standard reference/text book on copulas), as well as CHERUBINI et al. [32]. NELSON [91, 92] also wrote a couple of survey papers on copulas, which can be downloaded in the internet. A 'classical' book on dependence concepts among random variables is the book by JOE [58]. In this book, one also finds a general approach to mixture models playing a dominant role in credit risk modeling. A starting point of the re-discovery of copula functions and correlated default times came from LI [74].

Structured Credit and CDOs

A standard reference for credit derivative modeling is the book by SCHOENBUCHER [103]. Management and product-oriented books, which help in understanding structured credits are from CHOUDRY [33] and FELSENHEIMER et al. [46].

Besides books on CDOs, certain investment banks published surveys or manuals on structured finance and CDOs, e.g., J. P. MORGAN [64]. From the same bank, one also finds an introduction to base correlations in [65], identifying JPM as the 'inventor' of this now widely used quoting standard. Another example for an investment bank's survey on CDOs with a focus on market research (as of December 2005) is from LEHMAN [73]. However, every major investment bank has such annual reports and research brochures, the two mentioned papers are just examples for a general credit research culture.

Research on STCDOs includes AMATO and GYNTELBERG [8], ANDERSON et al. [9, 10] (where the second paper introduces a stochastic correlation model in order to match correlation smiles; it can be shown that their copula is a convex sum of two one-factor copulas, which can be written in a form showing that the correlation parameter is randomly chosen, leading to so called random factor loadings), BAXTER [17] (who works with a so-called Brownian variance gamma (BVG) APPs, which actually can be written as a time-changed Brownian motion based on which he develops a structural model for CDO evaluation and Basket pricing), BEINSTEIN et al. [18], HULL and WHITE [60] (for an approach to fit market spreads on tranches by means of an implicit copula approach), O'KANE and LIVESEY [95], FITCHRATINGS [49], and WILLEMANN [115]. Many other researchers could be named here.

Besides the just-mentioned papers, we want to draw the readers attention to the papers by FRIES and ROTT [52] (focussing on CDO sensitivities), WALKER [112] (discussing a risk-neutral term structure valuation model for CDO tranches), and LONGSTAFF and RAJAN [78] who introduce a 3-factor portfolio model by which market data can be explained very well.

Readers interested in an extensive discussion on iTraxx, including many examples from the market, should look into FELSENHEIMER et al. [45]. The authors introduce iTraxx as an index and all relevant iTraxx-related products like, e.g., index tranches.

Probability Theory

Last but not least we mention four books on probability theory, which provide the necessary basis for understanding concepts related to stochastics in this book. The first book [16] from BAUER is available in German and English and can be seen as a comprehensive course in mathematical (measure theoretic) probability. The second book is a classical and highly readable text book on probability theory from BREIMAN [31]. The third book is a more recent book from STEELE [109] on stochastic processes and stochastic calculus. The last book we want to mention is the well-known text from KARATZAS and SHREVE [67] on Brownian motion and stochastic calculus.

Chapter 6

Appendix

6.1 The Gamma Distribution

The gamma distribution is widely used in different appliied fields of probability theory. It has the probability density

$$\gamma_{\alpha,\beta}(t) = \frac{1}{\beta^\alpha \Gamma(\alpha)} \exp\left(-\frac{t}{\beta}\right) t^{\alpha-1}$$

for $t \geq 0$ and $\alpha, \beta > 0$. Basic summary statistics for a gamma distributed random variable X are

- $\mathbb{E}[X] = \alpha\beta$
- $\mathbb{V}[X] = \alpha\beta^2$
- $\gamma_{\alpha,\beta}(t_{\max}) \stackrel{!}{=} \max$ for $t_{\max} = \beta(\alpha - 1)$.

The parameter α can be interpreted as a *shape* parameter, whereas the second parameter β can be considered as a *scale* parameter. Special gamma distributions include the

- Erlang distribution in case of $\alpha \in \mathbb{N}$
- Exponential distribution in case of $\alpha = 1$
- Chi-squared distribution with n degrees of freedom in case of $\alpha = n/2$ and $n \in \mathbb{N}$.

The Laplace transform of the gamma distribution is given by

$$\mathbb{E}[\exp(-tX)] = (1 + \beta t)^{-\alpha} \qquad (X \sim \Gamma(\alpha, \beta);\ t \geq 0).$$

In case of the Clayton copula in Section 2.5.6.5, we use this fact for the special case $X \sim \Gamma(1/\eta, 1)$ to obtain

$$\mathbb{E}[\exp(-tX)] = (1 + t)^{-1/\eta}$$

which equals the inverse of the generator of the Clayton copula.

6.2 The Chi-Square Distribution

The χ^2-distribution is one out of several distributions constructed by means of normal random variables. Starting point is sample of independent normal variables $X_1, ..., X_d \sim N(0,1)$ where d is a positive integer. Then,

$$X_1^2 + \cdots + X_d^2$$

is said to be χ^2-*distributed with d degrees of freedom*. We write $X \sim \chi^2(d)$. The first and second moments X are

$$\mathbb{E}[X] = d \quad \text{and} \quad \mathbb{V}[X] = 2d.$$

The χ^2 distribution can be derived from the gamma distribution because any $\chi^2(d)$-density equals the gamma distribution density with parameters $\alpha = d/2$ and $\beta = 2$.

6.3 The Student-t Distribution

The Student-t distribution (or, in short: the t-distribution) has two main building blocks: a standard normal variable $Y \sim N(0,1)$ and a χ^2-distributed variable $X \sim \chi^2(d)$ where d is a positive integer. The two building blocks are assumed to be independent. Then, the random variable

$$Z = \frac{Y}{\sqrt{X/d}}$$

is said to follow a *t-distribution with d degrees of freedom*. As notation, we indicate the t-distribution by writing

$$Z \sim t(d).$$

The corresponding distribution function will be denoted by Θ_d, such that, e.g., the likelihood that Z exceeds some number z writes as

$$\mathbb{P}[Z > z] = 1 - \Theta_d[z].$$

The probability density of Z is given by

$$\theta_d(x) = \frac{\Gamma((d+1)/2)}{\sqrt{\pi d}\,\Gamma(d/2)} \left(1 + \frac{x^2}{d}\right)^{-(d+1)/2} \qquad (x \in \mathbb{R}).$$

The first and second moments of $Z \sim t(d)$ are given by

$$\mathbb{E}[Z] = 0 \quad (d \geq 2) \quad \text{and} \quad \mathbb{V}[Z] = \frac{d}{d-2} \quad (d \geq 3).$$

For large degrees of freedom d, the t-distribution is close to the normal distribution. In fact, we have convergence 'in distribution' to the standard normal distribution for $d \to \infty$. This can be verified by looking into any textbook on probability theory. In general, the t-distribution puts more mass in the tails than a normal distribution: the lower the degrees of freedom d, the fatter the tails of the distribution.

The t-distribution has a multi-variate version. Analogous to the one-dimensional t-distribution, the multi-variate t-distribution has two building blocks: a multi-variate Gaussian vector $\boldsymbol{Y} = (Y_1, ..., Y_m) \sim N(0, \Gamma)$ with a correlation matrix Γ and again a χ^2-distributed random variable $X \sim \chi^2(d)$. Again, it is assumed that $\boldsymbol{Y}$ and X are independent. The random vector

$$\boldsymbol{Z} = \frac{1}{\sqrt{X/d}} \boldsymbol{Y}$$

is said to be *multi-variate t-distributed with d degrees of freedom*. We denote the distribution function of $\boldsymbol{Z}$ by $\Theta_{m,\Gamma,d}$ and write

$$\boldsymbol{Z} \sim t(m, \Gamma, d).$$

Note that the dependence structure of $\boldsymbol{Z}$ is more complex than the dependence structure of $\boldsymbol{Y}$: in the multi-variate normal case, Γ rules the linear dependence of the variables $Y_1, ..., Y_m$. In the multi-variate t-distribution case, there are two sources of dependency between the marginal variables, namely

- The correlation matrix Γ such that $\boldsymbol{Z}$ 'inherits' the linear correlation structure from $\boldsymbol{Y}$,

$$\text{Corr}[\sqrt{d/X}\, Y_i, \sqrt{d/X}\, Y_j] = \text{Corr}[Y_i, Y_j]$$

  (this can be verified by a straightforward conditioning argument on X)

- The factor $\sqrt{d/X}$, which simultaneously is applied to all components in the random vector $\boldsymbol{Y}$

The latter-mentioned induces a certain dependence among the marginal components of a multi-variate t-distributed random vector even in case of zero linear correlations.

6.4 A Natural Clayton-Like Copula Example

Clayton copulae arise quite naturally in statistical problems. We found a nice example supporting this statement in NELSON [90]. For interested readers, the following example provides an opportunity to 'play' a little more with the copula concept.

Let $X_1, ..., X_m$ be a sequence of i.i.d. continuous random variables with (common) distribution function $\mathbb{F}$. Denote by

$$X_{(1)} = \min\{X_1, ..., X_m\}$$

the minimum of the sequence and by

$$X_{(m)} = \max\{X_1, ..., X_m\}$$

the maximum of the sequence, being the most left and right parts of the order statistics of $X_1, ..., X_m$. Denote by $\mathbb{F}_{(1)}$ and $\mathbb{F}_{(m)}$ the distribution functions of $X_{(1)}$ and $X_{(m)}$, respectively. We have

$$\begin{aligned} \mathbb{F}_{(m)}[x] &= \mathbb{P}[\max\{X_1, ..., X_m\} \leq x] \\ &= \mathbb{P}[X_1 \leq x, ..., X_m \leq x] \\ &= \mathbb{P}[X_1 \leq x] \cdots \mathbb{P}[X_m \leq x] \\ &= \mathbb{F}[x]^m. \end{aligned}$$

For the minimum $X_{(1)}$ we obtain

$$\begin{aligned} \mathbb{F}_{(1)}[x] &= \mathbb{P}[\min\{X_1, ..., X_m\} \leq x] \\ &= \mathbb{P}[-\max\{-X_1, ..., -X_m\} \leq x] \\ &= \mathbb{P}[\max\{-X_1, ..., -X_m\} \geq -x] \\ &= 1 - \mathbb{P}[\max\{-X_1, ..., -X_m\} < -x] \\ &= 1 - \mathbb{P}[-X_1 < -x, ..., -X_m < -x] \\ &= 1 - \mathbb{P}[X_1 > x] \cdots \mathbb{P}[X_m > x] \\ &= 1 - (1 - \mathbb{F}[x])^m. \end{aligned}$$

Then, the copula $C_{(-1,m)}$ of the joint distribution of $(-X_{(1)}, X_{(m)})$ has the form of a Clayton-like copula,

$$C_{(-1,m)}(u, v) = [\max\{u^{1/m} + v^{1/m} - 1, 0\}]^m. \tag{6.1}$$

For a proof, we consider

$$\begin{aligned}\mathbb{P}[-X_{(1)} \leq x, X_{(m)} \leq y] &= \mathbb{P}[-x \leq X_{(1)}, X_{(m)} \leq y] \\ &= \mathbb{P}[X_i \in [-x, y] \; \forall \; i = 1, ..., m].\end{aligned}$$

In other words, we have

$$\mathbb{P}[-X_{(1)} \leq x, X_{(m)} \leq y] = \begin{cases} (\mathbb{F}[y] - \mathbb{F}[-x])^m & \text{it } -x \leq y \\ 0 & \text{if } -x > y \end{cases}$$

such that altogether

$$\mathbb{P}[-X_{(1)} \leq x, X_{(m)} \leq y] = (\max\{\mathbb{F}[y] - \mathbb{F}[-x], 0\})^m. \quad (6.2)$$

Denoting by $\mathbb{F}_{(-1)}$ the distribution function of $-X_{(1)}$, we obtain

$$C_{(-1,m)}(u,v) = \mathbb{P}\big[-X_{(1)} \leq \mathbb{F}_{(-1)}^{-1}[u], X_{(m)} \leq \mathbb{F}_{(m)}^{-1}[v]\big]. \quad (6.3)$$

Based on $\mathbb{F}_{(1)}[x] = 1 - (1 - \mathbb{F}[x])^m$ we obtain

$$\mathbb{F}_{(-1)}[x] = (1 - \mathbb{F}[-x])^m.$$

Setting $u = \mathbb{F}_{(-1)}[x] = (1 - \mathbb{F}[-x])^m$ provides inversion of $\mathbb{F}_{(-1)}$ via

$$u^{1/m} - 1 = -\mathbb{F}[-x]$$

such that Equations (6.2) and (6.3) can be summarized by

$$C_{(-1,m)}(u,v) = (\max\{v^{1/m} + u^{1/m} - 1, 0\})^m$$

by additionally setting $v = \mathbb{F}_{(m)}[y] = \mathbb{F}[y]^m$ in order to find the inverse of $\mathbb{F}_{(m)}$ in Equation (6.3). This proves (6.1).

The exercise above shows that certain copulas may look exotic at first sight but nevertheless can arise quite naturally when considered in a certain context.

6.5 Entropy-Based Rationale for Gaussian and Exponential Distributions as Natural Standard Choices

In various places in this book, the Gaussian distribution (and copula function) and the exponential distribution occur as standard choices for distributions in credit risk modeling. In this section, we briefly outline an argument borrowed from information theory explaining to some extent why these choices are not as esoteric or arbitrary as they might look at first sight.

A common measure of *indeterminacy* or *degree of randomness* of a random variable or probability distribution is its *entropy*. The typical application context of entropy is information theory, see, e.g., COVER and THOMAS [34]. For a discrete random variable X with a (discrete) probability distribution $\boldsymbol{p}$,

$$\boldsymbol{p} = (p_1, ..., p_n), \quad p_k = \mathbb{P}[X = x_k] \ \forall \ k = 1, ..., n, \quad \sum_{k=1}^{n} p_k = 1,$$

the entropy of X is defined as

$$H(X) = -\sum_{k=1}^{n} p_k \ln p_k.$$

Before we consider the continuous analogon of H, we should think about a few simple properties of H.

- $H(X) \geq 0$ for all random variables X

- $H(X) = 0$ if and only if there exists one k_0 such that $p_{k_0} = 1$ and $p_k = 0$ for all other $k \neq k_0$

- $H(X)$ attains its maximum value for uniformly distributed X, i.e., $H(X)$ is largest in case of

$$p_k = \mathbb{P}[X = x_k] = 1/n \quad \forall \ k = 1, ..., n.$$

For a proof, convex function theory can help; see, e.g., [34]

Properties 2 and 3 in combination say that *the entropy can be used as a measure of the randomness* of random variables. A random variable X with $H(X) = 0$ attains the value x_{k_0} *with certainty* for some k_0. Therefore, it can be seen as a deterministic number *exhibiting no randomness at all*. In contrast, a random variable X, where $H(X)$ attains its maximum, *bears greatest possible uncertainty* because all events $\{X = x_k\}$ have the same likelihood, there is no event which is more likely than other events. *The higher the entropy of a random variable, the more random the variable behaves.* This is consistent with other contexts, e.g., thermodynamics, where the entropy is used as a measure for chaos or disorder in a physical system.

If X is a continuous random variable with a density function $f(x)$, then the definition of entropy is the natural extension[1] of discrete entropy to the continuous case. It is given by

$$H(X) = -\int_{-\infty}^{\infty} f(x)\ln(f(x))\,dx,$$

hereby assuming that X lives on $(-\infty, \infty)$. If not, the integration range can be restricted accordingly.

Given certain side conditions, one can find *maximum entropy distributions*; see [34] for the following statements.

- In the universe of all continuous distributions living on the real axis with a specified mean μ and standard deviation σ, the normally distributed random variable $X \sim N(\mu, \sigma^2)$ has maximum entropy.

- The exponential distribution with intensity λ has maximum entropy compared to all continuous distributions with support in $[0, \infty)$ and mean value $1/\lambda$.

- Among all continuous distributions with support in the interval $[a, b]$, the uniform distribution on $[a, b]$ has maximum entropy.

The results carry over to the multi-variate case. Therefore, *from the perspective of entropy, the normal distribution and, more general, the Gaussian copula function are natural choices if one wants to choose the most randomly behaving distribution with given side conditions* (like matching mean and vola, etc.). In other contexts, the exponential distribution is the most natural choice if one wants to choose the most randomly behaving distribution living on the time axis with a mean given by the reciprocal of an intensity, e.g., a default intensity. This concludes our brief excursion on entropy maximizing distributions.

---
[1]As usual, the sum is replaced by an integral w.r.t. the probability density.

6.6 Tail Orientation in Typical Latent Variable Credit Risk Models

Some remarks in the context of copulas and dependence measures associate lower tail dependence with downside risk and upper tail dependence with upside chances. In this section, we briefly recall why this makes sense in the context of latent variable models used in this book. For this purpose, let us use a simple one-factor model as described in Equation (2.3) as a generic to model the credit risk of a portfolio with m credit-risky assets,

$$\text{CWI}_i = \sqrt{\varrho}\, Y + \sqrt{1-\varrho}\, Z_i \qquad (i=1,...,m) \qquad (6.4)$$

with $Y, Z_1, ..., Z_m \sim N(0,1)$ i.i.d., such that defaults in the portfolio are modeled by Bernoulli variables

$$\mathbf{1}_{\{\text{CWI}_i < c_i\}}$$

with c_i referring to a default-critical threshold for asset i. Recall that ϱ refers to the underlying CWI correlation. Then, the default condition

$$\text{CWI}_i < c_i$$

already determines the 'direction' or risk and chances: starting with some CWI greater than the critical threshold means that movements of the CWI toward lower values brings the asset closer to default or actually into default, whereas movements of the CWI toward higher values increase the bankruptcy remoteness of the asset.

Figure 6.1 taken from [25] illustrates this mechanism graphically. It shows the conditional PD as calculated in Equations (2.12) and (3.23),

$$g_{p_i,\varrho}(y) = g(p_i, \varrho, ; y) = N\left[\frac{c_i - \sqrt{\varrho}\, y}{\sqrt{1-\varrho}}\right]$$

where p_i refers to the (unconditional) one-year PD of asset i given by

$$p_i = \mathbb{P}[\text{CWI}_i < c_i] = \int_0^\infty g_{p_i,\varrho}(y)\, dN(y)$$

as usual in Bernoulli mixture models, see, e.g., [25], Chapter 2.

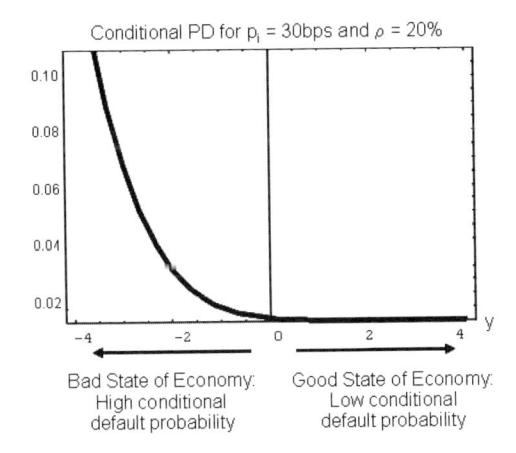

FIGURE 6.1: PD of assets conditional on realizations of the underlying systematic risk factor Y; see [25], Figure 2.3

Summarizing, the risk of movements of latent underlying factors toward lower values correspond to downside risk, whereas any movement toward higher values corresponds to upside chances, or, more precise, better chances of survival. Therefore, lower tail dependence can be associated with increased downside risk because assets will move in a, to some extent, coordinated way down to lower values, whereas upper tail dependence corresponds to better joint survival chances. Obviously, *no kind of general principle[2] forces us to see factor movements in specified directions in a specified way*, but the models considered in this book and many other publications by various authors behave in this way. If we want to change the assignment of factor moves to downside risk or upside chances, we have to change everything else in the model framework accordingly. The only really necessary condition is consistency.

---

[2]In fact, if we change the direction of underlying latent variables, the tail orientation in the induced model changes too. For instance, if spread would be used as an underlying latent indicator, then the tail orientation would be opposite to the orientation chosen in this book: increased spread means increased risk and decreased spread means a movement to a state with reduced risk.

6.7 The Vasicek Limit Distribution

The limit distribution for uniform portfolios with a uniform default probability p and a uniform CWI correlation ϱ goes back to a seminal paper by VASICEK [111] and constitutes one of the most famous and widely applied modeling 'tricks' in the world of credit models. The approach works as follows.

We start with m obligors, which are all assumed to have the same default probability p and reference the same exposure amount. We fix some time horizon T such that p is the default probability of obligors w.r.t. the evaluation horizon T. Introducing a Bernoulli mixture model for this situation (see, e.g., [25], Chapter 2), every obligor i defaults if and only if the indicator or Bernoulli variable

$$\mathbf{1}_{\{\text{CWI}_i < N^{-1}[p]\}}$$

equals 1 (indicating default), where the CWIs are parameterized by a uniform one-factor model

$$\text{CWI}_i = \sqrt{\varrho}\, Y + \sqrt{1-\varrho}\, \varepsilon_i \qquad (i=1,...,m).$$

We have seen this formula for one-factor models many times in this book, as well as the corresponding conditional PD for obligor i,

$$g_{p,\varrho}(y) = g(p,\varrho,;y) = N\left[\frac{N^{-1}[p] - \sqrt{\varrho}\, y}{\sqrt{1-\varrho}}\right]$$

see Section 2.5.4 and Proposition 2.2.3 as well as Equations (2.12), (2.31), and (2.32).

VASICEK's limit theorem says that in the setup described above the percentage portfolio loss

$$L_{PF(m)} = \frac{1}{m}\sum_{i=1}^{m} \mathbf{1}_{\{\text{CWI}_i < N^{-1}[p]\}}$$

admits almost surely a limit for $m \to \infty$ as follows:

$$\mathbb{P}\big[\lim_{m\to\infty}(L_{PF(m)} - g(p,\varrho;Y)) = 0\big] = 1 \qquad (6.5)$$

Appendix

Comparison of analytic loss distributions

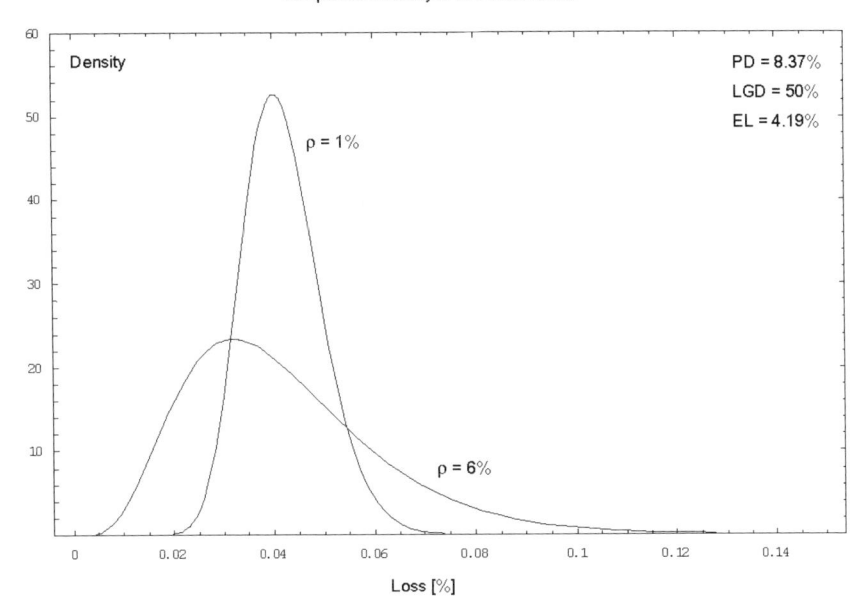

PD = 8.37%
LGD = 50%
EL = 4.19%

ρ = 1%

ρ = 6%

Loss [%]

FIGURE 6.2: Analytic loss densities w.r.t. the 5-year average PD from the reference portfolio described in Appendix 6.9 and applied in Section 3.3.6.1

see [111] or Proposition 2.5.4 in [25]. Equation (6.5) can be seen as a special case of Corollary 3.3.5 so that this book actually contains a proof of (6.5). The distribution function of this limit distribution has been applied in Proposition 3.3.2,

$$\mathbb{P}[L \leq x] = \mathbb{P}[g(p, \varrho; Y) \leq x]$$
$$= N\left[\frac{1}{\sqrt{\varrho}}\left(N^{-1}[x]\sqrt{1-\varrho} - N^{-1}[p]\right)\right]$$

where L denotes the limit random variable. Calculating the derivative w.r.t. the percentage portfolio loss x, we obtain the loss density shown in Equation (3.24). Figure 6.2 shows two examples.

As an interpretation of VASICEK's limit theorem one can say that for $m \to \infty$ the *fraction of defaulted assets* equals the conditional PD of the underlying assets with *exchangeable* risk characteristics (same PD, uniformly correlated, uniform exposures). The limit distribution is often taken as a proxy for the 'true' portfolio loss distribution for portfolios with many obligors. In Section 3.3.6.1 we applied the limit distribution in the sense of an *analytical approximation*.

6.8 One-Factor Versus Multi-Factor Models

In this section, we want to point out some facts regarding the comparison of one-factor versus multi-factor models. For this purpose we consider the two fundamentally different CWI factor representations used in different parts of this book:

$$\text{one-factor model:} \quad \text{CWI} = \sqrt{\varrho_i}\, Y + \sqrt{1-\varrho_i}\, \varepsilon_i$$
$$\text{multi-factor model:} \quad \text{CWI} = \beta_i \Phi_i + \sqrt{1-\beta_i^2}\, \varepsilon_i$$

where Φ_i is the composite factor of firm i defined in Equation (2.94) as

$$\Phi_i = \frac{\Psi_{n_{\text{region}}(i)} + \Psi_{5+n_{\text{industry}}(i)}}{\sqrt{2 + 2\text{Corr}[\Psi_{n_{\text{region}}(i)} + \Psi_{5+n_{\text{industry}}(i)}]}} \tag{6.6}$$

(see, e.g., Equation (2.3) for the one-factor model and Section 2.6.1 and Appendix 6.9 for the multi-factor model).

At first sight it seems that the two models can be compared 'component by component' associating ϱ_i with β_i^2 (which equals the R-squared of firm i) and Y with the composite factor Φ_i of name i, recalling that by construction via normalizing factors ν_i from Equations (2.92) and (2.93) both systematic factors have a variance of 1, and so on. However, as we will see in this section, *such a comparison is completely misleading*. Before we look at this more detailed, let us play a little bit with the multi-factor model to get a better feeling for the multi-variate setup.

For this purpose, we consider two cases as examples.

- Case 1: Firms i and j have the same region and same industry. The CWI correlation of the firms then calculates as

$$\text{Corr}[\text{CWI}_i, \text{CWI}_j] = \beta_i \beta_j \text{Corr}[\Phi_i, \Phi_j] = \beta_i \beta_j$$

because both firms have identical composite factors. This case actually coincides with the calculation of CWI correlations based on the one-factor model where we have

$$\text{Corr}[\text{CWI}_i, \text{CWI}_j] = \sqrt{\varrho_i}\sqrt{\varrho_j} \qquad (6.7)$$

based on the distributional assumptions made on involved random variables, see, e.g., (2.3). Recall that β_i corresponds to $\sqrt{\varrho_i}$.

- Case 2: Firms i and j have a different region or a different industry or are different w.r.t. both. Then,

$$\text{Corr}[\Phi_i, \Phi_j] = \nu_i \nu_j (\rho_1 + \rho_2 + \rho_3 + \rho_4) < 1 \qquad (6.8)$$

such that some diversification at systematic risk level takes place; see Equation (2.97). The numbers $\rho_1, \rho_2, \rho_3, \rho_4$ are given by

$$\begin{aligned}\rho_1 &= \text{Corr}[\Psi_{n_{\text{region}}(i)}, \Psi_{n_{\text{region}}(j)}] \\ \rho_2 &= \text{Corr}[\Psi_{n_{\text{region}}(i)}, \Psi_{5+n_{\text{industry}}(j)}] \\ \rho_3 &= \text{Corr}[\Psi_{5+n_{\text{industry}}(i)}, \Psi_{n_{\text{region}}(j)}] \\ \rho_4 &= \text{Corr}[\Psi_{5+n_{\text{industry}}(i)}, \Psi_{5+n_{\text{industry}}(j)}]\end{aligned}$$

and ν_i, ν_j are the variance normalizing factors defined in Equations (2.92) and (2.93), for instance for obligor i we have

$$\nu_i = \frac{1}{\sqrt{2 + 2\text{Corr}[\Psi_{n_{\text{region}}(i)} + \Psi_{5+n_{\text{industry}}(i)}]}}.$$

The main insight following from this case is that in such a setup *the composite factor of firms is subject to diversification effects based on its decomposition into different (further diversifying) systematic subcomponents like regions and industries.* Hereby, the degree of diversification at systematic level is controlled by the correlation matrix from Table 6.3, which shows the pairwise correlations between systematic indices.

Case 1, where all obligors are mapped to one region and one industry (making the composite factors of firms indeed equal to just one factor) is in fact the only case where the two factor representations from the beginning of this section are comparable. In Case 2, the potential for *diversification at systematic level* induced by the multi-factor setup of the factor parameterization of CWIs completely changes the picture. If we want to approximate the loss distribution based on the multi-factor model by a loss distribution based on the one-factor model we cannot expect to find a *functional relation between the parameters of the multi-factor model* and the vector of parameters $(\varrho_i)_{i=1,...,m}$ determining the one-factor model.

Let us calculate some explicit summary statistics figures in the context of our multi-factor model as described in Appendix 6.9 to illustrate the non-comparability of the two factor representations in Case 2 as described above. Based on the same calculations, which led to Figure 2.31 in Section 2.6.1, we can calculate the *mean value $\bar{\varrho}$ over all pairwise CWI correlations* in the the 100-names portfolio underlying different examples in this book; see Tables 6.1 and 6.2. We get

$$\bar{\varrho} = 16.9\% \tag{6.9}$$

as can roughly be guessed from Figure 2.31. For (6.8), we have the following mean values regarding components of $\mathrm{Corr}[\Phi_i, \Phi_j]$ when averaged over all pairwise combinations of names in the sample portfolio:

- $\bar{\rho}_1 = 48.7\%$
- $\bar{\rho}_2 = 32.9\%$
- $\bar{\rho}_3 = 30.5\%$
- $\bar{\rho}_4 = 23.7\%$

Table 6.3 combined with the distribution of regions (Figure 6.5) and industries (Figure 6.6) for names in the sample portfolio can be used to derive these average correlations. In addition, the average over all normalizing constants yields

$$\bar{\nu} = \frac{1}{m}\sum_{i=1}^{m}\frac{1}{\sqrt{2+2\mathrm{Corr}[\Psi_{n_{\mathrm{region}}(i)}+\Psi_{5+n_{\mathrm{industry}}(i)}]}} = 60.9\%.$$

As a proxy, we can now use these average figures to calculate the average correlation of composite factors. According to (6.8) we get

$$\overline{\mathrm{Corr}[\Phi_i,\Phi_j]} \approx (60.9\%)^2(48.7\% + 32.9\% + 30.5\% + 23.7\%)$$
$$= 52.1\%. \tag{6.10}$$

We also can calculate $\overline{\mathrm{Corr}[\Phi_i, \Phi_j]}$ directly and exact by calculating all pairwise composite factor correlations. We get

$$\overline{\mathrm{Corr}[\Phi_i, \Phi_j]} = 51.8\%$$

which is very close to the approximative number in Equation (6.10). Using the just-derived average numbers, we can re-calculate/approximate the average pairwise correlation in the portfolio. We obtain

$$\mathrm{Corr}[\mathrm{CWI}_i, \mathrm{CWI}_j] = \beta_i \beta_j \mathrm{Corr}[\Phi_i, \Phi_j] \approx \beta_i \beta_j \times 52.1\% \quad (6.11)$$

based on Equations (6.8) and (6.10). By explicitly calculating the portfolio average of the beta products $\beta_i \beta_j$, we arrive at

$$\mathrm{Corr}[\mathrm{CWI}_i, \mathrm{CWI}_j] \approx 32.4\% \times 52.1\% \approx 16.9\% \quad (6.12)$$

confirming the exactly calculated average correlation from (6.9).

So within the multi-factor setup, we can confirm summary statistics of correlation-related parameters by different approximative calculations. Now assume that we want to approximate the multi-factor model based loss distribution by a one-factor model based loss distribution, for the sake of simplicity with uniform R-squared parameters,

$$\varrho_i \stackrel{!}{=} \varrho$$

with a (uniform) CWI correlation ϱ,

$$\mathrm{Corr}[\mathrm{CWI}_i, \mathrm{CWI}_j] = \varrho \quad \forall\, i,j \text{ with } i \neq j.$$

As explained above, the multi-factor model allows, in contrast to the one-factor model, for diversification in terms of different systematic risks (regions and industries). Therefore, we *expect ϱ to be lower than the multi-factor model based average pairwise CWI correlation* $\overline{\varrho} = 16.9\%$. But how much lower is hard to say without further investigations. We are in the lucky position to exactly know the multi-factor model and the corresponding loss distribution. So we actually can find out what ϱ yields a good fit in our particular situation.

Typical practitioner's approaches to perform such a so-called *analytical approximation* are

- First and second moment matching, enforcing equal expected and unexpected[3] losses of the two loss distributions

---
[3] Standard deviation of the loss distribution.

- First moment and quantile matching, enforcing equal expected losses and equal quantiles w.r.t. a chosen level of confidence of the two loss distributions

Let us do this exercise in order to see what we get. The first moment or expected loss matching is the easy part. We generate a uniform *average PD term structure* for the portfolio via

$$\overline{p}^{(t)} = \frac{1}{m} \sum_{i=1}^{m} p_i^{(t)} \quad (t \geq 0). \tag{6.13}$$

Note that this is the best choice in case of equal exposures as it is the case in our examples. In general, it makes sense to calculate average PDs *weighted w.r.t. net exposure*. For the cumulative loss distribution at the 5-year horizon (term of the CSO in Section 3.3.1), we choose

$$p = \overline{p}^{(5)}$$

and calculate the portfolio loss according to the distribution

$$\mathbb{P}[L = k \times \text{LGD}] = \binom{m}{k} \int_{-\infty}^{\infty} g(p, \varrho; y)^k \left(1 - g(p, \varrho; y)\right)^{m-k} dN(y) \tag{6.14}$$

where $g(p, \varrho; y)$ is the conditional PD from Equation (3.23) (see also Appendix 6.7) and L denotes[4] the portfolio loss at the one-year horizon. The binomial coefficient in the formula expresses the *exchangeability of assets* in choosing k defaulters out of m assets without having to be bothered about *which* assets are defaulting. Equation (6.14) is true for a deterministic uniform LGD which is the case in our examples where we set LGD = 50%. Equation (6.14) yields an expected loss

$$\text{EL} = \sum_{k=0}^{m} k \times \text{LGD} \times \mathbb{P}[L = k \times \text{LGD}]$$

matching the EL from the multi-factor model by construction.

Regarding the determination of ϱ we do a standard deviation and a quantile matching and compare the so obtained uniform CWI correlations ϱ. We get the following results:

---
[4]For the sake of a simplified notation, we drop some subscripts in this appendix.

Appendix 327

- The standard deviation matching yields $\varrho = 6\%$ whereas

- The quantile matching yields $\varrho = 7.21\%$ for the 99.9%-quantile and $\varrho = 7.24\%$ for the 99.98%-quantile

In other words, *switching the factorization of CWIs from our multi-factor to a uniform one-factor model yields an average CWI correlation drop from roundabout 17% down to roundabout 6%-8%.* As always in this book, we ask if this is plausible. Figure 6.3 shows the results of the analytic approximations based on moment matching.

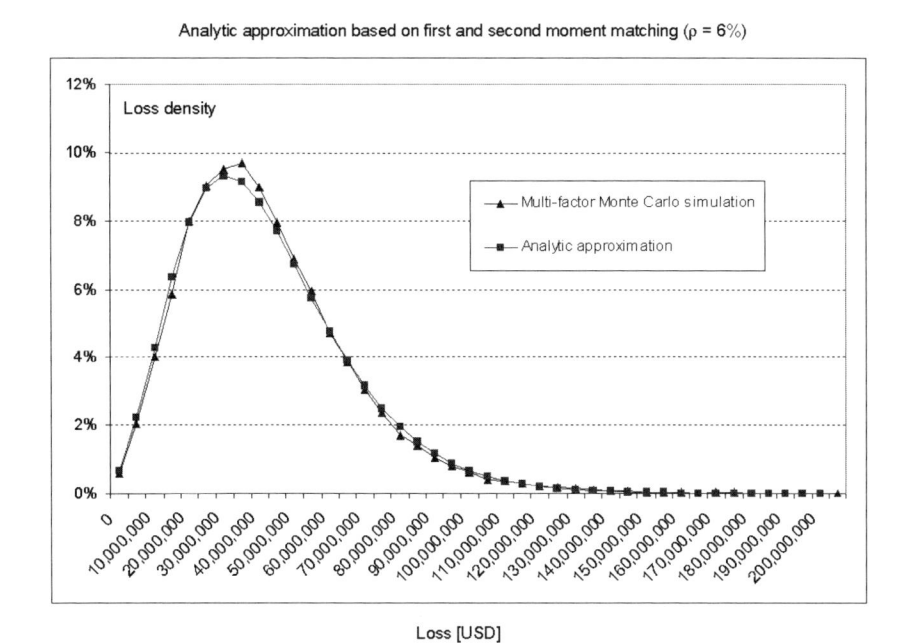

FIGURE 6.3: Illustration of analytic approximations

A simple exercise we can do in order to double-check if a ϱ of 6%-8% is a reasonable CWI correlation parameter - for the analytic approximation in our particular example, is *JDP matching*. The rationale underlying this calculation is the following. In a Gaussian[5] copula model,

---
[5]Recall that multi-variate Gauss distributions as well as Gauss processes are, from a distributional point of view, uniquely determined by their expectation function and

joint default events are exclusively driven by the correlation matrix of the names, no matter how we came up with these pairwise correlations. Our multi-factor model is one particular parameterization yielding the corresponding pairwise correlations pictured in Figure 2.31, but for the final loss distribution the correlations matter and not the underlying factor model *inducing* the correlations. JDP matching, based on joint default probabilities (see Remark 2.2.3), *takes the average CWI correlation $\bar{\varrho}$ based on all pairwise (multi-factor) CWI correlations as well as the rating/PD distribution in the reference portfolio into account.* The JDP matching implied CWI correlation is the ϱ satisfying the equation

$$\frac{1}{m(m-1)/2} \sum_{i=1}^{m} \sum_{j=i+1}^{m} N_2\big[N^{-1}[p_i^{(5)}], N^{-1}[p_j^{(5)}]; \bar{\varrho}\big] \stackrel{!}{=} \quad (6.15)$$

$$\stackrel{!}{=} N_2\big[N^{-1}[\bar{p}^{(5)}], N^{-1}[\bar{p}^{(5)}]; \varrho\big]$$

where $N_2[\cdot, \cdot\,; \varrho]$ denotes the standard bi-variate normal distribution function (see also Equations (2.5), (2.58), and (2.60)), $\bar{p}^{(5)}$ denotes the 5-year PD from the average portfolio PD term structure defined in (6.13), and ϱ is the JDP matching implied CWI correlation for a uniform reference portfolio. The correlation $\bar{\varrho}$ in Equation (6.15) is the average of the multi-factor based pairwise CWI correlations, which equals 16.9% according to Equation (6.9). In our example we get

$$\frac{1}{m(m-1)/2} \sum_{i=1}^{m} \sum_{j=i+1}^{m} N_2\big[N^{-1}[p_i^{(5)}], N^{-1}[p_j^{(5)}]; \bar{\varrho}\big] = 90 \text{ bps.}$$

The average 5-year PD in the portfolio is given by

$$\bar{p}^{(5)} = 8.37\%$$

such that the bivariate normal distribution can be applied to get

$$\varrho = 7.85\%$$

which is close to the quantile-based matching implied uniform correlations. So we finally find via JDP matching additional confirmation for a uniform CWI correlation somewhere in the range of 6% to 8% successfully applicable in an analytic approximation of the loss distribution of the reference portfolio described in Appendix 6.9.

---

covariance kernel.

6.9 Description of the Sample Portfolio

In this section, we describe the sample portfolio from which assets are chosen for inclusion in examples in this book. Note that the portfolio and its parameterization including the matrix of systematic index correlations is purely illustrative. It solely serves the purpose to generate examples with a realistic flavor.

Figures 6.1 and 6.2 exhibit 100 fictitious credit-risky names. Let us briefly describe the data fields used to characterize/parametrize the assets.

- **Asset No.** has no deeper meaning; its only purpose is to have a unique key in order to reference assets.

- **Rating** refers to the credit rating of assets. Figure 6.4 illustrates the rating distribution in the portfolio.

- **PD** specifies the default probability of assets. It reflects the mapping of rating letter combinations to default probabilities according to Table 1.1.

- **Region** is a systematic risk driver and refers to the broader geographic region companies in the portfolio have their main business in. Figure 6.5 reports on the regional distribution of names in the portfolio.

- **Industry** is the second driver of systematic risk in the portfolio. It describes the leading industry of the overall business of firms. Figure 6.6 illustrates the industrial distribution of names in the portfolio.

- **Beta** refers to the beta of firms. In our example, it is given by the square-root of the R-squared parameter of the considered firm. Figure 6.7 illustrates the distribution of R-squared parameters of names in the portfolio.

- **Exposure** shows the exposure aligned to each name. For reasons of simplicity, we assume uniform exposures as base case example. Variations of the scheme will be explicitly mentioned.

TABLE 6.1: Portfolio of sample assets, part I

Asset No.	Rating	PD	Region	Industry	R-squared	Exposure	LGD
1	AAA	0.002%	1	2	49%	10,000,000	50%
2	AAA	0.002%	1	5	50%	10,000,000	50%
3	AAA	0.002%	1	8	42%	10,000,000	50%
4	AAA	0.002%	2	6	48%	10,000,000	50%
5	AAA	0.002%	2	4	47%	10,000,000	50%
6	AA	0.01%	1	10	49%	10,000,000	50%
7	AA	0.01%	1	9	64%	10,000,000	50%
8	AA	0.01%	1	2	55%	10,000,000	50%
9	AA	0.01%	1	4	48%	10,000,000	50%
10	AA	0.01%	1	10	35%	10,000,000	50%
11	AA	0.01%	2	5	29%	10,000,000	50%
12	AA	0.01%	2	1	44%	10,000,000	50%
13	AA	0.01%	2	10	38%	10,000,000	50%
14	AA	0.01%	2	1	25%	10,000,000	50%
15	AA	0.01%	5	4	43%	10,000,000	50%
16	AA	0.01%	5	5	57%	10,000,000	50%
17	AA	0.01%	5	4	33%	10,000,000	50%
18	A	0.04%	1	1	51%	10,000,000	50%
19	A	0.04%	1	6	42%	10,000,000	50%
20	A	0.04%	1	5	47%	10,000,000	50%
21	A	0.04%	1	8	43%	10,000,000	50%
22	A	0.04%	1	5	34%	10,000,000	50%
23	A	0.04%	1	3	46%	10,000,000	50%
24	A	0.04%	1	8	30%	10,000,000	50%
25	A	0.04%	1	6	49%	10,000,000	50%
26	A	0.04%	1	6	29%	10,000,000	50%
27	A	0.04%	2	3	46%	10,000,000	50%
28	A	0.04%	2	5	49%	10,000,000	50%
29	A	0.04%	2	8	31%	10,000,000	50%
30	A	0.04%	2	3	40%	10,000,000	50%
31	A	0.04%	2	8	33%	10,000,000	50%
32	A	0.04%	2	4	42%	10,000,000	50%
33	A	0.04%	3	2	49%	10,000,000	50%
34	A	0.04%	3	2	20%	10,000,000	50%
35	A	0.04%	5	1	31%	10,000,000	50%
36	A	0.04%	5	4	46%	10,000,000	50%
37	A	0.04%	5	7	42%	10,000,000	50%
38	A	0.04%	5	5	31%	10,000,000	50%
39	A	0.04%	5	4	26%	10,000,000	50%
40	BBB	0.29%	1	7	34%	10,000,000	50%
41	BBB	0.29%	1	3	20%	10,000,000	50%
42	BBB	0.29%	1	10	40%	10,000,000	50%
43	BBB	0.29%	1	7	41%	10,000,000	50%
44	BBB	0.29%	1	8	43%	10,000,000	50%
45	BBB	0.29%	1	7	20%	10,000,000	50%
46	BBB	0.29%	1	7	15%	10,000,000	50%
47	BBB	0.29%	1	2	29%	10,000,000	50%
48	BBB	0.29%	2	1	43%	10,000,000	50%
49	BBB	0.29%	2	2	43%	10,000,000	50%
50	BBB	0.29%	2	8	32%	10,000,000	50%

TABLE 6.2: Portfolio of sample assets, part II

Asset No.	Rating	PD	Region	Industry	R-squared	Exposure	LGD
51	BBB	0.29%	2	1	20%	10,000,000	50%
52	BBB	0.29%	2	9	27%	10,000,000	50%
53	BBB	0.29%	2	8	41%	10,000,000	50%
54	BBB	0.29%	2	6	36%	10,000,000	50%
55	BBB	0.29%	2	7	32%	10,000,000	50%
56	BBB	0.29%	2	8	27%	10,000,000	50%
57	BBB	0.29%	2	4	31%	10,000,000	50%
58	BBB	0.29%	3	10	38%	10,000,000	50%
59	BBB	0.29%	3	1	31%	10,000,000	50%
60	BBB	0.29%	3	5	27%	10,000,000	50%
61	BBB	0.29%	3	10	34%	10,000,000	50%
62	BBB	0.29%	3	2	35%	10,000,000	50%
63	BBB	0.29%	4	9	19%	10,000,000	50%
64	BBB	0.29%	4	3	22%	10,000,000	50%
65	BBB	0.29%	4	7	23%	10,000,000	50%
66	BBB	0.29%	5	6	34%	10,000,000	50%
67	BBB	0.29%	5	8	38%	10,000,000	50%
68	BBB	0.29%	5	3	18%	10,000,000	50%
69	BBB	0.29%	5	6	25%	10,000,000	50%
70	BBB	0.29%	5	4	13%	10,000,000	50%
71	BBB	0.29%	5	8	43%	10,000,000	50%
72	BB	1.28%	1	10	27%	10,000,000	50%
73	BB	1.28%	1	4	24%	10,000,000	50%
74	BB	1.28%	1	1	18%	10,000,000	50%
75	BB	1.28%	1	4	29%	10,000,000	50%
76	BB	1.28%	2	7	21%	10,000,000	50%
77	BB	1.28%	2	3	18%	10,000,000	50%
78	BB	1.28%	2	4	30%	10,000,000	50%
79	BB	1.28%	3	4	19%	10,000,000	50%
80	BB	1.28%	3	7	29%	10,000,000	50%
81	BB	1.28%	3	1	21%	10,000,000	50%
82	BB	1.28%	3	7	27%	10,000,000	50%
83	BB	1.28%	4	4	27%	10,000,000	50%
84	BB	1.28%	4	1	24%	10,000,000	50%
85	BB	1.28%	5	4	22%	10,000,000	50%
86	BB	1.28%	5	1	16%	10,000,000	50%
87	BB	1.28%	5	4	19%	10,000,000	50%
88	BB	1.28%	5	3	16%	10,000,000	50%
89	B	6.24%	2	6	17%	10,000,000	50%
90	B	6.24%	3	7	25%	10,000,000	50%
91	B	6.24%	3	9	26%	10,000,000	50%
92	B	6.24%	4	7	17%	10,000,000	50%
93	B	6.24%	4	1	29%	10,000,000	50%
94	B	6.24%	4	7	42%	10,000,000	50%
95	B	6.24%	5	5	20%	10,000,000	50%
96	B	6.24%	5	6	23%	10,000,000	50%
97	CCC	32.35%	3	2	50%	10,000,000	50%
98	CCC	32.35%	4	3	27%	10,000,000	50%
99	CCC	32.35%	4	9	57%	10,000,000	50%
100	CCC	32.35%	4	3	35%	10,000,000	50%

- **LGD** is the loss given default of each name. In the same way as in case of exposures we assume a flat LGD equal to 50% as base case example. Deviations from the base case will be explicitly stated.

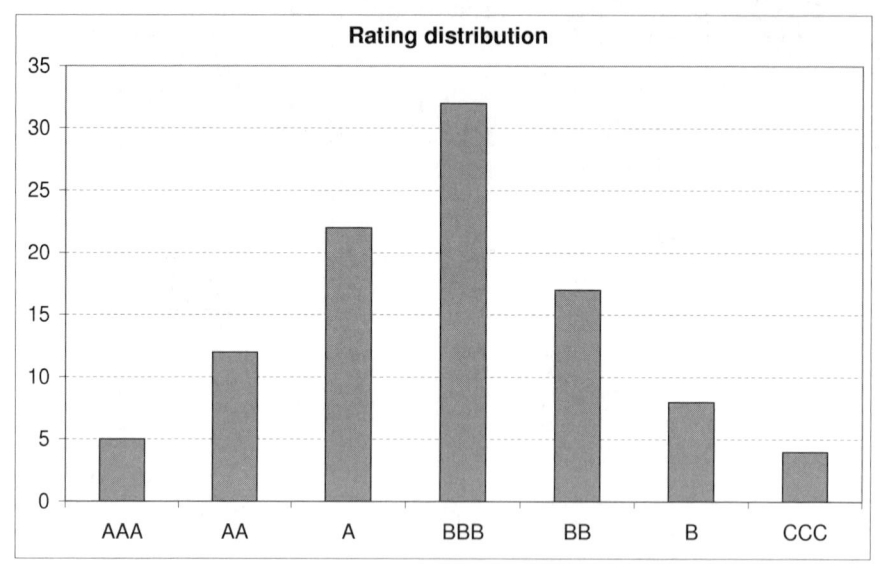

FIGURE 6.4: Rating distribution of names in the sample portfolio

Finally, we need to provide a correlation matrix for the systematic factors/indices. Table 6.3 shows the assumed correlations and Figure 6.8 illustrates the correlations graphically.

6.10 CDS Names in CDX.NA.IG and iTraxx Europe

In Section 3.4.4 we considered the two most liquid indices in the United States and Europe, namely CDX.NA.IG and iTraxx Europe. Tables 6.4 and 6.5 show a list of the names included in the Series 5 of the two indices (as of April 2006).

Appendix

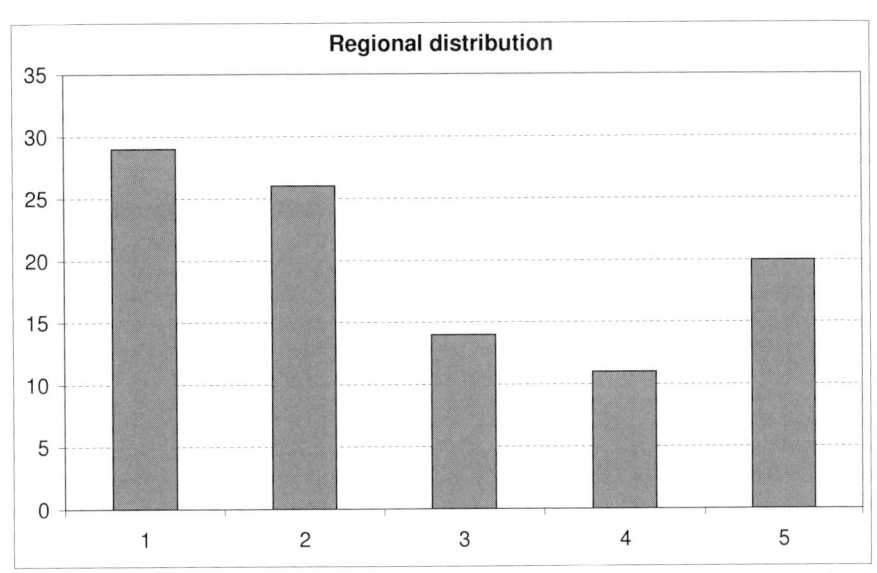

FIGURE 6.5: Regional distribution of names in the sample portfolio

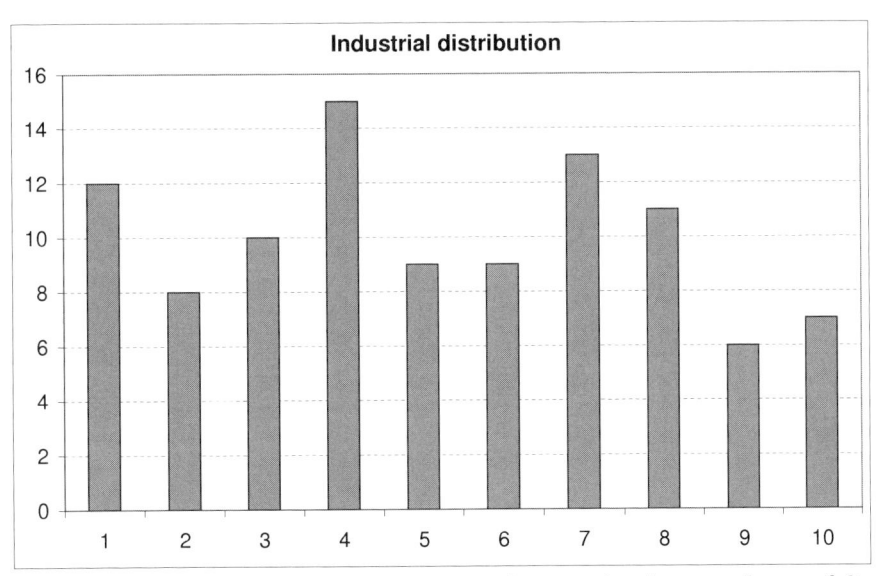

FIGURE 6.6: Industrial distribution of names in the sample portfolio

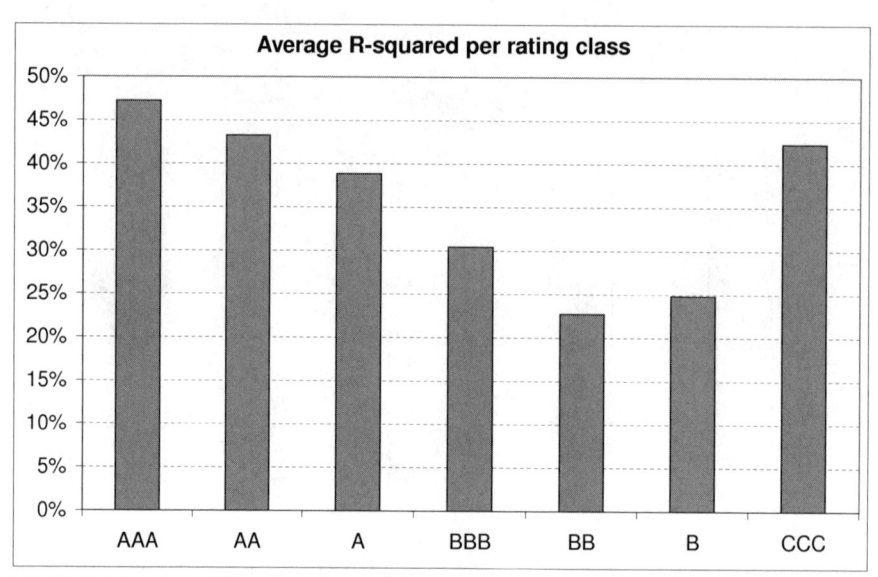

334   *Structured Credit Portfolio Analysis, Baskets & CDOs*

Average R-squared per rating class

[Bar chart showing: AAA ~47%, AA ~43%, A ~39%, BBB ~30%, BB ~22%, B ~24%, CCC ~42%]

FIGURE 6.7: Distribution of R-squared parameters of names in the sample portfolio; note that in our example we have $R_i^2 = \beta_i^2$ where the β_i's are listed in Tables 6.1 and 6.2

TABLE 6.3: Assumed correlation matrix of systematic factors (5 regions, 10 industries)

		Region					Industry									
		1	2	3	4	5	1	2	3	4	5	6	7	8	9	10
Region	1	100%	70%	50%	20%	30%	34%	27%	39%	47%	47%	48%	25%	35%	46%	28%
	2	70%	100%	20%	10%	40%	31%	26%	25%	29%	30%	25%	32%	33%	38%	33%
	3	50%	20%	100%	10%	10%	33%	11%	27%	30%	34%	21%	17%	28%	18%	29%
	4	20%	10%	10%	100%	10%	19%	18%	22%	21%	20%	17%	27%	30%	25%	16%
	5	30%	40%	10%	10%	100%	26%	39%	31%	33%	37%	37%	34%	49%	26%	40%
Industry	1	34%	31%	33%	19%	26%	100%	13%	17%	12%	21%	13%	13%	20%	15%	14%
	2	27%	26%	11%	18%	39%	13%	100%	14%	19%	10%	12%	19%	14%	22%	13%
	3	39%	25%	27%	22%	31%	17%	14%	100%	15%	19%	23%	11%	14%	16%	15%
	4	47%	29%	30%	21%	33%	12%	19%	15%	100%	13%	14%	15%	13%	22%	12%
	5	47%	30%	34%	20%	37%	21%	10%	19%	13%	100%	22%	12%	16%	19%	14%
	6	48%	25%	21%	17%	37%	13%	12%	23%	14%	22%	100%	12%	11%	18%	22%
	7	25%	32%	17%	27%	34%	13%	19%	11%	15%	12%	12%	100%	16%	17%	12%
	8	35%	33%	28%	30%	49%	20%	14%	14%	13%	16%	11%	16%	100%	16%	20%
	9	46%	38%	18%	25%	26%	15%	22%	16%	22%	19%	18%	17%	16%	100%	15%
	10	28%	33%	29%	16%	40%	14%	13%	15%	12%	14%	22%	12%	20%	15%	100%

Appendix 335

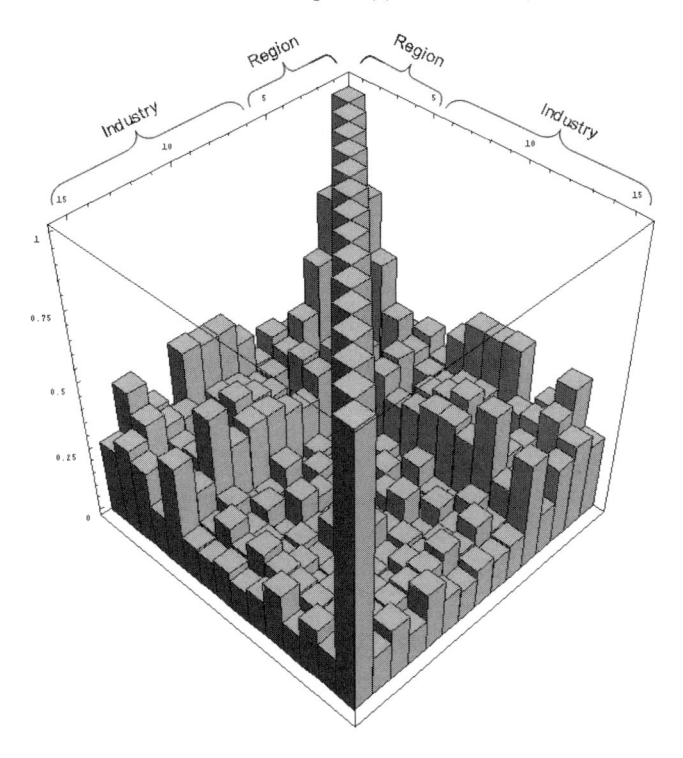

FIGURE 6.8: Graphical illustration of the assumed correlation matrix of systematic factors (5 regions, 10 industries)

TABLE 6.4: 125 names in the index CDX.NA.IG Series 5

	Company Name in CDX.NA.IG Series 5		Company Name in CDX.NA.IG Series 5
1	ACE Limited	63	Hewlett-Packard Company
2	Aetna Inc.	64	Hilton Hotels Corp.
3	Albertson's, Inc	65	Honeywell International Inc.
4	Alcan Inc.	66	IAC InterActive Corp.
5	Alcoa Inc.	67	Ingersoll-Rand Company
6	Allstate Corp. (The)	68	International Business Machines Corporation
7	ALLTEL Corporation	69	International Lease Finance Corp.
8	Altria Group, Inc.	70	International Paper Company
9	American Axle & Manufacturing Holdings	71	Jones Apparel Group Inc.
10	American Electric Power Co. Inc.	72	Knight Ridder Inc.
11	American Express Company	73	Kraft Foods Inc
12	American International Group	74	Kroger Co., The
13	Amgen Inc.	75	Lennar Corporation
14	Anadarko Petroleum Corporation	76	Limited Brands Inc.
15	Arrow Electronics, Inc.	77	Lockheed Martin Corporation
16	AutoZone, Inc.	78	Loews Corp.
17	Baxter International, Inc.	79	Marriott International Inc.
18	Bellsouth Corporation	80	March & McClennan Cos. Inc.
19	Boeing Capital Corporation	81	MBIA Insurance Corporation
20	Bristol-Myers Squibb Co	82	McDonald's Corp.
21	Burlington Northern Santa Fe Corporation	83	McKesson Corp.
22	Campbell Soup Co.	84	MeadWestvaco Corporation
23	Capital One Bank	85	Metlife Inc
24	Cardinal Health, Inc.	86	Motorola Inc.
25	Carnival Corporation	87	National Rural Utilities Cooperative Finance Corporation
26	Caterpillar Inc	88	Newell Rubbermaid Inc
27	Cendant Corporation	89	News America Inc
28	Centex Corporation	90	Nordstrom Inc
29	CenturyTel, Inc.	91	Norfolk Southern Corporation
30	Chubb Corporation	92	Northrop Grumman Corporation
31	Cigna Corp.	93	Omnicom Group Inc.
32	Cingular Wireless Llc	94	Progress Energy, Inc.
33	CIT Group Inc	95	Pulte Homes, Inc.
34	Clear Channel Communications Inc.	96	RadioShack Corp.
35	Comcast Cable Communications, Inc.	97	Raytheon Company
36	Computer Sciences Corporation	98	Rohm And Haas Company
37	Conagra Foods, Inc.	99	Sabre Holdings Corp.
38	Conocophillips	100	Safeway Inc.
39	Constellation Energy Group, Inc.	101	Sara Lee Corp.
40	Countrywide Home Loans, Inc.	102	SBC Communications Inc
41	Cox Communications, Inc.	103	Sempra Energy
42	CSX Corporation	104	Simon Property Group, L.P.
43	CVS Corp.	105	Southwest Airlines Co.
44	Deere & Company	106	Sprint Corporation
45	Devon Energy Corp	107	Supervalu
46	Dominion Resources, Inc.	108	Target Corp
47	Dow Chemical Company, The	109	Toll Bros. Inc.
48	Duke Energy Corporation	110	Textron Financial Corporation
49	Dupont, E.I. De Nemours & Co.	111	Time Warner Inc
50	Eastman Chemical Company	112	Transocean Inc.
51	EOP Operating Limited Partnership	113	Tyson Foods Inc.
52	Federal Home Loan Mortgage Corporation	114	Union Pacific Corporation
53	Federal National Mortgage Association	115	Valero Energy Corporation
54	Federated Department Stores, Inc.	116	Verizon Global Funding Corp.
55	FirstEnergy Corp	117	Viacom Inc
56	The Gap	118	Wal-Mart Stores Inc
57	General Electric Capital Corporation	119	Walt Disney Company, The
58	General Mills, Inc.	120	Washington Mutual Inc
59	Goodrich Corporation	121	Wells Fargo & Company
60	Halliburton Company	122	Weyerhaeuser Co.
61	Harrah's Operating Company, Inc.	123	Whirlpool Corporation
62	Hartford Financial Services Group	124	Wyeth
		125	XL Capital Ltd

TABLE 6.5: 125 names in the index iTraxx Europe Series 5

	Company Name in iTraxx Europe Series 5		Company Name in iTraxx Europe Series 5
1	ABN Amro Bank NV	63	GUS PLC
2	Accor SA	64	Hannover Rueckversicherung AG
3	Adecco SA	65	Hellenic Telecommunications Organization SA
4	Aegon NV	66	Henkel KGaA
5	Akzo Nobel NV	67	Iberdrola SA
6	Allianz AG	68	Imperial Chemical Industries plc
7	Altadis SA	69	Imperial Tobacco Group Plc
8	Arcelor Finance SCA	70	ITV Plc
9	Assicurazioni Generali SpA	71	Kingfisher PLC
10	Aviva Plc	72	Koninklijke Philips Electronics NV
11	AXA SA	73	Lafarge SA
12	BAA Plc	74	Linde AG
13	BAE Systems Plc	75	LVMH Moet Hennessy Louis Vuitton SA
14	Banca Intesa SpA	76	Marks & Spencer Group PLC
15	Banca Monte dei Paschi di Siena SpA	77	Metro AG
16	Banca Popolare Italiana	78	Muenchener Rueckversicherungs AG
17	Banco Bilbao Vizcaya Argentaria SA	79	National Grid PLC
18	Banco Comercial Portugues SA	80	Nestle SA
19	Banco Espirito Santo SA	81	Nokia OYJ
20	Banco Santander Central Hispano SA	82	O2 PLC
21	Barclays Bank PLC	83	Pearson PLC
22	Bayer AG	84	Peugeot SA
23	Bayerische Hypo-und Vereinsbank AG	85	Portugal Telecom SGPS SA
24	Bayerische Motoren Werke AG	86	PPR SA
25	Bertelsmann AG	87	Reed Elsevier PLC
26	Boots Group PLC	88	Renault SA
27	British American Tobacco PLC	89	Rentokil Initial PLC
28	British Telecommunications PLC	90	Repsol YPF SA
29	Cadbury Schweppes PLC	91	Reuters Group Plc
30	Capitalia SpA	92	Royal Bank of Scotland Plc
31	Carrefour SA	93	Royal KPN NV
32	Centrica PLC	94	RWE AG
33	Ciba Specialty Chemicals Holding Inc	95	Safeway Ltd
34	Cie de Saint-Gobain	96	Sanofi-Aventis
35	Commerzbank AG	97	Sanpaolo IMI SpA
36	Compagnie Financiere Michelin	98	Siemens AG
37	Compass Group PLC	99	Sodexho Alliance SA
38	Continental AG	100	Stora Enso Oyj
39	DaimlerChrysler AG	101	Suez SA
40	Degussa AG	102	Svenska Cellulosa AB
41	Deutsche Bank AG	103	Swiss Reinsurance
42	Deutsche Lufthansa AG	104	Tate & Lyle PLC
43	Deutsche Telekom AG	105	Technip SA
44	Diageo PLC	106	Telecom Italia SpA
45	DSG International PLC	107	Telefonica SA
46	E.ON AG	108	TeliaSonera AB
47	Edison SpA	109	Tesco PLC
48	Electricite de France	110	Thomson
49	Electrolux AB	111	UniCredito Italiano SpA
50	EnBW Energie Baden-Wuerttemberg AG	112	Unilever NV
51	Endesa SA	113	Union Fenosa SA
52	Enel SpA	114	United Utilities PLC
53	Energias de Portugal SA	115	UPM-Kymmene Oyj
54	European Aeronautic Defence and Space Co NV	116	Valeo SA
55	Finmeccanica SpA	117	Vattenfall AB
56	Fortum Oyj	118	Veolia Environnement
57	France Telecom SA	119	Vivendi Universal SA
58	Gallaher Group PLC	120	Vodafone Group PLC
59	Gas Natural SDG SA	121	Volkswagen AG
60	GKN PLC	122	Volvo AB
61	Glencore International AG	123	Wolters Kluwer NV
62	Groupe Auchan SA	124	WPP Group Plc
		125	Zurich Insurance Co

References

[1] *Artemus Strategic Asian Credit Fund Limited*; FitchRatings, Structured Finance, Synthetics/Asia New Issue Report, March 5 (2003)

[2] *Artemus Strategic Asian Credit Fund Limited; Synthetic Collateralized Debt Obligation*; Moody's Investors Service, Internation Structured Finance, New Issue Report, March 13 (2003)

[3] *New Issue: Artemus Strategic Asian Credit Fund Ltd.*; Standard & Poor's, Structured Finance, March 12 (2003)

[4] *Artemus Strategic Asian Credit Fund Limited II*; FitchRatings, Structured Finance, Credit Products/Hong Kong Presale Report, June 24 (2005)

[5] *Artemus Strategic Asian Credit Fund Ltd. II*; Standard & Poor's, RatingsDirect Presale, June 24 (2005)

[6] AKHAVEIN, J. D., KOCAGIL, A. E., NEUGEBAUER, M.; *A Comparative Empirical Study of Asset Correlations* Fitch Ratings, July (2005)

[7] ALBANESE, C., CAMPOLIETI, G., CHEN, O., ZAVIDONOV, A.; *Credit Barrier Models*; Risk **16** (6), 109-113 (2003)

[8] AMATO, J. D., GYNTELBERG, J.; *CDS Index Tranches and the Pricing of Credit Risk Correlations*; BIS Quarterly Review, March (2005)

[9] ANDERSEN, L., BAUM,D., KOLOGLU, B.; *Single Tranche CDOs. Tailored Investment Grade Portfolio Exposure*; Banc of America Securities, April (2003)

[10] ANDERSEN, L., SIDENIUS, J., BASU, S.; *All Your Hedges in One Basket*; Risk **X**, 67-72, November (2003)

[11] ARTZNER, P., DELBAEN, F., EBER, J., HEATH, D.; *Coherent Measures of Risk*; Mathematical Finance **9** (3), 203-228 (1999)

[12] AVELLANEDA, M., ZHU, J.; *Distance to Default*; Risk **14** (12), 125-129 (2001)

[13] AZIZ, J., CHARUPAT, N.; *Calculating Credit Exposure and Credit Loss: A Case Study*; Algo Research Quarterly, Vol. 1, No. 1, 3146, September (1998)

[14] BAHETI, P, MASHAL, R., NALDI, M., SCHLOEGL, L.; *Synthetic CDOs of CDOs: Squaring the Delta-Hedged Equity Trade*; Lehman Brothers, Fixed Income Quantitative Credit Research, June (2004)

[15] BASEL COMMITTEE ON BANKING SUPERVISION; *International Convergence of Capital Measurement and Capital Standards*; Bank for International Settlements, June (2004)

[16] BAUER, H.; *Probability Theory*; de Gruyter (1996)

[17] BAXTER, M.; *Dynamic modelling of single-name credits and CDO tranches*; Preprint, March (2006)

[18] BEINSTEIN, E., SBITYAKOV, A., ALLEN, P., MUENCH, D.; *Enhanced Base Correlation*; JP Morgan, Credit Derivatives Research, September (2005)

[19] BELKIN, B., SUCHOWER, S., FOREST, L. R. JR.; *The Effect of Systematic Credit Risk on Loan Portfolio Value-at-Risk and Loan Pricing*; CreditMetrics Monitor, Third Quarter (1998)

[20] BELKIN, B., SUCHOWER, S., FOREST, L. R. JR.; *A One-Parameter Representation of Credit Risk and Transition Matrices*; CreditMetrics^{TM} Monitor, Third Quarter (1998)

[21] BURTSCHELL, X., GREGORY, J., LAURENT, J.-P.; *A Comparative Analysis of CDO Pricing Models*; Preprint, April (2005)

[22] BIELECKI, T. R., RUTKOWSKI, M.; *Credit Risk*; Springer (2004)

[23] BLACK, F., SCHOLES, M.; *The Pricing of Options and Corporate Liabilities*; Journal of Political Economy **81**, 637-654 (1973)

[24] BLUHM, C.; *Applications of CDO Modeling Techniques in Credit Portfolio Management*; working paper, presented at RISK magazine's Quant Congress Europe, October (2006)

[25] BLUHM, C., OVERBECK, L., WAGNER, C.; *An Introduction to Credit Risk Modeling*; Chapman & Hall/CRC Financial Mathematics Series; 2nd Reprint; CRC Press (2003)

[26] BLUHM, C., OVERBECK, L.; *An Introduction to CDO Modelling and Applications*; appeared in: *Credit Risk: Models and Management*; edited by D. Shimko, 2nd edition, RISK Book (2004)

[27] BLUHM, C., OVERBECK, L.; *Semi-Analytic Approaches to CDO Modeling*; Economic Notes **33**, No. 2, 233-255 (2004)

[28] BLUHM, C., OVERBECK, L.; *Comonotonic Default Quote Paths for Basket Evaluation*; Risk **18** (8), 67-71, August (2005)

[29] BLUHM, C., OVERBECK, L.; *To Be Markov or Not to Be: Term Structures of Default Probabilities*; working paper, submitted (2006)

[30] BOUYÉ, E., DURRLEMAN, V., NIKHEGBALI, A., RIBOULET, G., RONCALLI, THIERRY; *Copulas for Finance. A Reading Guide and Some Applications*; July (2000)

[31] BREIMAN, L.; *Probability*; Siam (1992)

[32] CHERUBINI, U., LUCIANO, E., VECCIATO, W.; *Copula Methods in Finance*; John Wiley & Sons (2004)

[33] CHOUDRY. M.; *Structured Credit Products: Credit Derivatives and Synthetic Securitization*; John Wiley & Sons (2004)

[34] COVER, T. M., THOMAS, J. A.; *Elements of Information Theory*; John Wiley & Sons (1991)

[35] CREDIT SUISSE FINANCIAL PRODUCTS; *CreditRisk$^+$ - A Credit Risk Management Framework* (1997)

[36] CROSBIE, P.; *Modelling Default Risk*; KMV Corporation (1999) (www.kmv.com)

[37] CROUHY, M, GALAI, D., MARK, R.; *Risk Management*; McGraw-Hill (2000)

[38] DOW JONES INDEXES; *Guide to the Down Jones CDX Indexes*; September (2005); see http://www.djindexes.com

[39] DOW JONES INDEXES; *Dow Jones iTraxx - Product Descriptions*; see www.djindexes.com

[40] DUFFIE, D., SINGLETON, K. J.; *Credit Risk: Pricing, Measurement, and Management*; Princeton University Press (2003)

[41] EBMEYER, D., KLAAS, R., QUELL, P.; *The Role of Copulas in the CreditRisk$^+$ Framework*; appeared in: *Copulas: From Theory to Application in Finance*; edited by J. Rank, RISK Book (2006)

[42] EMBRECHTS, P., MCNEIL, A., STRAUMANN, D.; *Correlation and Dependence in Risk Management: Properties and Pitfalls*; appeared in: *Risk Management: Value at Risk and Beyond*; edited by M. Dempster, Cambridge University Press (2002)

[43] ENGELMANN, B., HAYDEN, E., TASCHE, D.; *Testing Rating Accuracy*; RISK, January (2003)

[44] ETHIER, S. N., KURTZ, T. G.; *Markov Processes Characterization and Convergence*; John Wiley and Sons (2005)

[45] FELSENHEIMER, J., GISDAKIS, P., ZAISER, M.; *DJ iTraxx: Credit at Its Best!*; Credit Derivatives Special, HypoVereinsbank Corporates & Markets, October (2004)

[46] FELSENHEIMER, J., GISDAKIS, P., ZAISER, M.; *Active Credit Portfolio Management*; Wiley-VCH (2006)

[47] FINGER, C. C.; *Conditional Approaches for CreditMetrics Portfolio Distributions*; CreditMetrics Monitor, April (1999)

[48] FINGER, C. C.; *A Comparison of Stochastic Default Rate Models*; RiskMetrics Journal, Volume **1**, November (2000)

[49] FITCHRATINGS; *Single-Tranche Synthetic CDOs*; Credit Products Special Report, June (2003)

[50] FITCHRATINGS; *Analysis of Synthetic CDOs of CDOs*; Global CDO Criteria Report, September (2004)

[51] FREY, R., MCNEIL, A. J.; *Modelling Dependent Defaults*; Preprint, March (2001)

[52] FRIES, C. P., ROTT, M. G.; *Fast and Robust Monte Carlo CDO Sensitivities and Their Efficient Object Oriented Implementation*; Preprint, May (2005)

[53] FRYDMAN, H., SCHUERMANN, T.; *Credit Rating Dynamics and Markov Mixture Models*; Working Paper, June (2005)

[54] FRYE, J.; *Collateral Damage* RISK **13** (4), 91-94 (2000)

[55] GOESSL, C.; *Predictions Based on Certain Uncertainties - A Bayesian Credit Portfolio Approach*; Preprint, July (2005)

[56] HAUSDORFF, F.; *Summationsmethoden und Momentfolgen I*; Math. Z. **9**, 74-109 (1921)

[57] HILLEBRAND, M.; *Modeling and Estimating Dependent Loss Given Default*; Working Paper, Munich University of Technology, October (2005)

[58] JOE, H.; *Multi-variate Models and Dependence Concepts*; Chapman & Hall (1997); first CRC reprint 2001

[59] HULL, J., WHITE, A.; *Valuing Credit Default Swaps II: Modeling Default Correlations*; The Journal of Derivatives, 12-21, Spring (2001)

[60] HULL, J., WHITE, A.; *The Perfect Copula*; Preprint, November (2005)

[61] ISRAEL, R., ROSENTHAL, J., WEI, J.; *Finding Generators for Markov Chains via Empirical Transition Matrices with Application to Credit Ratings*; Mathematical Finance **11** (2), 245-265 (2001)

[62] JARROW, R. A., LANDO, D., TURNBULL, S. M.; *A Markov Model for the Term Structure of Credit Risk Spreads*; Review of Financial Studies **10**, 481-523 (1997)

[63] JORION, PH.; *Value at Risk. The New Benchmark for Managing Financial Risk*; McGraw Hill (2000)

[64] J. P. MORGAN; *Structured Finance CDO Handbook*; JPMorgan Global Structured Finance Research, February (2004)

[65] J. P. MORGAN; *Introducing Base Correlations*; Credit Derivatives Strategy, March (2004)

[66] KADAM, A., LENK, P.; *Heterogeneity in Ratings Migration*; Working Paper, October (2005)

[67] KARATZAS, I., SHREVE, S. E.; *Brownian Motion and Stochastic Calculus*; Springer, 2nd Edition (2004)

[68] KEALHOFER, S.; *The Economics of the Bank and of the Loan Book*; Moody's KMV, www.mkmv.com, May (2002)

[69] KIMBERLING, C. H.; *A Probabilistic Interpretation of Complete Monotonicity*; Aequationes Mathematicae **10**, No.2-3, 152-164 (1974)

[70] KREININ, A., SIDELNIKOVA, M.; *Regularization Algorithms for Transition Matrices*; Algo Research Quarterly **4** (1/2), 25-40 (2001)

[71] LANDO, D.; *Credit Risk Modeling: Theory and Applications*; Princeton University Press (2004)

[72] LANDO, D., SKODEBERG, T. M.; *Analyzing Rating Transitions and Rating Drift with Continuous Observations*; Journal of Banking and Finance **26**, No. 2-3, 423-444 (2002)

[73] LEHMAN BROTHERS; *CDO/Structured Credit. Annual 2006*; Lehman Brothers, Structured Credit Research, December (2005)

[74] LI, D. X.; *The Valuation of Basket Credit Derivatives*; CreditMetrics Monitor, April (1999)

[75] LI, D. X.; *On Default Correlation: A Copula Function Approach*; The Journal of Fixed Income **6**, 43-54, March (2000)

[76] LINDSKOG, F.; *Modelling Dependence with Copulas and Applications to Risk Management*; Master Thesis, ETH Zurich, July (2000)

[77] LINDSKOG, F., MCNEIL, A., SCHMOCK, U.; *Kendall's Tau for Elliptical Distributions*; in: Credit Risk Measurement, Evaluation and Management; edited by Bol, N. et al., PhysicaVerlag, 149156 (2003)

[78] LONGSTAFF, F. A., RAJAN, A.; *An Empirical Analysis of the Pricing of Collateralized Debt Obligations*; Preprint, April (2006)

[79] LOTTER, H., OVERBECK, L.; *Ratings Based on Credit Portofolio Models*; appeared in: *Credit Rating: Methodologies, Rationale, and Default Risk*; edited by Michael K. Ong, RISK Book (2002)

[80] MARSHALL, A. W., OLKIN, I.; *Families of Multi-Variate Distributions*; Journal of the American Statistical Association **83**, 834841 (1988)

[81] MCNEIL, A. J., FREY, R., EMBRECHTS, P.; *Quantitative Risk Management. Concepts, Techniques and Tools*; Princeton University Press (2005)

[82] MCNEIL, A. J., WENDIN, J.; *Bayesian Inference for Generalized Linear Mixed Models of Portfolio Credit Risk*; Preprint, October (2005)

[83] MIKOSCH, T.; *Copulas: Tales and Facts*; Working Paper

[84] MILLER, K. S., SAMKO, S. G.; *Completely Monotonic Functions* Integr. Transf. and Spec. Funct. **12**, No. 4, 389-402 (2001)

[85] DWYER, D, KOCAGIL, A., STEIN, R.; *The Moody's KMV EDF RiskCalc v3.1 Model*; Moody's KMV, January (2004)

[86] MERTON, R.; *On the Pricing of Corporate Debt: The Risk Structure of Interest Rates*; The Journal of Finance **29**, 449-470 (1974)

[87] MORO, B; *The Full Monte*; RISK **8** (2), 57-58 (1995)

[88] MOROKOFF, W. J.; *Simulation Methods for Risk Analysis of Collateralized Debt Obligations*; Moody's KMV, August (2003)

[89] MOROKOFF, W. J.; *Simulation of Risk and Return Profiles for Portfolios of CDO Tranches*; Proceedings of the 2005 Winter Simulation Conference, edited by M. E. Kuhl, N. M. Steiger, F. B. Armstrong, and J. A. Joines (2005)

[90] NELSEN, R. B.; *An Introduction to Copulas*; Springer, New York (1999)

[91] NELSON, R. B.; *Dependence Modeling with Archimedean Copulas*; Working Paper

[92] NELSON, R. B.; *Properties and Applications of Copulas: A Brief Survey*; Working Paper

[93] NICKELL, P., PERRAUDIN, W., VAROTTO, S.; *Ratings- Versus Equity-Based Credit Risk Modeling: An Empirical Analysis*; Working Paper (1999)

[94] NORIS, J.; *Markov Chains*; Cambridge Series in Statistical and Probabilistic Mathematics; Cambridge University Press (1998)

[95] O'KANE, D., LIVESEY, M.; *Base Correlation Explained*; Lehman Brothers, Fixed Income Quantitative Credit Research, November (2004)

[96] ONG, M. K.; *Internal Credit Risk Models: Capital Allocation and Performance Measurement*; RISK Books (1999)

[97] OVERBECK, L., SCHMIDT, W.; *Modeling Default Dependence with Threshold Models*; Preprint, submitted for publication (2003)

[98] PAN, G.; *Equity to Credit Pricing*; Risk **14** (11), 99-102 (2001)

[99] PITT, A.; *Correlated Defaults: Let's Go Back to the Data*; Risk **17** (6), 75-79 (2004)

[100] PLUTO, K., TASCHE, D.; *Thinking Positively*; RISK **18** (8) (2005)

[101] SARFARAZ, A., COHEN, M., LIBREROS, S.; *Use of Transition Matrices in Risk Management and Valuation*; Fair Isaac White Paper, September (2004)

[102] SCHMIDT, W., WARD, I.; *Pricing Default Baskets*; RISK **15** (1), 111-114 (2002)

[103] SCHOENBUCHER, P.; *Credit Derivatives Pricing Models*; John Wiley (2003)

[104] SCHUERMANN, T., JAFRY, Y.; *Measurement and Estimation of Credit Migration Matrices*; Wharton School Center for Financial Institutions, University of Pennsylvania, Working Paper 03-08 (2003)

[105] SCHUERMANN, T., JAFRY, Y.; *Metrics for Comparing Credit Migration Matrices*; Wharton School Center for Financial Institutions, University of Pennsylvania, Working Paper 03-09 (2003)

[106] SKLAR, A.; *Fonction de Repartition à n Dimension et Leur Marges*; Publications de l'Insitute Statistique de l'Université de Paris **8**, 229-231 (1959)

[107] SKLAR, A.; *Random Variables, Joint Distribution Functions and Copulas*; Kybernetika **9**, 449-460 (1973)

[108] STANDARD & POOR'S; *Annual Global Corporate Default Study: Corporate Defaults Poised to Rise in 2005*; S&P Global Fixed Income Research, January (2005)

[109] STEELE, J. M.; *Stochastic Calculus and Financial Applications*; Applications of Mathematics **45**, Springer (2001)

[110] TRUECK, S., OEZTURKMEN, E.; *Adjustment and Application of Transition Matrices in Credit Risk Models*; Working Paper, University of Karlsruhe, September (2003)

[111] VASICEK, O. A.; *The Loan Loss Distribution*; KMV Corporation

[112] WALKER, M. B.; *CDO Models - Towards the Next Generation: Incomplete Markets and Term Structure*; Preprint, March (2006)

[113] WIDDER, D. V.; *The Laplace Transform*; Princeton University Press (1941)

[114] WILDE, T., JACKSON, L.; *Low Default Portfolios Without Simulation*; RISK **19** (8) (2006)

[115] WILLEMANN, S.; *An Evaluation of the Base Correlation Framework for Synthetic CDOs*; Preprint, December (2004)

[116] ZENG, B., ZHANG, J.; *An Empirical Assessment of Asset Correlation Models*; Moody's KMV (2001)

[117] ZHOU, C.; *Default Correlation: An Analytical Result*; Federal Reserve Board, Washington, D.C., Finance and Economics Division (1997)

Index

1st-order generator approximation, 44

ability to pay as a function of assets and liabilities, 9
ability to pay process (APP), 9, 194, 216
accuracy versus practicability, 24
additivity of expected shortfalls, 288
analytic approximation, 226
analytic CDO evaluation, 220
analytic CDO models, 220
analytic CSO approximation, 225, 227, 228
analytic expected loss for CDOs, 224
analytic loss density, 226
application of semi-analytic default time modeling, 236
approximate Q-matrix, 46
approximative generator (Q-matrix) for M, 45
Archimedean copula simulation, 96
Archimedean copulas, 88
Artemus cash flows, 181
Artemus transactional diagram, 180
asset correlation, 20
asset picking, 294
asset scenario, 191
asset selection, 294
asset side, 190
asset side of a CDO, 167
asset side of Artemus, 180
asset value log-return, 214
assets and liabilities, 10
attachment point, 172, 273
AUROC, 6, 7

back-of-the-envelope evaluation of CDO, 201
backloaded defaults, 183
balance sheet scoring, 4
base correlation, 280, 281, 284, 286

Basel II, 3, 13–15
Basel II versus Basel I in securitizations, 187
basket, 27
basket reference portfolio, 121
basket simulation study: first-to-default time distribution, 132
basket simulation study: second-to-default time distribution, 132
basket simulation study: third-to-default time distribution, 134
basket-linked note, 147
benefit of improved predictive power in rating or scoring systems, 7
Bernoulli variable, 8, 20, 21, 28
bespoke, 252
best case scenario, 141
bi-variate normal distribution function, 30
bid/ask spread, 279
Black & Scholes, 8
bottom-up approach, 300
bottom-up cash flows, 170
breakeven spread, 138
Brownian first passage time, 216
Brownian motion, 8, 214
Brownian motion based models, 214
Brownian motion time changed, 216

calibration of default probabilities, 5
capital accord, 3
capital structure, 169, 172, 174
capital structure in STCDOs, 252
capped EL, 175
cash flow evaluation in default baskets, 136
cash flow illustration in CLN transaction, 151
cash flow illustration in first-to-default basket, 136

349

cash flow scenario, 191
cash flow waterfall, 169, 198
cash payment priorities, 210
cash settlement, 130, 147
causal model, 8, 12, 14
causal rating, 4
CBO, 168
CDO modeling approaches, 194
CDO modeling principles, 190
CDO modeling scheme, 190
CDO modeling with dependent default times, 194
CDO of CDOs, 168, 299
CDO scenarios, 191
CDO-squared, 168, 299
CDS index, 253
CDS index tranches, 250, 255, 271
CDS indices, 253, 256
CDX, 253, 256
CDX.NA.IG, 254–256
Clayton copula, 91, 97
Clayton copula for nth-to-defaults, 127
Clayton copula for stressed simulations, 112
Clayton copula for stressed tail behavior, 118
CLN, 147
CLN transactional diagram, 148
CLO, 167
CMBS, 168
code space, 156
collateral value volatility, 15
collateralized bond obligation, 168
collateralized loan obligation, 167
collateralized synthetic obligation, 168
combination of tranches, 174
combined approach, 19
commercial mortgage-backed securities, 168
committed undrawn line, 11
comonotonic approximation, 242, 245, 248
comonotonic CDO evaluation, 220, 236
comonotonic coincidence, 240
comonotonic copula, 92, 237
comonotonic default quote path simulation, 240
comonotonic duo basket simulation, 244

comonotonicity in default quote paths, 238
competitive advantage by discriminatory power, 6
completely monotonic, 89, 90
complexity premium, 29
composite factor, 19
compound correlation, 280, 282, 283, 286
concordant pairs, 105
condition for application of semi-analytic CDO evaluation, 230
conditional default probability, 35, 64, 222
conditional default timing, 233
conditions for application of analytic CDO models, 221
copula for default times, 82
copula function, 85
copula functions, 24
copula impact on dependent default times, 114
copulas in practice, 93
copulas with Gaussian marginals, 100
copulas with Student-t marginals, 101
correlation, 17, 20, 28, 32
correlation as dependence measure, 102
correlation cat, 135
correlation desks, 278
correlation matrix for systematic sample indices, 334
correlation products, 17, 20
correlation skew, 283, 285, 286
correlation smile, 283, 285, 286
correlation study from Fitch, 20
correlation study from Moody's KMV, 20
correlations inter-industry, 20
correlations intra-industry, 20
cost of carry, 276
cost-to-securitize, 294, 295, 297
cost-to-securitize illustration, 298
coupon, 147
credit conversion, 13
credit conversion factor, 13
credit curves, 39
credit event, 130, 253, 254
credit exposure, 10
credit worthiness index (CWI), 8

credit-linked note, 147, 148
CreditMetrics, 8
CreditRisk$^+$, 91
critical barrier, 8
critical threshold, 8, 64
cross dependencies, 299
crystal ball rating or
    scoring system, 7
CSO, 168
CSO example, 194
CSO simulation results, 204
CSO simulation results with excess
    spread trigger, 212
CSO transactional diagram, 195
CWI correlation, 20, 28, 31, 32
CWI correlation matrix for the CLN
    reference pool, 158
CWI correlation matrix of the basket
    reference portfolio, 124

default basket, 27, 129
default correlation, 20, 30–32
default correlation in dependence on
    CWI correlation, 31, 32
default correlation over time, 60
default frequencies from rating agencies, 41
default indication, 80
default indicator, 8, 18, 21, 28
default leg, 259
default leg in index tranche, 282, 284
default point, 9, 10, 28, 64, 79, 80
default point standardization, 79
default probability, 2
default quote paths, 238
default time copula, 82
default time densities, 69, 72
default time density, 68, 69
default time distribution, 68
default time distributions for the CLN
    example, 153
default time expectation, 73
default time simulation, 83, 84
default time standard deviation, 73
default time summary statistics, 74
default times, 21
default times and PD term structures,
    67
default timing, 292

delta, 274, 275, 301
delta exchange index tranche trading,
    279
delta exchange trading, 278, 279
delta hedging, 271, 276, 277
density of analytic loss distribution,
    223
dependence measure: correlation, 102
dependence measure: Kendall's rank
    correlation, 104
dependence measure: Spearman's rank
    correlation, 104
dependence measure: tail dependence,
    107
dependence measures, 99, 102
dependence measures applied to copula examples, 108
dependence modeling via copulas, 85
dependent default times, 67
dependent default times simulation,
    125
detachment point, 172, 273
deviation of default times, 73
diagonal adjustment, 40
diagonal dominant migration matrix,
    42
discordant pairs, 105
discriminatory power, 4, 6
distributional scenario, 120
diversification, 30, 32
Dow Jones CDX Indices, 256
Dow Jones indices, 256
duo basket, 27
duo basket transaction, 243
duo baskets for multi-year horizons,
    59

economic capital (EC), 22
economic capital (EL), 22
effective risk transfer, 175, 176
effective spread w.r.t. delta exchange
    trading, 279
embedding of time-discrete into time-continuous Markov chain, 46
entropy, 315
entropy maximizing distribution, 87,
    317
equity correlation, 20
equity performance, 209, 213

equity piece, 169, 175
equity tranche, 169, 229
error of first kind, 5
error of second kind, 5
ESFC, 290
excess cash, 183
excess cash trap, 210
excess cash triggered, 211
excess spread, 182
excess spread and pool losses, 209, 213
exchangeable exposures, 242
expected cash flow to protection seller, 137
expected default time, 73
expected discounted cash flow, 139
expected loss (EL), 21, 22
expected loss of tranche, 282
expected loss of tranches, 203
expected shortfall, 22
expected shortfall (ESF), 23, 288, 289
expected shortfall based on quantile based economic capital matching, 23
expected shortfall contributions, 288, 290, 291
exponential distribution, 266, 317
exponential distribution term structure, 266
exponential matrix expansion, 44
Exponential of 1st-order generator approximation, 44
exposure at default, 12
exposure at default (EAD), 10
exposure measurement, 10
exposure uncertainty, 11, 13
extrapolation of credit curves, 57
extrapolation of PD term structures, 57
extreme correlation cases, 37

factor decomposition, 93, 95, 96
factor model, 93, 95, 96
fair index spread, 265, 270
fair spread, 139
false alarm rate, 6
Felix Hausdorff, 90
financial ratios, 4
finding a best-matching parameterization for $\varphi_{\alpha,\beta}$, 54

first loss piece, 169
first loss tranche, 169
first passage time, 8, 215, 216
first passage time models, 213
first-order Q-matrix, 43
first-to-default, 34
first-to-default basket cash flows, 136
first-to-default example, 132
first-to-default probabilities w.r.t. different copulas, 116
first-to-default probability, 34, 64
first-to-default probability in dependence on correlation, 37
first-to-default probability over time, 66
fixed recovery transaction, 131
fixed time horizon model, 8
fixed-time horizon model, 21
forward default probabilities, 49
forward default probability, 48
frontloaded defaults, 183
funded and unfunded STCDO, 252
funded basket credit swap, 148
funded transaction, 130, 147
funding benefits, 184

gamma, 301
gap risk, 301
Gaussian copula, 81, 87, 317
Gaussian copula for nth-to-defaults, 121
Gaussian copula with Gaussian marginals, 93
Gaussian copula with Student-t marginals, 95
generator of a Markov chain, 43
generator of Archimedean copula, 89
geometric Brownian motion, 78
geometric Brownian motion mean value, 79
geometric Brownian motion variance, 79
Gumbel copula, 113

hazard rate, 68, 69, 267
hazard rate interpretation, 69
hazard rates, 69, 77
HCTMC-based credit curves, 47

Index

HCTMC-based PD term structures, 40
hedge ratio, 275
hedging against single names, 277
Historic iTraxx Europe tranche spreads, 273
historical (real-life) default probabilities, 146
historical default probability, 268
hit contributions to tranches, 292
hit rate, 6
hitting probability, 34
hitting probability of tranches, 202
hitting probability of tranches in analytic CDO models, 224
hybrid rating, 5

iBoxx, 256
idiosyncratic risk, 17–19
imperfect hedge, 276
implied correlation, 255, 280–283
implied correlation: conclusion, 287
independence copula, 86, 88
index, 147
index administration, 256
index spread, 261, 270
index spread: mean versus weighted mean, 263
index spreads for iTraxx Europe, 272
index tranches, 250, 271
indices, 19
industry indices, 121
insurance paradigm, 175
interest coverage (I/C) tests, 182
interest stream, 12, 150, 169, 198
interest stream stability, 209
Internal Ratings Based (IRB) approach, 3
International Index Company, 256
interpolation of credit curves, 62
ISDA agreements, 130, 254
iTraxx, 253, 257
iTraxx Europe, 258, 259, 271
iTraxx Europe index spread, 261
iTraxx Europe spreads, 260
iTraxx Europe tranches, 274
iTraxx Europe tranching, 271
iTraxx Europe: fair index spread, 265

iTraxx Europe: historic index spreads, 272
iTraxx tranches, 258

JDP, 36
JDP matrix for the CLN reference pool, 158
joint default probability, 81, 118
joint default probability (JDP), 36
joint default probability in dependence on correlation, 39, 65
joint default probability over time, 63
jump-to-default, 301

Kendall's rank correlation, 104
Kendall's rank correlation of Gumbel copula, 113
Kendall's tau of Gaussian copula, 108
Kendall's tau of Student-t copula, 108
Kimberling's theorem, 90

Laplace transform representation of Archimedean copula, 97
Laplace transform representation of Archimedean generator, 91
Laplace transform representation of nArchimedean generator, 96
latent variables, 9
latent variables approach and default times, 78
leverage ratio of an index tranche, 276
LGD and workout process, 17
LGD determination, 15, 16
liability side, 167, 191
liability side of Artemus, 180
liability side scenario, 191
liability spread, 297
LIBOR, 149, 150
limit breachers, 295
linear dependence, 103
linearity of expected shortfall, 288
liquidity drivers, 253
liquidity premium, 29
log-expansion of one-year migration matrix, 45
loss distribution, 20, 21
loss distribution of tranche, 206
loss given default (LGD), 10, 14
loss given default of tranche, 203

loss given default of tranches, 224
loss severity, 14
loss trigger, 210
lower strike of a tranche, 273
lower tail dependence, 107

managed STCDOs, 252
management of CDO and mixed portfolios, 299
margin, 29
marginal default quote likelihoods, 237, 246, 249
mark-to-market changes, 275
mark-to-market of illiquid assets, 296
mark-to-model, 296
mark-to-model of illiquid but securitizable credit risks, 298
Markov chain embedding, 46
Marshall and Olkin's theorem, 96
MBS, 168
Merton model, 8
mezzanine tranche, 169
minimum variance portfolio, 32
Monte Carlo simulation of CDOs, 191, 229
Monte Carlo simulation of dependent default times, 125
Monte Carlo simultion of CSO, 202
Moody's KMV, 20
Moody's KMV's PortfolioManager, 8
mortgage-backed securities, 168
most randomly behaving, 87, 317
multi-period models, 213
multi-step models, 213, 214
multi-year default frequencies from S&P, 58
multi-year horizons, 59

Names in the CDX.NA.IG Series 5, 336
Names in the iTraxx Europe Series 5, 337
negative carry, 276
NHCTMC-based credit curves, 55
NHCTMC-based PD term structure, 52
non-exchangeable exposures, 245
non-homogeneous continuous-time Markov chain, 52

notional exposure, 10
nth default time, 240
nth-to-default modeling, 120
nth-to-default simulation study, 129

offering circular, 11
one-factor model, 28, 222
one-year default probabilities, 3
one-year migration matrix based on S&P, 42
order statistics, 129
overcollateralization (O/C) tests, 182

pairwise correlations histogram, 126
parameterization of NHCTMC-based PD term structures, 53
partially funded, 169
payment in kind (PIKing), 12
PD calibration, 5
PD term structure, 39, 40, 53, 266
PD term structure (risk-neutral), 269
PD term structure example, 47
PD term structure interpolation, 62
perfect correlation case, 66
perfect credit score, 9
physical settlement, 130, 147
pool EL, 152
pool PD, 154
portfolio credit risk, 17
portfolio in risk/return space, 33
Portfolio of sample assets, part I, 330
Portfolio of sample assets, part II, 331
portfolio profit, 29
portfolio variance, 30
portfolio-referenced CLN, 147
practicability versus accuracy, 24
premium, 169
premium leg, 130, 259
premium leg in index tranche, 282, 284
prepayment, 11, 13
pricing floor, 295
principal, 147
principal stream, 12, 150, 169
principles of CDO modeling, 190
protection buyer, 129, 147
protection seller, 130
PV01, 275

Index

Q-matrix, 43
quantile function, 10

R-squared, 18
rank correlation, 109
rating, 2
rating arbitrage, 178
rating system optimization, 5
receiver operating characteristic ROC, 6, 7
recovery, 16
reference pool, 167
regional indices, 121
regulatory arbitrage, 186
regulatory capital relief, 186
replenishment, 11
residential mortgage-backed securities, 168
residential mortgage-backed securities (RMBS), 11, 12
residual effect, 123
return and risk, 29
risk and return, 29
risk measure contributions, 290
risk measures, 300
risk measures for tranches, 287
risk transfer, 172, 174
risk-neutral default probabilities, 146, 264, 265, 267, 269
risk-neutral default probability, 268
risk-neutral PD, 264
risk-neutral PD term structure, 267
risk-neutral survival probabilities, 265, 269
risk-neutral uniform PD term structure, 282
risk/return of portfolio, 30, 33
RiskMetrics Group, 8, 20
risky duration, 270, 271
risky duration of tranche, 251
RMBS, 168, 236
running spread, 272, 274
RWA formula in Basel II, 187

sample CSO, 194
scenario analysis, 140, 142, 145
scenario transformation in CDO modeling, 192
scenarios, 288, 290

scenarios matching ESF conditions, 289
scoring, 4
scoring system optimization, 5
second-to-default, 34
second-to-default example, 132
second-to-default probabilities w.r.t. different copulas, 117
second-to-default probability, 34, 35, 65
second-to-default probability in dependence on correlation, 38
securitization, 175
securitization cost versus benefit, 176
securitization effect on bank portfolio, 177
securitization effect on economic capital, 177
securitizations in Basel II (remarks), 189
semi-analytic CDO evaluation, 220, 230
semi-analytic default time modeling, 235
semi-analytic default time simulation, 233
senior tranche, 169
sensitivities, 300
separation approach, 19
sequence space, 156
settlement of credit events in Artemus, 181
shortfall measures, 22
simulation efficiency, 242
simulation of dependent times, 84
single-name credit risk, 2
single-name default times, 67
single-name hit contributions to tranches, 292
single-name shortfall contributions, 291
single-tranche CDOs (STCDO), 250
single-tranche loss distribution, 206
skew, 283
Sklar's theorem, 85
smile, 283
Spearman's rank correlation, 104
Spearman's rank correlation of Gaussian copula, 108
Spearman's rank correlation of Student-t copula, 108
special purpose vehicle, 167

spread arbitrage, 178, 183
spread delta, 275
spread leverage in index tranches, 274
spread performance, 247
spread stream, 169
spread-implied risk-neutral default probabilities, 266
spreads for CDS in iTraxx Europe, 260
spreads in CSO example, 197
static model, 213
STCDO negotiation process, 251
STCDO quotes based on iTraxx Europe, 274
STCDO: choice of capital structure, 251
STCDO: determination of maturity, 251
STCDO: selection of reference names, 251
stochastic matrix, 43
stochastic process as default trigger, 9
Student-t-copula, 88
Student-t-copula for nth-to-defaults, 127
Student-t-copula with Gaussian marginals, 96
Student-t-copula with Student-t marginals, 95
subordinated, 170
subordinated capital, 173, 183
subordination, 170, 210
super senior swaps, 184
survival function, 68
survival function and hazard rates, 69
survival probabilities (risk-neutral, 269
synthetic CDO, 169
synthetic transaction, 130
systematic factor, 19
systematic indices, 19
systematic risk, 17–19, 299

t-copula, 88
tail dependence, 107
tail dependence of Clayton copula, 111
tail dependence of Gaussian copula, 109
tail dependence of Gumbel copula, 113

tail dependence of Student-t copula, 109, 110
tail risk measures, 23
tailor-made CDO tranches, 251
term structure function, 68
term structure of default probabilities, 10, 39
third-to-default example, 134
tight spreads for mezzanine tranches, 285
time scale transformation, 216
time-changed Brownian motion, 216
time-homogeneous Markov chain approach to PD term structure, 40
time-scale transformation, 218, 219
top-down cash flows, 170
Trac-x, 256
traded delta, 274, 275
tranche decomposition into difference of two equity tranches, 284
tranche delta, 301
tranche expected loss, 203
Tranche expected loss contributions (TELCs), 294
tranche expected loss contributions (TELCs), 293
tranche expected loss in analytic CDO models, 224
tranche gamma, 301
tranche greeks, 302
tranche hit contribution, 292
tranche hit contributions of single names, 292, 293
tranche hitting probability, 202, 224
tranche loss distribution, 206
tranche loss given default, 203, 224, 229, 249
tranche loss profile, 173, 223, 273, 282
tranche loss profile function, 259, 273
tranche loss severity, 203
tranche PV delta, 275
tranche PV01, 275
tranche risk measures, 287
tranche spreads for iTraxx Europe, 273
tranched loss distribution, 172, 173
tranching of CDX, 254
tranching of iTraxx, 258
trigger for excess cash redirection, 210

Index 357

unexpected loss (UL), 23
unfunded CDO, 169
unfunded transaction, 130
uniform sample, 93
upfront payment for equity index
    tranches, 272, 274
Upper cumulative PD term structure
    for scenario analysis, 143
upper Frechet copula, 92, 237
upper strike of a tranche, 273
upper tail dependence, 107

variance reduction, 242
Vasicek density, 223
Venn diagram, 155
visualization of copulas, 99

waterfall, 198
weighting of index spreads, 266
wide spreads for senior tranches, 285
wipe-out, 152
wipe-out probability, 34, 154, 156
wipe-out scenario, 172
worst case expected loss, 152
worst case scenario, 141